ABSOLUTE BEAUTY

ABSOLUTE BEAUTY

RADIANT SKIN AND INNER HARMONY
THROUGH THE ANCIENT SECRETS
OF AYURVEDA

PRATIMA RAICHUR

WITH MARIAN COHN

HARPER

NEW YORK • LONDON • TORONTO • SYDNEY

Grateful acknowledgment is made for permission to reprint portions of
the following copyrighted material:

Touching (3rd Edition) © 1971, 1978, 1986 by Ashley Montagu reprinted with
permission from HarperCollins Publishers.
Care of the Soul © 1992 by Thomas Moore reprinted with permission from
HarperCollins Publishers.
The Face of Love © 1995 by Ellen Zetzel Lambert courtesy of Beacon Press.
Conversations with God © 1995 by Neale Donald Walsch reprinted by permission
of The Putnam Publishing Group.
The Book of Indian Beauty © 1981 by Mulk Raj Anand and Krishna Nehru
Hutheesing reprinted by permission of the Charles E. Tuttle Company.
The Principles of Anatomy as Seen in the Hand (2E) © 1942 by F. Wood Jones
courtesy of Williams & Wilkins.
The Miracle of Mindfulness © 1987 by Thich Nhat Hanh courtesy of Beacon Press.

A hardcover edition of this book was published in 1997 by
HarperCollins Publishers.

First HarperPerennial edition published 1999.

Designed by Barbara Balch
Illustrations by Charlene Rendeiro

The Library of Congress has catalogued the hardcover edition as follows:

Raichur, Pratima. 1938–
Absolute beauty : radiant skin and inner harmony through the ancient
secrets of ayurveda / by Pratima Raichur with Marian Cohn. — 1st ed.
p. cm.
Includes bibliographical references.
ISBN 0-06-270172-X
1. Skin—Care and hygiene. 2. Medicine, Ayurvedic. 3. Beauty, Personal.
I. Cohn, Marian. II. Title.
RL87.R22 1997
646.7'26—dc21   97-1908

ISBN 0-06-092910-3 (pbk.)

12 13 ❖RRD 20 19 18 17 16 15 14

TO MY DEAR HUSBAND, VENKATESH,

who has been there for me for the last thirty-two years
and has given me a stable, pleasant home,
and unconditional love

CONTENTS

PART IV.
BEYOND THE SENSES
TO THE SOUL: BREATH,
MIND, AND SOUL PURIFICATION

PRANA, MANAS, AND ATMA SHUDHI

ACKNOWLEDGMENTS

My gratitude goes first to Shanta Dandekar, my spiritual mentor, who thirty years ago helped me to know myself and my purpose clearly; to my first teacher, the late Vaidya Rele, and to my present teacher, Dr. Vasant Lad, who answers my calls at any time of day or night with graciousness and wisdom; and to my beloved parents and family—my husband, Venkatesh, and my son, Sandeep, who have helped me with many details of my work, and my dear daughter, Leena, who did all my typing for this book.

I also wish to thank:

My agent, Lynn Franklin, and my friends Deepak Chopra, Bipin Shah, and Anita Raj, who brought Lynn into my life.

My friend, Sunita Vase, who helped to find and translate rare Sanskrit books for me.

The numerous beauty editors who wrote about my work at Tej Skin Care Clinic even before most American readers had heard of Ayurveda—especially Felicia Milewicz at *Mademoiselle,* who was the very first to break this ground, and who has been an advocate and a friend ever since.

My fellow teachers Karen and Blair Lewis, David Frawley, and Bri Maya Tiwari, who have helped to expand knowledge of Ayurvedic science in the United States.

My excellent staff at Tej and my business partner in the manufacture of my Bindi product line, Ramesh Sarva, for keeping everything going while I worked on this book. Without their devoted efforts and support, I would not have been able to complete it.

All my clients, who have inspired and encouraged me, and have become my friends through the years.

Peternelle van Arsdale and the many people at HarperCollins, who worked so diligently and enthusiastically to produce *Absolute Beauty*.

Barbara Balch, for her elegant book design.

And most of all, my co-author, Marian Cohn, for being my voice, lending her eloquence and insight, and sharing my deep appreciation for the wisdom of Ayurveda.

PRATIMA RAICHUR

I would like to add my thanks to the following people:

Lynn Franklin, without whom I would not have had the opportunity to meet Pratima Raichur or the privilege of collaborating with her.

Marty and Bea Gross and Nancy Gross Belok for their constant love and support, and for generously giving me space, both literal and figurative, to write this book.

Karen Lucic, friend and scholar, for her invaluable and astute editorial input from first page to last, and for her encouragement from start to finish.

Richard LaMarita, friend, writer, and chef, for sharing his Ayurvedic knowledge with me whenever I asked, for pitching in during the final crunch to help with Chapter 11, and for making me some delicious Ayurvedic meals.

Doug Winblad, friend and philosopher, for his research assistance and his uncannily lucid explanations of the most abstruse philosophical ideas.

Mark Hardesty and my sister, Regine Urbach, who provided essential computer support and who rushed to my aid, technical and personal, in numerous emergencies.

Doug Beube, Lenore French, and Joanne Rhinehart for cheering me on.

Susan Peerless, sister-in-spirit and "first reader," who patiently coached me—body, mind, and soul—through the long months of work and helped me keep the text on track with keen pointers on every draft. She eagerly tried out Pratima's advice as quickly as I could write it down, and kept me inspired by her own enthusiasm for its results. Her unconditional support throughout was the most extraordinary act of friendship.

Maharishi Mahesh Yogi, who gave me not only the profound knowledge and experience of consciousness, but also the precious skill to teach it to others.

And my wonderful parents, Sidney and Marlene Cohn, who gave me a love of words as well as the opportunity and encouragement to develop my abilities. In giving me life and love, they have given me all the important things.

MARIAN COHN

INTRODUCTION

In India, where I was born in 1939, physicians have known the secrets to flawless skin and ageless beauty for six thousand years. Those secrets, which you will learn in this book, are contained in one of the world's oldest systems of health care and healing, named *Ayurveda,* which means knowledge of life or longevity.

My informal introduction to this ancient science came during my childhood at home, where everyone in my extended family lived in accord with Ayurvedic traditions. Our meals were prepared in Ayurvedic style to maintain physical health, and our daily routine included meditation and other practices to balance the mind and spirit. It was a loving, happy family life, and like a typical child, I was unaware that our way of living was different from any other. However, I did recognize at a young age that there seemed to be something special about my mother and her mother: They were striking women whose inner poise and outer radiance never went unnoticed, no matter where we were.

My formal education in Ayurveda began, I confess, somewhat less happily. It was in my early adolescence, under the tutelage of a famous physician who happened to be my family's neighbor. At eighty, he had come to need some assistance with his work, and I became the reluctant recruit whom he called in to help. Every evening after school, I spent a couple of hours in his home, either reading to him or helping him to make various pills and remedies. Not surprisingly, at age thirteen, I was not very pleased to spend my time with this elderly man or to learn his strange formulas, but he was a relentless taskmaster and insisted that I record everything he said in notebooks. "Today you don't understand the importance of what we are doing," he told me, "but one day you will have a use for all this in your life."

Years later, when I embarked on the career that has culminated in this book, I remembered the doctor's words and realized how prescient he was. Those long-forgotten notebooks were filled with Ayurvedic prescriptions for the care of the skin. By then, I had a science degree from Bombay University and was married, with an infant daughter. I began to work as a chemist in England while my husband completed his medical training, and later I returned to India to work at a cancer research hospital.

One day, two of my hospital colleagues came into the laboratory very distraught. Both had acne problems and had gone the day before to get facials. Their treatments had left black patches all over their skin, which looked worse than the acne, and now they were in tears. To everyone's surprise, including my own, I said to them, "You should have told me before you went for the facials. I could have prepared something to help you."

That night, for the first time in over a decade, I reread the Ayurvedic recipes I had so dutifully transcribed as a young girl. Since Ayurveda does not include a specific investigation of skin disease, I had never thought to apply its principles and techniques to the problem before this incident. By then, however, I knew a lot about cosmetic chemistry, and with my combined knowledge of ancient and modern science, I created a special mix of herbs and oils, which I delivered to my friends the next morning. Within days, their complexions cleared up completely. Word spread, and soon I was inundated with requests for help from people with problem skin. Each time I prepared a new remedy, everyone could see the positive results on the person who used it.

Meanwhile, the efficacy of the Ayurvedic preparations aroused my curiosity as a scientist. I wondered how these age-old formulas worked from a biochemical standpoint. The more I thought about it, the more questions I had: How do these preparations activate healing? What causes skin problems in the first place? And why do some of us have them while others don't? I decided to pursue my research where I could make the greatest contribution—by bridging my understanding of modern chemical science with my knowledge of the use of herbs and oils according to India's oldest science. With my baby at home to care for as well, I left my job at the hospital and began to study Ayurveda in earnest.

That was twenty-five years ago. Today, I am founding director of the Tej Skin Care Clinic in New York City and the creator of three lines of Ayurvedic beauty products, including the Bindi, Tej, and Ojas labels, which are sold in health stores and used in spas nationwide. Over the years, I have continued to develop my skills as an aesthetician, scientist, and researcher, testing my findings both in and out of the laboratory. Since emigrating to the United States in 1977, I have expanded my training to include a certificate in acupuncture and a doctorate in naturopathy. Using my unique system of Ayurvedic products and techniques, I have by now treated ten thousand men and women with skin problems ranging from acne, eczema, and psoriasis to the common symptoms of stress and aging. In many cases, I have cleared up ailments that defied years of treatment by top dermatologists.

Why do my treatments succeed where other beauty regimens and even modern medicine fail? They work because Ayurveda provides the key to health and healing that Western approaches lack: knowledge of the *individual*, not just the illness.

When most Western physicians and skin care experts see a condition like my friends', they are trained to examine the parts—the disorder and its various physical symptoms. Generally speaking, oily skin is treated with astringents, and dry skin is treated with oils; that is, the outward effect dictates the cure. In the case of acne, a doctor will diagnose the presence of a certain type of bacterium as the cause of the problem, and will treat the infected skin with the appropriate drugs or topical products. In a certain percentage of cases, this approach will help reduce or eliminate the acne, at least until the next breakout. But in a significant portion, the treatment will provide little or no permanent relief—or worse yet, it will actually exacerbate the condition, as it did with my friends.

Why are the results so haphazard? Because modern medicine works only on the level of molecules—of matter—without reference to the sentient human being, the unique and complex person who actually *feels* ill. Life is the totality of experience, not merely a collection of physical parts, and human experience happens fundamentally through the filter of the mind and senses on the level of *consciousness.* How we view the world and how we feel about things affects our experience, and experience *changes* the body. If that weren't so, there would be no happy smiles, no sad tears, no embarrassed blushes, no worried brows, no angry looks, no surprised gasps. When modern medicine asks *where* the body is diseased but not *why* the patient is ill, it ignores the basic truth of our experience. Most disease results from a breakdown in the immune system; immune breakdown results from stress; stress is due to perception; and perception derives from consciousness. In other words, the classical Western approach ultimately fails because it disregards the body's network of intelligence—the mind factor—which is the level of life where sickness and healing actually begin.

Of course, in the last two decades, this material bias of modern medicine has slowly begun to erode in the face of new scientific evidence of the biochemical links between our psychological experience and the action of the neuroendocrine and immune systems. These findings, which describe the physiological processes that enable insubstantial thoughts and emotions to affect bodily functions, point to an essential mind-body unity. This has led to the creation of a new Western science known as *psychoneuroimmunology,* or what we commonly call mind-body medicine. Undoubtedly, many of you are familiar with its concepts from the numerous books and articles that have appeared on the subject just within the past ten years.

Nevertheless, for the past three hundred years, most Western minds, scientists and nonscientists alike, have been steeped in a dualistic worldview that splits the universe into the mutually exclusive realms of reason and nature, or mind and matter. Shortly after this idea was proposed by Descartes in the early seventeenth century, it attained the seemingly inviolable status of scientific truth when Isaac Newton published his revolutionary work on gravitation and motion in 1687. Typically, Newton's model of the universe is likened to the game of billiards, in which solid objects, acted upon by outside forces, move, collide, repel, and ultimately come to rest in mathematically predictable cause-and-effect ways. Newton's predictions of objective behavior proved so accurate that his laws have been undisputed ever since, at least in terms of the visible—that is, the large-scale—universe. Spurred by Newton's triumph, the Age of Reason dawned in the West, and with it the common belief that it is possible to describe any activity in terms of fixed rational principles. Like every other field of human endeavor in the eighteenth century, medicine fell under the spell of science. Sad to say, the ancient art of healing, which had been practiced and preached by the likes of Hippocrates, slowly fell to the wayside.

Ironically, as an art, Western medicine had placed a premium on the personal dimensions of disease and had viewed the patient in terms of her whole experience, not just her illness. As a precise science, however, medicine began to place greater and greater importance on events in the laboratory and lesser importance on events in a patient's life. In this century, the mechanistic view of the body has become so overriding that medicine itself has become compartmentalized into dozens of subspecialities, fragmenting the body into progressively smaller and more isolated parts. Clearly, modern science and medicine have advanced the human condition in countless ways; yet progress has not come without a cost. Given the increasing levels of stress in contemporary life, Westerners pay a high price for ignoring the consequences of everyday experience on health. Fortunately, a more holistic approach is slowly taking root in the mainstream medical community. Nevertheless, the mind-body split is still so deeply ingrained in modern scientific thought, and so largely unquestioned, that most doctors trained in the West are as yet far more conversant in the behavior of molecules than in the behavior of human beings.

In contrast to this material view, Ayurveda is based on the premise that the mind and body are unified on the level of consciousness, and through this unifying "field" have direct reciprocal effects. Indeed, as the *first* psychoneuroimmunological science, Ayurveda has emphasized the causative role of thought and behavior in health and disease for thousands of years. As a result, when Ayurvedic practitioners see a skin problem, we observe the person who has it, not just the symptoms themselves. A line or a blemish on the face—like any sign of stress or illness—is only one puzzle piece in a diagnostic picture that encompasses the entire range of an individual's life, from the innermost aspects of mind and emotions to the outermost aspects of lifestyle and environment.

When someone comes to me with a case of acne, as did my friends, I don't try to eliminate the infection that produced the symptoms, although that will happen in the course of treatment. Instead, my ultimate goal is to eliminate the physical or emotional *imbalance* that weakened the body's immunity and enabled the bacteria—which are present even in a healthy person—to negatively affect the skin. Ayurvedic treatments accomplish this, as you shall see, by harnessing the body's inherent ability to heal and balance itself in accord with its own nature.

When I speak to Western doctors and beauty specialists, they inevitably respond to these statements with the same question: "What do you mean by *balance*? What do you mean by the body's *nature*?" Immersed in the knowledge of fragments, they are understandably skeptical of holistic truths. Indeed, *Tufts University Diet and Nutrition Letter* recently advised its readers that any product promising to "balance" the body, "bring it into harmony with nature," or " 'stimulate' [its] power to heal itself" cannot live up to the claim because "no one can prove" you are out of balance in the first place. "After all," the experts challenged, "what does it even mean to be out of harmony with nature?"

Millennia ago, the *rishis*—the "knowers of reality" who gave us Ayurveda—answered this question. They said that you and I, along with everything else in the universe, are made up of the same five constituents: *space, air, fire, water,* and *earth.* Although these elements are present in everyone, each of us has them in a different proportion, a different *balance.* Just like your genetic imprint, your particular mixture of elements is set at conception and remains a constant throughout life—and it determines your basic characteristics, including your type of skin.

Although we refer to it in terms of proportions and combinations, the balance of elements is not strictly a physical phenomenon, like a cake mix, because our basic ingredients are not essentially *things.* Indeed, when the rishis said we are made of space, air, fire, water, and earth, they did not mean we are warm mud—though in the dust-to-dust sense, of course, that's more or less what the material body is. Rather, from their highly developed state of awareness, they realized that the fundamental component of existence is not a speck of matter or even the energy locked inside it. It is *unbounded intelligence.* This intelligence, the rishis said, is beyond direct sensory experience and even beyond the scope of objective science, because it exists on the level of human consciousness, which is subjective by its very nature. Nevertheless, the effects of this intelligence are evident everywhere: in the rhythms of nature, the motion of planets and galaxies, the complex structures of matter—and in the evolution of life itself and the "genius" of DNA. Without it, both energy and matter would be awash in chaos, incapable of the cosmic organization that clearly underlies existence.

Quantum mechanics, the twentieth-century science that describes the realm of the "smaller than the small," where Newtonian law breaks down, puts a modern spin on this ancient theory of five elements. As physicist John Hagelin suggests, the basic constituents of the Ayurvedic universe are different *vibrational modes* within the virtual energy field that underlies subatomic matter. It is called a *virtual* field because it is so abstract, it cannot be detected directly, even by our most powerful technology. Western science infers its existence from clues left behind by force-carrying particles that instantaneously appear and disappear out of this seeming void. Sixty years ago, physicists named this invisible layer of reality the *quantum field*. Sixty *centuries* ago, Ayurvedic scientists named it the *field of pure consciousness*. By either name, Ayurveda considers this unseen, omnipresent continuum of intelligence to be the ultimate source of mind and matter. The five elements are simply particular *patterns of intelligence*—that is, vibratory patterns—within that field, which shape individual and material existence.

Depending upon which elements—which vibratory patterns—are most prevalent in an individual—and usually one or two will dominate—Ayurveda classifies everyone according to three universal *natures,* or constitutional types, known as the *prakriti.* Later, we will discuss these different natures and how they determine the characteristics of your skin. But generally speaking, your prakriti is like a personal wellness norm that describes your overall appearance, your emotional disposition, and your mental aptitudes when you are balanced; and also predicts specific ailments—skin problems included—that you tend to develop when you are not balanced. A fundamental principle of Ayurveda is that all illness results from deviations in your prakriti—your ideal constitutional formula—due to physical, mental, behavioral, or environmental factors. In other words, whenever our unique configuration of elements becomes imbalanced—whenever the innate patterns of intelligence are disturbed—disorder or disease ensues. Accordingly, all Ayurvedic treatments work, as you will see, by restoring balance to the elements—that is, by restoring the flow of intelligence.

If you and I had an identical balance of elements, we would look exactly alike. Because our "formulas" are different, however, not only do we look different, but our bodies and senses *respond* differently to everything in the environment, including the factors that cause aging and disease, and the methods to treat them. Healthy, glowing skin is a natural condition when mind and body are balanced. But Ayurveda shows us that the formula to achieve balance is different for each person, depending upon his or her innate body type and temperament. As a result, *there is no single type of treatment*—not a soap, moisturizer, age cream, doctor's prescription, natural remedy, fitness program, diet, or lifestyle change—that can work for everyone, *because not everyone is born with the same type of constitution or the same type of skin.* To find the right beauty program to reduce the signs of aging and enhance your skin, you must know what your Ayurvedic skin type is. This book will tell you how.

There are three skin types, as there are three basic constitutions. If you were a new client coming into my clinic, the first thing we would do is have a personal consultation to determine your type. I do that by "examining" you for a specific set of *outward* physical, mental, and behavioral characteristics that give me information about your internal state of being. I begin by looking you over from head to toe: your height, weight, body frame, facial structure, body language, mental attitude, hair, nails, complexion, and skin. I also ask about your lifestyle, including work, eating habits, exercise, family life, sleeping patterns, and skin care regimen—and about your day-to-day moods and feelings, and the general way you handle stress in your life. In other words, I look at the whole person. On the basis of this information, I can determine which elements are dominant in your constitution, and therefore which type of skin you have and what type of treatment will balance your complexion.

Once you know what characteristics to look for, you can easily determine your skin type for yourself. In Chapter 2, you will have the chance to do so by observing yourself in a mirror and answering a simple quiz. Once you know a little more about Ayurveda and how the elements determine your attributes, you can even make an entertaining game of observing friends and strangers and guessing their type. You will be quite amazed at the insights Ayurveda provides into the natures of people you barely know. And you may also find it helpful in understanding what inner forces drive the people you love—not to mention yourself. Used wisely, Ayurvedic "typing" can be a useful tool in human relations.

My own clients are sometimes so startled to hear how much I know about them just by looking at their face and skin, they ask if I am psychic. Of course, I am not. But what I say in response—that their skin *speaks* to me—often surprises them more. If I am not clairvoyant or imagining things, what do I mean by this statement?

Modern scientists describe the skin as the body's protective tissue and its largest organ, responsible for a remarkable range of functions, including regulation of waste, water, and temperature. It is also the major producer of endocrine hormones, which control most physiological functions, as well as the organ of touch and sensation connected to every other organ and cell of the body through its vast web of cutaneous nerves.

Ayurvedic scientists, however, who gained their insights by means different from their modern-day counterparts, looked beyond the mechanical processes of the body, as we said, to the "quantum" level of existence where the boundary between mind and matter disappears. On the basis of direct cognition of this quantum field, the rishis declared long ago that consciousness is the essential "stuff" of the universe. It took contemporary scientists quite a bit longer to discover what Ayurvedic practitioners have known for millennia, that there is no purely objective reality, no solid "stuff"—no animal, vegetable, or mineral—that we can point to definitively as "the world out there."

Einstein struck the first blow to this familiar and dependable reality with his proof of the equivalence of energy and matter and his work on relativity. In one stroke, he drove a deep wedge into cherished notions of duality, and eventually demolished our belief in the "absolute" dimension of time. He proved that time is not the same for everyone under all conditions; because of space's curvature, it passes faster or slower for each observer depending upon her traveling speed and distance from the center of gravity. In 1926, Werner Heisenberg struck the final blow to materialism when he postulated his famous "uncertainty principle," which introduces the determining role of human consciousness into the so-called objective realm of science. In broad terms, this principle states that because of the intrinsic properties of light and matter, it is not possible to know both the exact position of a particle and its exact velocity, because in the very act of measuring one, we necessarily alter the other. Hence, there is always a degree of uncertainty in our present picture of reality.

We live in a world of probability rather than predictability, and at the finest level of existence, our own consciousness is a deciding factor in the outcome of every phenomenon we observe. The inescapable implication of this quantum view is that the world looks as it does because we are looking at it. That is, the world is the way it is because we *think* it is. As physical participants in this world, *we ourselves are what we think.*

This has been Ayurveda's message for six thousand years: That every fluctuation in thought—in consciousness—produces a corresponding change in the body. At the fundamental level of existence, mind structures matter, not the other way around. Knowing this, the rishis understood that skin is the physical reflection of our inner being. Skin not only feels sensations, but *expresses* what we feel. Through its nerve endings and endocrine glands, the skin takes chemical messages to and from all parts of the body and transcribes every single event into a language whose words we can read in the "angry" rashes, "weeping" eczema, "worry" lines, and other marks and blemishes on the surface tissue. To someone like myself, trained to recognize each nuance of this remarkable code, the skin literally does speak volumes. In fact, skin problems are not problems of the skin at all, but signals of specific imbalances deep within the body and mind, far out of reach of the creams and lotions we apply to our body and face.

Ayurveda teaches that to correct these imbalances at the source, we have to work on four levels of life: body, breath, mind, and spirit. To do this, I start with every client on the external level, because that is what we see. Once I have diagnosed a person's skin type, the first thing I do in the clinic is cleanse, nourish, and moisturize the face and body using pure herbal extracts and essential oils. Then I prescribe a personalized daily skin care routine to do at home to counter the effects of stress and pollution and stimulate new cell growth. Within days, these measures will help to alleviate surface symptoms and enhance the appearance of the complexion.

But I must tell you in all honesty that there is no lasting solution for aging or disease to be found in a bottle, not even if that bottle comes from me. By themselves, no external treatments can permanently restore balance, because they fail to affect the deep structure of the cells where all disorder originates. In order to reach that level of life, we go within, using all five senses to harness the body's intelligence. Therefore, my complete Ayurvedic beauty regimen includes a program of diet, rhythmic breathing, massage, sense therapies, and meditation tailored to your skin type.

Ayurveda does not stop, however, even when the symptoms of imbalance are gone. The final goal of this life science is not just freedom from disease. It is nothing short of wholeness, which is a state of perfect inner harmony mediated by the body's built-in intelligence. When we live life on this basis, balanced in body and mind, the face and skin look naturally vital and blemish-free, and we literally radiate a fresh, happy, peaceful feeling from within. This is what I mean by beauty from the inside out. It is the source of that unmistakable grace and elegance I saw in my mother and grandmother as a child—their ancient beauty secret from me to you.

As I hope you have begun to see, my book is about a very different kind of beauty and skin care than anything you have known before. At my Tej clinic in New York, where I have only a few people trained in my techniques, there are just so many individuals we can treat in a day. I have written this book for the millions of women and men like you around the country whom I am unable to see personally but who are seeking more balance, radiance, and beauty in their lives through the ancient and compassionate wisdom of Ayurveda.

When I ask my clients which benefits they think my readers should know about, many of them say to me, "Oh, Pratima, tell them about the *feeling* of this place! I leave here feeling so happy and peaceful." Even if you cannot come to my clinic, you can create the same experience for yourself at home using the simple and profound ideas you will find in the pages to come.

The book itself is organized to bring you this peace and radiance in a step-by-step manner. Part I talks about the Ayurvedic notion of beauty and the contributions of contemporary mind-body science to our understanding of the skin. It also goes further into the ideas of constitutional balance and *types,* which are central to Ayurvedic skin care, and includes the Skin Type Quiz. Part II describes the personalized daily skin care routine for each type. Part III gives dozens of other Ayurvedic techniques to balance the skin and body, inside and out, using all five senses. Part IV goes beyond the body and senses to describe the techniques, including meditation, for balancing breath, mind, and soul. The final chapter will inspire you, I hope, to seek the highest joy in every experience. It discusses the role of consciousness in everyday life, which goes to the heart of Ayurveda, and offers some simple principles gleaned from the rishis and from personal experience to help you achieve the inner sense of purpose, poise, harmony, and bliss that is the essence of grace and the true secret of ageless beauty.

I have tried throughout the book to focus on the practical applications of Ayurveda to skin care, confining my discussion of theory to the ideas you will need to understand why skin problems develop and how treatments work. If you are interested in a more comprehensive study of this ancient science of health and healing, I recommend general books in the Bibliography.

My Ayurvedic approach to beauty is *not* something you will find in these texts or any others, however, because the ancient seers gave no special attention to skin care beyond general health concerns. In fact, Ayurveda formally addresses only eight branches of medicine, and dermatology is not among them. What Ayurveda does reveal is the remarkable interconnectedness of skin, the body's largest organ, with all other organs and life processes—physical, mental, emotional, and spiritual. It is this revelation—that the health and appearance of our complexion are an actual expression of our total well-being—that drew me to look deeper into Ayurvedic principles in order to develop an effective beauty system. The classification of three skin types and many of the treatments described in this book are the product of my personal insights, based upon a comparative study of this ancient mind-body medicine with its modern counterpart. As such, this book offers a totally new and unique way to achieve beautiful, youthful skin that incorporates the best wisdom of scientists and seers.

Before you begin on this rewarding journey, I want to share one simple story from the rishis, who often used the method of allegory to unfold their knowledge of life. The rishis recognized that all the manifest universe is an expression of pure intelligence—consciousness—and therefore we can find this intelligence everywhere in nature—even in the actions of an animal. Below is my favorite parable about the restless deer who seems to forage the woods in a ceaseless search.

In this story, the deer is attracted by a beautiful aroma that is always in the air. But though he hunts for it far and wide, he never discovers its source. The sad-eyed deer doesn't know that the irresistible scent is musk from his own belly. Only when he looks within will he find the perfection he seeks.

Now, *you* are wiser than the deer—and luckier. The beauty you desire is attainable, it resides within you, and all the directions you need to get there are contained within this book.

PART I

BEAUTY
AND
AYURVEDA

Beauty is the translucence, through the material phenomenon, of the eternal splendor of the "one."

WERNER HEISENBERG

CHAPTER 1

WHAT IS BEAUTY?

*If a woman, it is said in a Tantra, abandons herself often
enough to the dreams that spring from her heart, the mood
that arises will color the whole of her person. Is it not one of
the most common of commonplaces in conversation that in
moments of intellectual or emotional excitement the features
of the plainest person assume an aspect of exquisite beauty?*
MULK RAJ ANAND AND KRISHNA NEHRU HUTHEESING

Every single human being wants to be beautiful. It doesn't matter whether we are
young or old, female or male. The desire for physical beauty—and the capacity
to recognize it—seems deeply ingrained in the human psyche. A common index
of beauty is harmony or proportion, and developmental scientists believe that
our innate ability to perceive the symmetry of the human face is a mechanism for
survival. With no understanding why, infants instinctively light up at a friendly,
pleasant face, and cry at ugly or distorted expressions, providing a built-in signal of
potential danger. By age four or five, children are well aware of subtle physical differences among people and will judge others on the basis of appearance. They are
also conscious of their own looks, and love to experiment—the more elaborate or
glamorous they can make themselves, the better. Just watch any youngsters playing
dress-up and notice the delight when they see their reflection in the mirror.

No civilization on earth has existed without some standards of beauty and
dress, even if those standards have differed radically from our own. As anthropologist Ashley Montagu observes, "Each society has found its own ways of decorating,
and thus celebrating, the human form." Indeed, the history of art and culture is, in
large measure, a testament to the universal allure of beauty and humanity's quest
for perfection.

VANITY OF VANITIES:
THE QUEST
FOR BEAUTY IN A BOTTLE

Unfortunately, this age-old quest has become in contemporary American life little more than a fixation on images fueled by media and advertising, and compounded by public attitudes towards health and aging. Historical ideals of beauty, which stressed the perfectability of our deepest nature, have eroded in mass culture into something more aptly called "good looks," which we achieve with the right makeup, the right wardrobe, the right personal trainer, and if all else fails, the right plastic surgeon. In recent years, for instance, fashion and rock video joined forces to popularize a "look" and a dance form epitomized by the exaggerated styles, postures, and attitudes of runway mannequins; Madonna branded it "voguing," in apparent homage to the beauty magazine. The notion of a beauty that—like fine art—takes time to create and bring to the surface in all its subtle and varied shades, is virtually lost from the common visual lexicon. What we see instead is the cover-girl glance, the Hollywood pose, the MTV clip, the commercial spot—all visual equivalents of sound bites. Our unprecedented capacity to endlessly reproduce and instantly flash these glossy images around the world creates an infinitely distorted reflection of ourselves, not unlike a fun-house hall of mirrors. The effect might be humorous if the supermodel look itself were not so extreme, and if its proliferation were not a significant factor in the rise of eating disorders, depression, and other self-esteem problems among women and teens.

Our point here is not to disparage the role of cosmetics, fashion, or entertainment. On one level, these glamour industries are just playing out for the collective psyche the same sorts of fantasies we enacted as children dressing up. At any age, dressing up is, as Montagu suggested, an act of self-affirmation, not to mention fun. However, these highly stylized, homogenized images, by their very ubiquity and form, have fixed our vision of physical perfection in two dimensions, reinforcing the misguided belief that beauty is only skin deep. As a result, this society has literally lost sight of what it means—and what it takes—to *be* beautiful.

At the same time, modern medical advances have led us to hope that we can find eternal youth in a bottle, and freedom from disease in a pill. Americans today, both in our personal lifestyles and public policies, exhibit a blind and blinding faith in the power of science to cure all ills, no matter what we do to cause them. Many people are happier to take drugs with toxic side effects, or even to go under the knife, than they are to change their diet or cut out harmful habits. Health insurers themselves will cover the high costs of lung disease treatment, for example, but not necessarily the low cost of an aid to stop smoking, even if it has been medically prescribed. Moreover, most physicians here are trained in *allopathic* practices, which focus on the treatment of acute disease, not on how to prevent it. In fact, despite its victories over polio, smallpox, and other terrible sicknesses, allopathic medicine has little to do with wellness. Rather, by chemically suppressing the symptoms of illness, or surgically removing diseased parts, it allows us to achieve an appearance of good health without actually having to be healthy. This treatment strategy is not so beneficent as it may seem. It masks—not heals—the fundamental physiologic imbalance; and temporarily out of view, the disease process can establish new strongholds in previously healthy tissue while surviving strains of infectious agents grow ever more resistant to treatment, as we are witnessing in some diseases treated with antibiotics.

Modern medicine's "quick fix" approach to health has a direct bearing upon popular attitudes towards physical beauty. At a recent cosmetics convention in New York, I saw thousands of new "wonder drugs" claiming to cure skin problems, remove wrinkles, stop aging, and make us look great. That means we put dozens of new formulas on the market *weekly*! It also means that whatever we have already in the jars and tubes that fill our makeup bags and line our bathroom shelves is *not* working. This includes many so-called natural products, whose recent popularity has only added to consumer confusion. Advertisers have latched onto the "natural" concept as an effective marketing strategy, but the truth is that federal standards do not restrict companies from printing the word on labels, as you will see in Chapter 4, even when products contain many synthetic and chemical additives in addition to some natural ingredients. Meanwhile, the purest products, although they are not likely to cause harm, are not cure-alls, either.

The bottom line is this: For decades, Americans have spent billions of dollars annually looking for the next "magic" skin care ingredient, and we still have not found one. I believe we never will find the solution in a bottle because the secret to lasting beauty is not outside us at all, but comes from deep within the body and mind.

BEAUTY: THE HIGHEST PLEASURE

"The essence of all beauty, I call love. . . . "
ELIZABETH BARRETT BROWNING

"Beauty is not a veneer upon things; it is not skin deep; it is not something added to make an ugly thing acceptable. It belongs to the nature of the thing made."
UNKNOWN

Why are physical appearances so important to us? What is this quality of beauty that we want so much? In the anthropologist's terms, the desire for beauty is a primal one: We want to be beautiful because beauty, to borrow Ashley Montagu's phrase, confers basic survival benefits on the individual who has it. Long before a child develops the rudiments of language and can conceive of beauty as a cultural ideal, the senses perceive pleasure and pain—alerting the body to comfort or harm, friend or foe, harmony or discord. Just as we are, in infancy, naturally attracted to the sight of a pleasant demeanor, we are soothed by a caress or a lullaby. Conversely, we are upset by a rough touch or sudden, loud noise. In this manner, solely through sensory-motor awareness, we begin in the first weeks of life to differentiate the qualities of experience that we come to understand in cognitive terms as loving and fearsome, good and bad, beautiful and ugly.

In fact, *Webster's* dictionary calls the quality of beauty "that which gives the highest degree of pleasure to the senses or the mind. . . . " Its opposite, ugliness, is rooted in a word meaning dreadful or fearful. In other words, what threatens or harms us is *not beautiful*; it literally displeases the senses, triggering a series of neurochemical reactions that spell danger to the body and stimulate the self-protective mechanisms that we experience as the fight-or-flight impulse. By the same token, what is naturally pleasurable to the senses—and thus, lovely—sets off a completely different neurochemical response that we experience as feelings of calm and well-being. As Dr. Andrew Weil remarks in *Spontaneous Healing*, "Beauty in any form has a salutary effect on spirit."

Although beauty has no meaning in itself to a one-month-old, the tone of a mother's voice, the arrangement of the features on her face, the quality of her touch, her behavioral response to physical and emotional needs all do, as we said, convey information that the infant's body comprehends instinctively. The basic intelligence that enables us to recognize certain patterns in the environment is encoded from conception in the DNA. The senses act in effect like supermarket scanners of nature's bar codes, identifying the inherent mark of each item—each stimulant—and with feedback from the central nervous system, assessing its immediate cost in life-or-death terms.

Through the vehicles of pleasure and pain—that is, through body and senses—the newborn also builds the foundation of identity. By the natural process of trial and error, we literally feel our way through our new surroundings outside the womb, discovering as we do that—within the seemingly undifferentiated universe to which we are born—there is, in fact, an "I" and a "not I," a "self" and "other." Perceiving our physical separateness, we begin to develop an ego, and with it, our first relationships, which give us, in turn, our earliest lessons in self-worth. Depending upon the measure of pleasure or pain such primal relationships provide, we begin to know ourselves as loved, lovely, and lovable, or not.

Consequently, our self-image—whether or not we are beautiful or lovable *in our own estimation*—is inextricably tied not only to the body, but more specifically to the *skin*, which constitutes the physical boundary of our person and provides, through touch, the primary mode of communication we have at birth. This intimate relationship of the skin, senses, "feelings," and "self" is at the basis of Ayurvedic beauty, since our emotions—how we feel in and about ourselves—directly affects our outward facial appearance.

Ellen Zetzel Lambert, a feminist writer who examines this primal experience in her own early development, captures its essence when she defines beauty as "the face of love." In her book of the same title, she recalls how, as an adult, she looks at pictures of herself taken in the few short years before her mother's death. She wonders what eventually happened to the delightful, pretty child in the photos; how such obvious physical charm became transformed into the plain—even "scowling" and "ugly"—face she sees in her later portraits. Reflecting deeply on the question, Lambert concludes that the pain of her loss and subsequent lack of affection from her stepmother had redrawn the "dazzling smile" created by her mother's love. "[T]he beautiful face," she writes, "is the one itself informed, or animated, *by* love. Thus as a small child I gave back to the world the reflection of the love I had received, and in so doing I was beautiful."

Our desire for beauty, then, is not just a legacy of women's conditioned power-lessness, as many have contended. To the contrary, it is an innate impulse and the natural expression of the self *empowered* by love. Thus, as Lambert believes, "appearances not only do but *should* matter," and we needn't apologize for caring about them. "In fact," she states, "beauty matters to me precisely *as a feminist issue*. It matters just because outward beauty is the expression of the inner self, because it is the bearer of identity. I believe it is a very basic need for an adult, as for a child, to be loved *in the body*; and as feminists we are mistaken to deny the validity of that need. . . . To do so is to deny (for a man or a woman) our wholeness."

What is wrong with mass culture is not that it asks us to be beautiful, but that the objectified, depersonalized image of beauty so prevalent in the media belies the very essence of our desire for it—the wish to fully realize our selfhood. In its ideal form, physical beauty is not the means to an end; it is neither a tool for seduction nor a lucky charm to win love. It is itself, as Lambert suggests, the end product of love—of *well-being*—the ultimate and full expression of individuality and inner power.

In Ayurvedic terms, beauty is the face of the *Self* unbounded—the pure energy of consciousness reflected in subtlest physical form through the body. In the words of Werner Heisenberg, the renowned quantum theorist, "Beauty is the translucence, through the material phenomenon, of the eternal splendor of the 'one.'" We know it in ourselves as a profound experience of wholeness, and we see it in others as an effortless poise, grace, and vibrance: the individual totally at ease from *deep within the skin* and radiant from without.

Ayurveda teaches that anyone can achieve this state of inner wholeness, *regardless* of upbringing or place in the world, by learning to balance all levels of life in accord with our constitutional makeup. This ranges from the outermost physical aspects—behavior, body, and breath—to the subtle subjective aspects—emotions, thought, and ego—to the unmanifest source of existence—the underlying field of consciousness. When every level of existence, from the way we eat and sleep to the way we think and play, is concordant with our innate energy patterns, then we are living as nature designed us to be. As you will see, there is no higher pleasure than such harmony of being.

RADIANCE: THE TEJ FACTOR

"The glowing blush that mantles the cheek,
The dazzling fire that sparkles from the eyes,
The soft, shining sheen of the wavy hair,
Are all mere expressions of good health."

TANTRA

"At the very instant that you think, 'I am happy,' a chemical messenger translates your emotion, which has no solid existence whatever in the material world, into a bit of matter so perfectly attuned to your desire that literally every cell in your body learns of your happiness and joins in."

DEEPAK CHOPRA

It is natural to desire beauty—and it is natural and normal to be vibrant and beautiful, according to Ayurveda—but it is limiting and ultimately detrimental to think of beauty as something dependent upon the shape and features of our anatomy. The body itself alters as we mature. Skin and muscles lose firmness and density, our hair changes color and texture, our ears and nose grow measurably longer, and our stature, shorter. And as we said, the popular picture of attractiveness is ever shifting anyway, in the tides of economic and social change. At one time, for example, milky skin was the mark of an aristocrat; and tanned skin, the sign of a laborer. Now a suntan is a badge of privilege; and pallor, an occupational hazard. Similarly, the voluptuous shape that once made Marilyn Monroe a generation's sex symbol today would appear ridiculous in a Victoria's Secret catalog. Kate Moss's starved look, *en vogue* right now, would be anything but glamorous seen through the lens of the Great Depression.

The picture of perfection also changes from culture to culture, as we mentioned. Indian poets describe the ideal beauty as "moonfaced," "elephant-hipped," "serpent-necked," "swan-waisted," and "lotus-eyed." Yet someone fitting this well-rounded profile is not apt to be voted Miss America. Any objective ideal we may achieve is ephemeral at best, since nothing in the physical universe, least of all flesh or fashion, ever remains the same. That is the law of nature. If we base our identity on such changing values of perception, then we will be caught in a wild goose chase, doomed to empty, unhappy, even unhealthy lives.

Ironically, many women think that being beautiful will *make* them happy, but the truth is the other way around. Without happiness, lasting beauty is an unattainable goal. Even the most perfectly drawn features lose their attractiveness on a person who is deeply distressed or dissatisfied. Beauty does not take form from the parts of the body, but from the whole of our existence. Our face and complexion are, as we said, the physical manifestation of all that we think and do—an exacting mirror of the soul—and as long as any corner of our mind or heart is unfulfilled, beauty will be elusive.

Modern mind-body science has shown that when we are relaxed and happy, we live in a body that is significantly different biochemically than the one we live in when we are tense, angry, or sad. "To think is to practice brain chemistry," writes physician and mind-body expert Deepak Chopra in *Quantum Healing*, ". . . the mind and the body are like parallel universes. Anything that happens in the mental universe must leave tracks in the physical one." Every thought and emotion takes form in the body as a molecular messenger that travels through the bloodstream to deliver its particular "marching orders"—its intelligence—to every cell, altering our appearance from the inside out. When we are afraid, for example, chemical communicators signal the adrenal glands to release adrenaline. This, in turn, affects the kidneys, resulting in dehydration throughout the body, and also causes the heart to pound as it pumps an increased blood supply to the limbs in readiness for fight or flight. These changes result in the characteristic effects of fear: a dry mouth and blanched complexion. Usually, when the stimulus is gone, the fear subsides and the biochemistry normalizes. However, when we are under constant stress, as most of us are in the 1990s, the body cannot resettle from this state of overdrive, and physical breakdown begins: Constant dryness leads to wrinkles, elevated blood pressure leads to heart disease, and so on. So if you want to change your appearance, you must first change the thoughts, emotions, and habits where stress and aging originate. And if you want to be beautiful, you must first create a whole and happy inner life, so that "every cell in your body learns of your happiness and joins in."

This state of pure unbounded happiness is the source and essence of Vedic beauty. It is known in Sanskrit, the language of Ayurvedic literature, as *sat chit ananda*, or pure bliss consciousness. It is *pure* in that it is unalloyed, self-contained, and nonchanging; it is *unbounded* in that it is not the consequence of any particular event or thing, but exists free of all conditions and limitations in time or space, a self-generated constant of our experience. It is happiness of the highest order, and as you will see, it is the spontaneous effect of a balanced life.

In mythology, which illuminates the universal phenomena of human existence, the archetypal beauties of many cultures share one trait in common: They are portrayed as *radiant*, and this characterization is often key to their role in the cosmic order. David Kinsley cites numerous examples in *The Goddesses' Mirror*. In the sacred poems of Tantra, the beautiful goddess Lakshmi is likened to "the fat that keeps a lamp burning." When she is absent, "the worlds become dull and lusterless and begin to wither away." The Greek goddess Aphrodite, along with her Roman counterpart Venus, embodies womanly perfection in Western tradition. Described as "golden" with "flashing eyes" and a "smiling . . . sunlit sexuality," Aphrodite's supreme beauty, like Lakshmi's, is depicted not as a composite of her features but as a subtle and luminous dimension of her being, an inner warmth that seems to emanate as light from her pores. In one myth, Aphrodite disguises herself as an old woman, but her "eyes that were full of shining" give her away. Her radiance *is* her identity—it is the beauty for which she is known. In an African myth, beauty is a water spirit who incarnates as a "dazzling" woman, a mermaid with "glistening" eyes and a "smooth, glowing forehead." As art historian Sylvia Boone describes her in *Radiance from the Waters*, the spirit is ethereal and sublime—part of an unseen world. However, humans can see her essence in what the culture calls "the *new leaf of* a leaf," the first tender, vivid green sprouting of a plant that is "the visible sign of the surging power of new life." According to common belief, it is this same quality of "brightness and heightened visibility" in a woman's appearance that first draws one's attention to her and ultimately "satisfies the sight."

This brightness, this *radiance*, is absolute beauty—the *nonchanging* value of physical perfection. It is an inner vitality so compelling, and a complexion so lustrous, that our attractiveness transcends all modes of fashion and all popular ideals. It is unmistakable and unforgettable. It is not imaginary and it is not something we can fake for very long, even with the best makeup or plastic surgeon. Once we achieve it, not age or time alone can diminish it. I call this quality the *Tej* factor—the "inner fire" that gives the face a fresh, happy, serene look and the skin, a visible glow. According to Ayurveda, this luster is a normal phenomenon produced by the presence of a very subtle substance called *ojas,* which is a by-product of all healthy body tissue.

Actually, ojas is one of three vital essences. Along with the forces of *prana* and *tejas,* ojas arises out of what Robert Keith Wallace in *The Physiology of Consciousness* calls the "self-interacting dynamics of pure consciousness." Together, these three essences generate and sustain human life (and, as we describe later, also have a role in determining our constitutional type). Prana is the life "breath"—the equivalent of *chi* in Chinese medicine, or what many cultures refer to as *spirit*. Tejas is the metabolic "fire," or transformative energy, from which the Tej factor gets its name. Ojas, which literally means "vigor," is the sustaining and protective life force and the essential substance of the body.

In Western terms, ojas is associated with the protoplasm, which biologists regard as the basic living substance of all cells. Ojas is not protoplasmic matter itself, however, but a subtler essence of it. In Ayurvedic terms, it is the *unifying* power of consciousness, the cohesive life force that fuses matter and intelligence to create the first biological substance out of inert molecules. In other words, ojas is consciousness in its subtlest biological form; it is the sap of life, and according to Ayurveda, the source of the body's immunity and strength.

The biosynthesis of ojas depends upon the balanced functioning of the other two forces. Prana, the moving life force, is literally the pulse of creation. Like all pulsating energy, it naturally releases heat and light. This radiant energy is tejas, which is the source of the body's intrinsic warmth (body temperature), as well as metabolism and digestion. In its form as the digestive fire, tejas is also called *agni*, and it is responsible for breaking down the raw nutrients in our diet. According to Ayurveda, these nutrients include not just food substances, but the emotions and perceptions that constitute our mental "diet." In this sense, tejas is both the transformative power of consciousness and the "digestive" power of the senses and intellect. When prana and agni are in balance, the body tissue produces the superfine essence of ojas. If prana and agni are too high or too low, however, ojas is "burned" up or depleted. In other words, the body's life-sustaining intelligence becomes disturbed and the tissues lose their vigor. On the other hand, the radiant quality of the skin—the Tej factor—is a perceptible sign that these three forces are in dynamic equilibrium—prana is in good supply, the fires of tejas are strong, and ojas is circulating adequately throughout the body. This "frictionless flow" of unifying consciousness—of ojas—is the essence of life; its stagnation is death.

Just as the outer effect of flowing ojas is radiance, the inner effect is the sensation of pure bliss. When raw ojas becomes "refined" on the level of consciousness, it assumes the form of *soma*, or what the rishis called "the nectar of the gods." When the "taste" of this nectar is "digested" in consciousness, the mind experiences bliss. The charismatic look of ease and contentment that is the hallmark of a truly beautiful face is not an affectation or an attitude, but the direct reflection of this blissful state. Vedic texts describe soma as a heavy, cool, soft, clear, sticky substance with the pale yellow color of clarified butter—the color of "brilliance"—the taste of honey, and the sweet smell of *laja*, a type of prepared rice. As we've seen, these qualities of clear, golden, sweet, lubricating, liquid brilliance are similar to those appearing in classical descriptions of beauty around the globe.

Although soma cannot be found in a test tube yet (its form may be too subtle for detection by the strictly material means of modern biological science), contemporary researchers in Ayurvedic medicine theorize that it may be related to two hormones: *seratonin*, well known for its calming effects on the body and mind, and *melatonin*, a secretion of the pineal gland, located at the brain's center. Before the recent discovery of melatonin, scientists knew little about this small gland, or even whether it had more than a vestigial role in the body. Now scientists link the pineal secretion with a remarkable range of functions, including the regulation of biorhythms, improved immune efficiency, and increased resilience under stress. The only hormone known to have antioxidant properties, melatonin also seems to be a key agent in the body's fight against free radicals, the molecular scavengers that break down healthy cells and start the process of aging and disease. Scientific studies have also shown that the level of these hormones tends to increase in the body during the practice of meditation. As you will read in Chapter 13, Ayurveda has used meditation techniques for six thousand years to strengthen biological ojas and cultivate the experience of bliss, or soma, in consciousness.

Ojas in its form as soma is also known as the "glue" of the universe, the "natural affinity of forces," as Vedic scholar David Frawley explains, without which elementary particles would not bond to form the building blocks of matter. "This cohesiveness [is] seen not only as a chemical property; it also reveals a conscious intent. It manifests the power of love. . . . the real force that holds all things together." This Ayurvedic principle is evident in the glowing face of every bride and groom: Soma is the biological substance of love. As Frawley describes the principles of prana, tejas, and ojas, "The energetic principle (life), possesses radiance (light), which in turn has a bonding power (love). We must ever seek greater life, light and love as this is the nature of the universe itself."

In this sense, the desire for beauty is not merely a primal urge for self-preservation, but an expression of our highest evolutionary impulse: the wish to know our own universal nature, the absolute value of life, light, and love that resides in individual awareness. Although this is the source of all life, human beings are the only creatures conscious of our innate consciousness; we are the only creatures who know that we know. This unique self-reflective capacity therefore gives us a sublime purpose, even a responsibility, in the natural scheme: that is, the goal of understanding our own nature—the nature of consciousness—which Ayurveda calls *Self-realization*.

As the essential "stuff" of *all* nature, consciousness is the essence of *our* nature; it is the "I" in "I am," the "whole" of our whole*ness*, the basic "stuff" of our identity, our *Self*. As the unifying element of the universe, it is one*ness*. As the source of all things at all times, it is infinite and eternal—the field of all possibilities. Ayurveda says: The Self is unbounded. Lacking nothing, it is complete and utter *fullness*, and the inner experience of fullness is *pure bliss*. The direct experience of consciousness, then, is the ultimate source of radiance and beauty. All of Ayurvedic practice is directed at enabling each of us to achieve this supreme level of being through the daily routines of life.

BLISS: THE PRESCRIPTION FOR HEALTH AND BEAUTY

"There is no cosmetic for beauty like happiness."
LADY BLESSINGTON

"From Bliss, indeed, all these beings originate;
by Bliss they are sustained; towards Bliss they
move; into it they merge."
TAITTIRIYA UPANISHAD

As we've seen, the flow of ojas in the body and mind gives rise to the radiance and bliss that are the basis of ageless beauty. But how does the body produce ojas and how do we keep it properly flowing?

Ojas is the end product of healthy development of the body tissues, known in Ayurveda as the seven *dhatus*: plasma, blood, muscle, fat, bone, bone marrow and nerve tissue, and reproductive tissue. These bodily structures develop sequentially, one from the other, so that each dhatu, starting with the first, provides the raw materials, including raw ojas, for the transformation of the next—plasma gives rise to blood tissue, blood to muscle tissue, and so forth. That means the health of each tissue depends upon the health of the one that precedes it. Once a problem develops along this production line, the functioning of all subsequent dhatus, and hence the production of ojas, is compromised. Such problems develop when either prana or tejas is imbalanced, as we said. When metabolism is normal, however, ojas is properly distilled at every stage of tissue formation until it finally appears as the fine essence of the final dhatu, the reproductive tissue and its vital fluids. In this refined form, ojas circulates via the heart and other subtle energy channels to maintain the life span of cells and to strengthen the body's resistance to disease. Thus, ojas is not just the source of radiance and bliss; it is, according to Ayurveda, the source of our natural immunity—the cause and effect of perfect health. For this reason, beauty and health—or beauty and *wholeness*—necessarily go hand in hand.

We measure the presence of ojas by observing specific functions and effects in the mind and body. As the essential energy of the immune system, it is naturally implicated in the etiology of all disease, and Ayurveda diagnoses and treats many specific disorders on the basis of its condition. Problems can result from either tissue dysfunction, improper distribution of ojas (obstructed or dislocated flow), or loss of ojas due to poor diet or routine. A person whose skin is pale, wasted, rough, prematurely aged, or unusually prone to allergic reactions has obvious symptoms of low ojas—since a weak immune system means weak ojas, the complexion naturally looks dull when we are sick. Other physical signs of impaired ojas include (but are not limited to) loss of strength, hypertension, swelling, emaciation, dehydration, osteoporosis, liver problems, mononucleosis, rheumatic fever, lupus, Epstein-Barr syndrome, and many skin diseases, including eczema, acne, psoriasis, and dermatitis. Mental symptoms of impaired ojas include fear, worry, hopelessness, lack of patience, and disturbances of the senses.

From a Western medical perspective, this symptomology looks like a meaningless catch-all of problems and complaints. In fact, many ailments that are inexplicable by Western standards—including chronic pain, depression, and fatigue; mysterious, lingering infections and nervous disorders; and degenerative diseases—frequently are due to impaired ojas. In the Ayurvedic view, depleted ojas is even the cause of autoimmune diseases, including AIDS. The difference in views stems from the fact that allopathic medicine concentrates on fighting acute forms of illness and does not have a well-defined standard of individual health. In contrast, Ayurveda is a science of longevity and immunity whose first aim is to maintain balance and overall well-being. As such, it is concerned with *any* condition, mental or physical, that falls short of perfect health, and it has a diagnostic system far broader and more refined than its modern counterpart.

This system breaks down the disease process into six distinct stages, and classifies in terms of their origins even such "vague" symptoms as fleeting moods and sensations, which most Western doctors would overlook altogether, or dismiss as mild stress or even hypochondria. Indeed, while allopathic medicine typically goes into gear only in the last three stages, when disease is already full-blown, Ayurveda detects and remedies problems even in the first three stages, when symptoms are still invisible to the Western eye. Consequently, modern medicine has little to offer the patient who has simply lost her glow. Ayurveda, on the other hand, pays careful attention to signs like the luster of the skin as a measure of health or imbalance, and offers specific treatments, as you will read, to replenish ojas and restore radiance.

Any factor that disrupts the forces of prana and tejas—of life energy and metabolism—or otherwise causes imbalance to our innate constitution will hinder tissue development and thus impair ojas. A diet or routine incompatible with our makeup; alcohol; cigarette smoke; drugs of all kinds; canned, processed, and preserved foods (which are low in prana); poor sleep; poor digestion; improper breathing; physical trauma; intense and prolonged grief, anger, frustration, or anxiety; overexposure to bacteria, viruses, parasites, sun, wind, or cold—all destroy cellular intelligence, and as a result, cause ojas to become depleted or degraded.

Ojas is naturally highest at birth and diminishes with age, which is why the very young look so effortlessly vibrant, even though their individual lifestyles already may be imbalanced. It takes time and considerable abuse of the body to completely destroy our innate reserve of ojas. On the other hand, unless an imbalance is of a serious order and has persisted for a long while, we can replenish ojas by using specific Ayurvedic remedies, including oil massage, meditation, herbal tonics, milk, ghee (clarified butter), almonds, and other dietary aids to improve digestion (agni) and elimination. Maintaining constitutional balance through right diet and routine, right breathing, right thinking, and right purpose in life also strengthens ojas, produces wholeness and bliss, and thereby creates radiant beauty. This, of course, is the subject of the rest of this book.

BEAUTY IS AN
EIGHT-ARMED GODDESS:
AN ANCIENT IMAGE FOR A
NEW MILLENNIUM

*"To see beauty as the face of love rather than the arbitrary
gift of fortune is . . . to enlarge our sense of life's possibilities."*

ELLEN ZETZEL LAMBERT

*"Give me beauty in the inward soul; and may the inner and
the outer be at one."*

SOCRATES

Despite the plethora of celebrity icons and the cult of the supermodel, we have in the West no common symbol that captures the all-encompassing perfection of radiance—the beauty that starts with the soul. So before we continue, I want to offer as a possibility one beautiful image that I have carried with me since childhood. It is an image I received from my mother, who taught me the most important beauty lesson of my life when I was a young girl and just starting to develop a personal interest in my physical appearance.

To my mother's consternation at the time, I spent long periods each day looking at myself in the mirror, and an equal amount of time observing others. I loved to watch the women around me perform their daily skin care rituals, and I was as intrigued by the natural herbs and oils that each one used as I was by their individual beauty. I would question everyone to find out what they mixed in their concoctions—store-bought soaps and creams are rare in Ayurvedic households—and then experiment with their formulas, adding or subtracting various ingredients in the hope of achieving new effects. Of course, I would then insist that everybody try the preparations so I could study the results. Even for a teenage girl, my fascination was unusual, but it was genuine scientific curiosity as much as vanity that prompted it. No one was surprised years later when I took my college degree in botany and chemistry.

I was still in that self-conscious stage, however, when my mother taught me the real meaning of beauty. As I said, I was always staring in a mirror to examine the changes in myself as childhood gave way to puberty. My feelings about my emerging femininity went to extremes. Pride and excitement one instant crumbled into awkwardness and embarrassment the next. The only constant in my mind was the desire to be gorgeous—and for me, as for most girls, wearing makeup seemed the epitome of beauty.

When I finally got permission to use cosmetics, I was thrilled. As my mother looked on, I carefully applied my first lipstick and blusher, then stared at the results. With just a few strokes of color added to my features, I seemed, in my own eyes, quite sophisticated and lovely—and my conceit must have been apparent. At first, my mother complimented my new look. Then she turned to me and said, "So, Pratima, you are growing up. You are becoming a woman now and realizing the powers that you have. But this is only the beginning. There is much more to being beautiful than just your physical appearance."

These were surprising words coming from someone as stunning as my mother. As I watched her leave the room, I wondered what she meant. In a few minutes she returned with my answer. In her hand was a portrait of an eight-armed goddess who was adorned in jewels and sitting serenely atop a fierce tiger, as three other goddesses stood in miniature at her feet. I recognized the image immediately from the altar where my mother prayed each day, but I knew little of what it symbolized, except that the goddess, like her, was obviously radiant.

"This is Chymunda," she explained as she handed me the picture. "She has eight arms because she embodies within her own form three other goddesses, three aspects of a woman—Saraswati, who represents knowledge; Kali, destroyer of evil, who represents courage; and Lakshmi, who represents wealth and prosperity. It takes wisdom, courage, and power to ride the back of a tiger. Whatever you choose to do in life, you must have knowledge to do it well, and courage to overcome the challenges in your path. With knowledge and courage, you have the power to achieve abundance in all things. These qualities together are the foundation of bliss.

"If you really want to be beautiful to the full extent of your womanhood," my mother told me with great love, "then you must develop all these aspects of your Self, Pratima, not just your appearance."

You do not have to worship Chymunda or believe in any deity at all to benefit from my mother's advice. What is essential is to recognize the universal feminine principles embodied by this goddess. In the coming millennium, her multidimensional beauty is a much more fitting and complete symbol of womanhood than the cover girls and superstars we have been conditioned to idolize in twentieth century Western culture.

Without an effective method to achieve this quality of life, however, the lesson is utterly meaningless. Like my mother, you can fulfill its promise through the practice of Ayurvedic beauty.

CHAPTER 2

BALANCE:
THE PATH TO HEALTH
AND BEAUTY

*No one can be happy or healthy in an unbalanced
state, because it is just not natural.*

DEEPAK CHOPRA

As we've said, Ayurveda states that everybody and everything is composed of five fundamental elements—five vibrational modes—called space, air, fire, water, and earth. Each of us, like a microcosmic universe, is born with all five present. However, everyone has them in a unique proportion that determines our individual features, including our skin type, and our inherent mental and physical tendencies. The secret of absolute beauty is to live in harmony with nature's design for us as expressed by our personal mix of elements—that is, our constitutional type.

This is easy to do once you know the basic theory of balance and your Ayurvedic skin type, which you will learn in this chapter; and also the factors that cause imbalance, which you will learn in Chapter 3. We begin here with a look at Ayurvedic "psychophysiology" and the basic principles of balance, including:

- consciousness
- the forces of evolution, or *gunas*
- the elements
- the five senses
- the biological forces, or *doshas*
- the constitutional types, or *prakritis,* including the skin types
- the law of balance known as *like increases like*
- the Ayurvedic "anatomy"—body, breath, mind, ego, and consciousness

To bring these ideas to life, we conclude this introduction to Ayurveda with a personality portrait of each prakriti. With this understanding, you will be ready to take a simple quiz to find out your basic constitution and skin type, and whether it is balanced or imbalanced at this time. If there is an imbalance, you will have an opportunity in Chapter 3 to find out what it is and how Ayurveda works to correct it. If there is no imbalance, you will discover how Ayurveda works to prevent it. Then you will be prepared to select the appropriate daily regimen for your skin, and begin your own journey to absolute beauty using the many treatments and ideas you will find in the rest of the book.

Given the full scope and complexity of this ancient science, however, we are not suggesting that you will be ready to cure every ailment with what you are about to learn. The whole of Ayurveda covers eight medical fields: internal medicine; general surgery; surgery of the head, neck, eye, ear, nose, and throat; pediatrics; toxicology; psychiatry; obstetrics, gynecology and fertility; and gerontology. Unbeknownst to most Westerners, a number of surgical procedures used today in American hospitals, including rhinoplasty—the common "nose job"—were first described in these ancient writings. On the other hand, both Ayurvedic gerontology, which is primarily the science of rejuvenation (including techniques to reverse aging), and the Ayurvedic science of fertility, which focuses on maintaining sexual vigor, have no exact Western medical equivalents. Indeed, the complete collection of Ayurvedic texts rivals any library of contemporary medicine. There is a huge volume dedicated solely to the uses of the neem plant, for example, which is one medicinal herb among hundreds in Ayurveda's natural pharmacy—and which, by the way, has a tested track record many thousands of years older than any drug approved by the FDA. Furthermore, all Ayurvedic physicians take seven years of medical training at the university level; and for some doctors, Ayurvedic study is a life work that begins in childhood. In traditional Indian families, the knowledge is often passed down in a formal manner from generation to generation—not as a "folk medicine" but as the family duty, or *dharma*—much as other great arts have been kept alive throughout history.

Ayurveda functions on two levels: as a system of prevention and self-care for the lay person, as well as a medical science for trained physicians. It is, after all, the "science of everyday living," and its basic principles are inherently simple and easy to apply to ordinary experience. As you will see, all the essential data needed to determine your skin type or imbalance can be gathered directly by the senses, so you don't need any special tools to make an accurate assessment of your general condition. Moreover, Ayurveda is concerned with identifying the underlying nature of the imbalance rather than with diagnosing a specific illness, so you don't need any special knowledge of anatomy, biology, virology, or other areas of study.

To diagnose and treat most common complaints through Ayurveda, you only need to know yourself—that is, you need to know your basic body or skin type and to pay attention to the characteristic signals that your physical or mental state has deviated from this innate norm. Once you know these basics, you can easily observe even the early subtle signs of disorder and often correct your personal health and beauty problems—or prevent them altogether—by using a few home-made remedies and by moderating your daily routine according to the simple guidelines set forth here.

Of course, many of the concepts that follow will be totally new to you; and some of them, such as consciousness and the quantum field (with which we begin our discussion below), may seem surprisingly complex, especially for a beauty book. We come back to these ideas in later chapters, so you do not have to grasp them all at once. In addition, we include a glossary of Ayurvedic terms for easy reference at all times. Nevertheless, in this chapter you can hope to learn not only a simple system for determining your skin type and keeping it balanced, but also the principles that underly this system. You will learn as well the overarching philosophy of Ayurvedic medicine, which is fundamentally different from the classical Western view. This philosophy is built upon the premise that consciousness is the unifying essence of mind and body, and in this chapter you will discover its preeminent role in health and experience, according to Ayurveda. To provide a contemporary context for this ancient knowledge, we also include some key insights into consciousness provided by modern quantum science. If you give these ideas a little time and live with them, so to speak, as you read the book, you will discover a whole new way of seeing your skin, yourself, and the world. This "quantum" perspective of life is the secret of absolute beauty, as you will see in Part IV. What is important here, however, is just to grasp Ayurveda's central notion of the mind-body connection in the process of healing and balance, and to realize that Ayurvedic skin care encompasses more than herbal remedies.

CONSCIOUSNESS

Earlier we said that the physical universe comes from the five elements. But where do the elements come from? According to Ayurveda, the source of all creation—all that we know and all that we do not yet know—is the unseen, ever-present, infinite field of pure consciousness. This is the ultimate reality, and any system of knowledge that does not include the study of consciousness is, in the Ayurvedic view, an incomplete system. Moreover, an individual life that does not include direct experience of this fundamental field is an incomplete life. In this sense, consciousness is not just the source of creation; it is also the goal of knowledge and experience—the goal of human existence. The ultimate purpose of Ayurveda is to provide everyone the means to this fulfillment, which is also the essence of absolute beauty.

This principle—of consciousness as the source of all creation and experience—is the most fundamental of Ayurveda. Yet consciousness itself is so misunderstood, and the term so loosely used in Western culture, that it is worth spending some time here to clarify what it means in the context of this book. We will look at the idea from three points of view: from common experience, from the experience of the ancient rishis, and from the understanding of quantum science.

CONSCIOUSNESS AND THE INDIVIDUAL

In our own experience, pure consciousness is found at the source of thought. Normally, however, we do not notice it because our attention is absorbed in our ideas, perceptions, and feelings. We see a bird outside the window or an attractive person on the street. A sound pricks our ears, and we register the sensation. Or we become engrossed in conversation or lost in an inner world of daydreams or memories. Sometimes, we fall in love and become lost in another person. Yet, in fact, *we* are not ever lost. We do persist. We continue to follow the daily routine, and a part of ourselves is constant throughout. No matter what changes, there is an unseen bed upon which the stream of events flows, and which connects all the separate points into one whole experience that we recognize as "my life." That unbroken aspect of existence—the "always" and "ever" aspects of the *always* present, *ever*-changing stream—is what we call consciousness.

Although it is abstract—indeed, it is pure abstraction—the experience of consciousness itself is more familiar than you may realize. Consciousness is literally what makes the difference between your sense of self when you are in deep sleep—which, of course, is *no* sense at all—and when you are wide awake. In that fleeting gap between the last moment of dull slumber and the first glimmer of a conscious thought is a direct taste of conscious*ness*—the experience of inner wakefulness, or lively attention, *absent the boundary of any object of experience*. It is the state of pure knowing*ness* in which your mind recognizes its own unbounded nature and remembers, "I *am!*" In other words, with the attention unengaged, the *knower* simply knows herself. For this reason, the sages also refer to consciousness as Pure Existence, or Being, or simply the Self.

Of course, consciousness underlies not only the first thought of the day, but all mental activity, all perception, and all sensation—whether conscious, subconscious, or unconscious. Therefore, we can experience it in the gap—in the silence—that separates each and every impulse of thought we have. And, as you will see later, Ayurvedic meditation provides a technique to let us slip through this crack, so to speak, and experience the absolute silence of the mind directly and systematically. As the source of all matter and energy, consciousness is also the source of all intelligence in the universe—the orderliness and regularity we see in the cosmic architecture. Consequently, the direct experience of consciousness in meditation is a way to realign our intelligence with nature's organizing power.

Naturally, the question arises, if consciousness is the basis of all experience—if it is ever-present—why don't we notice it all the time? Why do we need to do anything special to experience it? Ayurveda answers in this way: In ordinary activity, as we saw, the Self is masked by a perpetual stream of mental images just as, in the cinema, the white light from the motion picture projector is masked by the film images. We constantly project our awareness—our inner "light"—outward through the senses, and with our attention thus focused on external objects and events, we lose the experience of pure Being. In truth, we could not have life without consciousness any more than we could have the movies without light; yet in both cases the *source* of reality is literally overshadowed by the object of perception—the unbounded value of awareness is obscured by the countless individual sights, sounds, smells, tastes, and feelings that capture the attention *as long as we only project it outward*. In this way, the infinite nature of the Self is lost in a thought!

Ayurveda refers to this phenomenon as *prajnaparadha*—the "mistake of the intellect." Indeed, logic tells us that something finite cannot obscure the infinite. The very idea is as comical as a game of peekaboo with a toddler who truly believes he becomes invisible when he hides his face in his hands. Yet we accept this illusion as the reality of life, and in so doing, we "forget" our Self. This illusion is not simply a mistake in logic or discrimination, however; it is the loss of *intelligence* itself, or what Frawley calls our "natural wisdom." In fact, Ayurveda states that all imbalance—all disease and disorder—is nothing but the loss of this natural wisdom, which is part and parcel of consciousness. Ayurvedic therapies work to correct the "mistake" by reawakening the mind to its unbounded value of existence in consciousness, and by creating a lifestyle in tune with the body's own intelligence.

The knowledge of how to do this originates in *Veda*.

CONSCIOUSNESS AND VEDA

Veda is the Sanskrit word meaning knowledge. Although Ayurveda is described in Vedic texts, Veda itself cannot be bound within a book because it is not the knowledge of facts, things, or ideas. Rather, it is knowledge through direct experience of the abstract and all-inclusive field of consciousness itself. Veda is *pure* knowledge.

The Self-realized men and women known as the ancient Vedic rishis experienced pure consciousness not only as their own subtle awareness, but also as the essence of everyone and everything around them. As a result, these seers lived in a state of *unity*—in every aspect of their existence, in every thought, perception, and action, there was no separation between their individual nature and Nature itself. This *unity consciousness* is the source of the often-quoted phrase, "I am That, thou art That, all this is That"—*That* being consciousness. The expression, which comes from Veda, is a subjective statement, because the rishis were speaking from their direct experience. They perceived all creation as an unbounded wholeness, indivisible from their own Being. And because no part of the universe exists outside of this wholeness—that is, no part of life exists outside of this unified field of consciousness—therefore nothing in the universe was unknowable to these enlightened individuals. Their own awareness permanently established in this sublime state, the rishis were able to cognize the abstract laws of nature—the patterns of intelligence within consciousness—that underlie and give dimension, direction, and order to the created universe. This includes everything from the most durable forms of matter to the most elusive forms of energy, including the fleeting thoughts and indelible memories of the human mind.

The classical Indian texts known collectively as the Vedic literature are the written expression of their cognitions. The texts themselves comprise the formula, so to speak, for the unfolding of consciousness from its unmanifest form through every stratum of creation. In the same way that the intelligence that structures human life is encoded in a strand of DNA, the total intelligence of the universe is encoded in Veda. No amount of objective analysis will yield the ultimate "secret" of creation, however, because consciousness itself is pure subjectivity—it is the essence of "I am." To decipher the Vedic code, we must, like the rishis, go beyond the intellect, mind, and ego directly to its source—which is our source—in consciousness.

Quantum physics, which is concerned with the existence and behavior of the fundamental particles of matter, appears to confirm this truth. As you will see, the closer we come through the objective means of modern science to knowing exactly what the world is made of, the more evidence we find that it is not made of anything (any "thing") *but* consciousness. The missing piece in the Western scientific puzzle—how abstract "mind" can affect or become physical matter—is found in Veda.

CONSCIOUSNESS AND
THE QUANTUM FIELD

Quantum mechanics is the branch of physics born at the dawn of the twentieth century, the brainchild of a family of great minds that included among others Einstein, Heisenberg, and Max Planck. In contrast to "classical" physics (the science of Newton, which describes the motion of bodies in the large-scale universe), quantum mechanics describes the interactions of force and matter in the ultra-small world, where an atom is astronomically large compared to the electrons, protons, neutrons, quarks, bosons, and other elementary particles that inhabit this realm. To get a sense of the quantum scale, consider the fact that a single quantum particle is tens of millions of times *smaller* than one small atom—and that it takes a few million atoms to add up to the thickness of a page in this book. To get an idea of the forces within this microcosm, consider the fact that the attraction between two quantum particles is *many millions of millions* of times stronger than the pull of gravity.

Obviously, this is not the world of our daily experience. At the borders of the small-scale universe, Newton's laws—the laws of the sensate world—break down. We cannot play billiards in the quantum realm because there are no dependably solid objects there—only packets of energy, or *quanta,* that flash in and out of existence and sometimes behave like particles and sometimes like waves, demonstrating the truth of Einstein's famous theorem that matter and energy are equivalent. The *quantum field* is the underlying force field—the unmanifest and omnipresent continuum—from which these quanta come and go. *It is the source of all energy, intelligence, and matter in the universe.* We described it in the introduction as a "virtual" field because we cannot directly detect it by the senses or by any technological means. Indeed, physicists have never actually "observed" it, but hypothesize its existence based upon certain effects they do observe in the subatomic world.

As you will see below, in the past half century, scientists have made many interesting discoveries about the behavior of quanta. However, none is more important to the discussion of Ayurveda and consciousness than Heisenberg's finding in 1926 that our measurement—that is, *our perception*—of quantum events influences their results. "The uncertainty principle had profound implications for the way in which we view the world," says noted physicist Stephen Hawking of this discovery of the effects of consciousness on physical phenomena. "Even after more than fifty years they have not been fully appreciated by many philosophers, and are still the subject of much controversy." As physicist and philosopher David Bohm poses the problem for Western science: "The question is whether matter is rather crude and mechanical or whether it gets more and more subtle and becomes indistinguishable from what people have called mind."

Of course, for the Vedic scientist, the answer has never been in doubt. The field of pure consciousness and the quantum field are the same thing: the fundamental reality from which all life is formed. As such, this virtual field is the source of all the forces and intelligence of nature. It is called the "home of natural law" and the "field of all possibilities." As far as the rishis are concerned, this is not hyperbole, but simple truth. When we live the full potential of consciousness, *nothing is impossible.*

Now Western science apparently is coming face-to-face with the same truth. Indeed, in their quest to find the fundamental constituent of matter, physicists have chased Einstein's theories through Newton's universe only to fall down a rabbit hole and land in a quantum wonderland. In this universe, they have found that elementary particles can pass through solid objects with the apparent ease of a ghost through a wall, for example. They have also shown that two particles separated by astronomical distances can instantaneously respond to each other's experiences, demonstrating the quantum equivalent of telepathy. And in May 1996, *The New York Times* reported that physicists at the National Institute of Standards and Technology in Boulder, Colorado "proved that an entire atom can simultaneously exist in two widely separated places." That is, science has outdone science fiction by putting the same object in two places at once—and then making, as the *Times* described, "a kind of movie" of the process!

Clearly, the world becomes curiouser and curiouser the deeper physicists reach into the small-scale universe. Vedic scientists, who have already mastered the field of all possibilities, know that one day their Western counterparts will reach the limits of objective analysis and come around to the Vedic point of view—that the ultimate constituent of the universe is found not outside us at all, but deep within our own consciousness.

THE FORCES OF CREATION AND EVOLUTION

Within the field of consciousness are three forces that set creation in motion. They are known in Ayurveda as the three *gunas*, or the fundamental attributes of the universe. *Sattva* is the pure essence of reality, what Chopra has called "the impulse to evolve"; *tamas* is the force of inertia, or "the impulse to stay the same"; and *rajas* is the force of change or movement, or "action for its own sake." The gunas are also referred to respectively as the creative, maintaining, and destructive principles of nature. Their interaction within the field of consciousness is what generates and sustains *evolution*—that is, the progressive development of life. All progress—all growth—requires activity, which is the function of the gunas. As Maharishi Mahesh Yogi explains in his commentary on the Vedic text, *The Bhagavad Gita,* rajas is the "spur" to action; sattva and tamas "uphold the direction of the movement;" and all together they create the process of growth. He writes:

> For any process to continue, there have to be stages in that process, and each stage, however small in time and space, needs a force to maintain it and another force to develop it into a new stage. The force that develops it into a new stage is [sattva], while [tamas] is that which checks or retards the process in order to maintain the state already produced so that it may form the basis for the next stage.

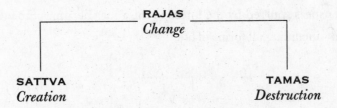

RAJAS
Change

SATTVA
Creation

TAMAS
Destruction

Everything that exists necessarily expresses all three attributes, since no guna can be active apart from the others. However, the influence of one guna may dominate at a given time. Thus, Ayurveda sometimes describes the subtle characteristics of a thing in terms of a particular guna. A food, or someone's mental state or activity, for example, is referred to as *sattvic* when it *predominantly* reflects the qualities of purity and clarity and has a life-giving and harmonious influence. It is *rajasic* when it reflects the qualities of change and disintegration and has a stimulating, distracting, or depleting influence. And it is *tamasic* when it reflects the qualities of inertia and heaviness, and has a dulling, clouding influence. Likewise, the sattvic mind acts in order to progress and create; the tamasic mind tends to be inactive and likes the status quo; the rajasic mind enjoys action for its own sake.

Nevertheless, the whole of creation in its infinite variety arises from the dynamics of the three gunas *together*. When the *one* (the field of consciousness) becomes the *many* (the permutations and combinations of the manifest world), the gunas are the forces that make it happen. In this sense, creation is an ever-evolving expression of the same essential energies—each layer of existence is nothing but consciousness creating out of itself by virtue of the gunas. As you read, keep this idea in mind. It will help you to understand why the principles of Ayurveda often seem to be different variations on the same basic theme. In fact, as you can see, this is true—every phase of creation recapitulates the primal creation of the manifest universe from unmanifest consciousness through the action of the gunas. Despite the differences in the form and function of all things, their ultimate constituent (consciousness) is always the same, as is the force (the action of the gunas) that creates them. In similar fashion, we understand that ice, water, and vapor are the same basic constituent (H_2O) changing expression according to one basic principle (thermal motion). Just as one molecule becomes three different forms of matter by moving at different speeds and collecting in different densities, so consciousness spurred by the forces of the gunas becomes the infinite variety of creation—including all forms of life.

CONSCIOUSNESS TAKES FORM:
THE FIVE ELEMENTS AND
THEIR EFFECTS

What is the body made of? Modern chemistry states that we are composed of 110 basic constituents defined by atomic structure. The average person is a mixture of oxygen, carbon, hydrogen, calcium, and nitrogen, plus fractional amounts of phosphorous, chlorine, sulfur, sodium, potassium, magnesium, and iron, and traces of the rest. The modern biologist states it a little differently: Human beings are roughly fifty-five parts water, nineteen parts lipid, eighteen parts protein, five parts ash, and about one part each carbohydrate, calcium, and phosphorous. The cognitive psychologist says that we are a body, a nervous system, and a brain, which together create the neurochemistry that creates the phenomenon of mind. The quantum physicist, who has come closest to describing all of nature in a single unified theory, states that we are 99.99 percent empty space filled out with a few trillion protons, neutrons, electrons, and other invisible subatomic particles.

None of these Western descriptions, however, accounts for the ingredients of mind and consciousness, or for the life force itself. Moreover, the rules of Western science have been broken, and their factual bases have changed many times over in the course of its unsuccessful search for the ultimate reality. In contrast, Ayurveda's simple classification of the five elements plus the three biological principles and five strata of existence, as you will see, leaves out nothing in creation, not mind or matter, and has provided a complete and consistent framework for understanding life for at least six thousand years.

The five elements are the first objective expressions of consciousness resulting from the interaction of the gunas. They arise sequentially at the switching point, so to speak, between unmanifest and manifest reality. In this process of manifestation, space is the first element to emerge. Although it is physically the "vacuum" state, space is not empty at all; rather, it is filled with the virtual energy of the quantum field. The dynamics of the gunas within space produces air, which is the first state of matter. As air moves, it creates friction, giving rise to heat and light, or fire. Fire liquefies matter, creating the element of water. When the liquid evaporates, the solid matter of earth remains.

In quantum terms, the elements appear as abstract patterns of intelligence— vibratory patterns within the quantum field, as we have said—that create the elementary forces and give order and form to the world we see. In Ayurvedic terms, the elements have an abstract form within the *ultimate* and *least* subtle layer of consciousness, or "mind," and a concrete form within the *first* and *most* subtle layer of matter. As such, they have a subjective as well as an objective role in human experience. On the level of individual consciousness, the elements give rise to the five senses; on the level of the body, they combine to create all the forms and functions of the psychophysiology.

According to Ayurveda, we can understand all experience in terms of the interaction of the elements. As the basic energetic forms of the universe, they have definite qualities and exert a predictable influence upon the way we look and think, as well as upon the way we are affected by the environment. We perceive, act upon, and experience the external world by virtue of the elements. They are the silent vocabulary we share with the universe. As you will see, understanding their energetic qualities and effects is the key to understanding our own inner nature as well as our place within nature. This, in turn, is the basis for achieving balance and beauty.

THE ELEMENTS IN MIND AND MATTER

In their subjective aspect—that is, on the level of individual consciousness— the five elements give rise to the five senses. Space is the essence of sound. Ayurveda describes the first faint stirring from the field of consciousness—the first emanation of creation—as *primordial sound*. Sound is vibrating energy—energy moving in waves. Without space, energy would have no place to move, and physical existence would be impossible. Air is the essence of touch; its existence makes vibrating energy palpable. When we feel vibrations from a bass instrument, for example, what we are feeling is the movement of air displaced by sound energy. Fire is the essence of sight. Radiant energy—light—is what makes matter visible to the eye. Water is the essence of taste; without it, the tastebuds do not work. Earth is the essence of smell.

Through the five senses, each element is also associated with specific body organs and functions. Obviously, we react to the elemental world through the ears, skin, eyes, tongue, and nose. However, the active organs and their sense-related functions are not so self-evident: The throat, mouth, and vocal cords give voice to sound, so they are related to space. The hands hold and touch things, linking them to air. The feet move us in a known direction by virtue of sight, linking them to fire. The genitals are known in Ayurveda as the "lower tongue," linking them to taste, and thus water. And through the excretory functions, the anus is linked to smell, and thus earth.

	SPACE	AIR	FIRE	WATER	EARTH
SENSE OBJECT	Sound	Touch	Vision	Taste	Smell
SENSE ORGAN	Ear	Skin	Eyes	Tongue	Nose
ACTIVE SENSE	Speech	Holding	Walking	Procreation	Excretion
ORGAN OF ACTION	Throat, mouth, vocal cords	Hands	Feet	Genitals	Anus

In their objective aspect, the elements give rise to all forms and functions of the animate and inanimate world. For example, air, water, and earth correspond respectively to the three states of matter: gas, liquid, and solid; fire is the power that transforms matter from one state to another; space is the "ethereal" element, the hollowness in which all matter exists. Air, fire, and water are the active, or mobile, elements; space and earth are inert. Each element also is associated with specific aspects of human anatomy and physiology. Space creates the body's hollow spaces; air creates all bodily movement; fire, heat and body temperature; water, all fluids and secretions; and earth, the solid structures like skin and bones.

The significance of these associations between the various elements and the specific aspects of the body, mind, and environment will become more apparent when we discuss the causes and treatment of imbalance and disease in later chapters. In the meantime, consider as an example the links between the element of fire, the sense of sight, and the phenomena of heat and light. Ayurveda teaches that a person whose innate constitution is primarily fire will tend to have intolerances to heat and bright light. Likewise, too much heat and sunlight can cause an imbalance of fire, and fire imbalance is a leading factor in most vision problems.

THE ELEMENTAL ATTRIBUTES

The properties of each element are just what you might guess from common experience: Space is expansive; air is light, moving, cold, and rough; fire is hot, slightly oily, mobile, light, and penetrating; water is soft, moist, and fluid; earth is heavy, oily, cold, and dense. These are just a few of their manifestations (a complete list appears below). In fact, anything we experience through our senses takes its form and qualities from the elements. As you become familiar with their individual attributes, you will begin to recognize their pervasive effects everywhere, even in your own feelings and behavior. How strongly each element reflects in your life, however, will depend upon your own unique balance of them. As you will see, you can determine your basic constitution and skin type by which elemental qualities are most evident in your body and temperament.

SPACE	AIR	FIRE	WATER	EARTH
Expansive	Thin	Hot	Moist	Heavy
Dry	Dry	Slightly oily	Fluid	Oily
Light	Moving	Penetrating	Adjustable	Dense
	Cold	Sharp	Soft	Cold
	Rough	Light	Heavy	Moist

THE DOSHAS
AND THEIR EFFECTS

The elements themselves are inanimate, but in combination they give rise to three biological forces, or principles, called *doshas*. Together, space and air create *Vata* dosha, or the "air" principle; fire and water create *Pitta* dosha, or the "fire" principle; and water and earth create *Kapha* dosha, or the "earth" principle. Thus, the five basic elements form three doshas:

Each of us naturally has a unique proportion of the three doshas, just as we have a unique proportion of the five elements. This proportion determines our body and skin type, as you will see.

THE BIOLOGICAL FUNCTIONS OF THE DOSHAS

The doshas are not physical by any Western understanding of the term. Like the elements that compose them, the doshas are patterns of intelligence that exist in the gap between mind and matter—that is, in the underlying field of consciousness—and consequently they govern mental as well as physical functions. As you will see, our personality, temperament, and intellectual capacities owe themselves to the effects of the doshas, just as our physical traits do.

The literal meaning of dosha is *impurity*. The doshas are impurities in the sense that, as the first biological forces arising in the manifest field of creation, they are no longer the pure essence of consciousness, but a dilution of it. In fact, the doshas are a grosser—that is, more expressed—aspect of the three fundamental life forces of prana, tejas, and ojas, which we described in Chapter 1. The essence of Vata is prana, the life pulse and moving force. The essence of Pitta is tejas, the transformative, radiant energy. The essence of Kapha is ojas, the cohesive force. As the expression of these forces on the level of individual existence, the doshas naturally have similar properties.

That which moves things. Vata exists as *movement* in the body and is visible in the action of the breath and lungs, the nerve impulses, the heartbeat, the expansion and contraction of muscles, the transport of nutrients and wastes, the microscopic movement of cells, and even the natural urges such as blinking, sneezing, spitting, and belching. In fact, Vata controls the movements of the doshas themselves—we say that it "leads" Pitta and Kapha, which are immobile in themselves—and thus controls all the functions and activity of life. Vata is the flow of life, and if it is blocked, the process of decay begins. On the other hand, if activity becomes frenetic, we become "spaced-out" and fatigued. Like the element of air which it contains, Vata expands into empty space and fills up the hollow cavities and energy channels of the body. It is the source of inspiration, positivity, and freshness, as well as fear, nervousness, gas, spasms, tremors, and pain.

That which digests things. Pitta exists as *metabolism* in the body and is responsible for body heat and temperature, skin coloration and luster, and the transformative actions of digestion, absorption, assimilation, metabolism, and all biochemical reactions. It is found in all acid forms in the body, including enzymes and hormones. Pitta is also responsible for the processing functions of mind, intellect, and senses. Good health depends upon Pitta's capacity to fully metabolize the nutritional, emotional, and sensory information we ingest. If its fires become too hot or too cool, normal digestion is disrupted, toxins are created, and the skin and eyes lose their glow. Pitta gives us warmth, intelligence, perception, and understanding, but it also provokes anger, jealousy, frustration, hate, burning sensations, rashes, allergies, ulcers, and heart disease.

That which holds things together. Kapha is the force of *cohesiveness* in the body, cementing together all the elements to make the material structures of life. Kapha heals wounds, enables physical growth, imparts strength and stability, and maintains the body's internal environment. Also known as "biological water" (water is the main chemical constituent of the physical body), Kapha is the source of all bodily fluids, including phlegm, plasma, and cytoplasm. It also provides energy to the heart and lungs, and natural resistance to disease. If these "waters of life" are depleted, the immune system becomes weak; if the "cement" becomes too thick (if the mix of earth and water is out of balance), we experience heaviness and blockage. Kapha creates the capacity for love, forgiveness, calmness, and wisdom, but it also creates greed, envy, attachment, laziness, depression, bloating, and obesity.

VATA	PITTA	KAPHA
(Prana)	(Tejas)	(Ojas)
Movement	*Metabolism*	*Structure*
Breathing	Body heat	All tissue substance
Heartbeat & blood flow	Skin coloration	Healing
Muscle action	Luster	Growth
Nerve impulses & blinking	Digestion	Vigor
Cellular motion	Assimilation	Stability
Flow of nutrients & waste	Absorption	Immunity
Sensory & motor functions	Understanding	Resistance
Natural urges	Perception	Lubrication & moisture

Note that we commonly refer to Vata, Pitta, and Kapha in terms of both their biological functions and their distinguishing elements. We speak of energy flow in the body as a quality of Vata, or air; heat as a quality of Pitta, or fire; and substance or solidity as an aspect of Kapha, or earth. We follow this custom throughout the book and use these English and Sanskrit terms interchangeably.

THE SEATS OF THE DOSHAS

Each dosha has bodily *seats*—that is, organs where its energy and action are naturally most concentrated. The large intestine and colon, which produce bodily *gas*—or "air"—are the primary sites of Vata. Vata is also concentrated in the skin (the organ of touch and elimination), kidneys, bones, thighs, and ears. The small intestine and liver, which produce bodily *acid*—or "fire"—are the primary sites of Pitta. Pitta is also concentrated in the gall bladder, the spleen, the heart, the sebaceous glands, the blood, and the eyes. The lungs and stomach, which produce bodily *mucus*—or "water"—are the primary sites of Kapha. Kapha is also concentrated in the head, sinuses, nose, throat and tongue, the lymph glands, pancreas, fat tissue, and joints.

Ayurveda speaks a different language from modern medicine, but in fact the two sciences are not so far apart in their meaning. To compare terms, we could say, for example, that oxygen, carbon, and other gases we breathe are Vata "molecules."

Enzymes, hormones, and energy liberated by chemical reactions in the body are Pitta "molecules." And body fluids, proteins, lipids, sugars, and carbohydrates are Kapha "molecules." What one science describes as too little Vata, Pitta, or Kapha, the other might label "low blood gases," "low estrogen levels," or "low blood sugar." The jargon differs but the message is the same: The natural balance of biological elements must be maintained in order to stay youthful and healthy.

All three doshas are necessary to life, and all disease and disorder result from their imbalanced interaction. This may be due either to the *increase* or *decrease* in any or all of these forces, causing a change in their innate proportions. As you will see, keeping their natural dynamics in balance through a lifestyle in harmony with your makeup is the basis of radiant beauty.

THE ATTRIBUTES OF THE DOSHAS

Doshas themselves are invisible (despite our analogy, you will not find a Vata "molecule" under any existing microscope), but our senses can detect their distinctive effects, just as we detect the elements. As we said, each dosha arises out of a pairing of elements. Naturally, then, each dosha has the combined attributes of its two constituents (see below). Whenever you perceive any of these qualities in yourself, in others, or in the environment, you are seeing the doshas in action.

VATA	PITTA	KAPHA
Dry	Slightly oily	Oily
Cold	*Hot*	Cold
Light	Light	*Heavy*
Mobile	Mobile	Steady
Bitter	Sour-smelling	Sweet
Irregular	Sharp	Soft
Quick	Fluid	Slow
Thin	Pungent	Dense
Clear		Dull
Dispersing		Sticky
Rough		Lubricating

Note that the Vata, Pitta, and Kapha doshas have one attribute each (see italicized words in the chart above)—respectively, dry, hot, and heavy—that belongs *only* to itself. Moreover, each of these one-of-a-kind effects finds its opposite quality—oily, cold, and light—in the two other doshas: *dry* Vata is balanced by *oily* Pitta and Kapha; *hot* Pitta is balanced by *cold* Vata and Kapha; *heavy* Kapha is balanced by *light* Vata and Pitta. These six attributes—three pairs of opposites—are key indicators of the doshas and play an important role in the process of maintaining health and beauty.

OPPOSITES BALANCE:

Dry	*Oily*
Vata	Pitta & Kapha
Hot	*Cold*
Pitta	Vata & Kapha
Heavy	*Light*
Kapha	Vata & Pitta

In fact, if you look down the list on page 40 again, you will see that *every* quality has an opposite effect: thin and dense, quick and slow, and so forth. This phenomenon reflects the dualistic nature of the sensate world. Ayurveda holds that creation manifests out of the unified field of consciousness through the interplay of opposing forces. We referred to this earlier in terms of sattva and tamas, the fundamental forces of creativity and inertia, which are both necessary to direct the force of change, rajas, toward evolution. According to Ayurveda, not just health and beauty, but all physical existence is manifested and maintained through the interaction and balance of unlike energies. The attributes of the doshas, therefore, are the perceptible expressions of these basic forces.

THE DOSHAS AND YOUR NATURE

The doshas are the key to your psychophysiological nature and to your skin type. Since everyone has a different balance of elements, we naturally have a different balance of doshas, too. Depending upon whether your innate constitution has more space and air, fire and water, or water and earth—one pair typically predominates—then we say that Vata, Pitta, or Kapha dosha *leads* the other two in the government of your psychophysiology. Your leading dosha determines your *prakriti*—your "nature"—and therefore all your basic inward and outward characteristics.

The three basic prakritis, or constitutional types, are known as Vata, Pitta, and Kapha, after the leading dosha. As we've said, your prakriti is the formula that describes the natural relationship of your doshas—that is, your natural balance. All three doshas are active in every individual, but what makes each life unique is the particular balance of doshas with which we are born. When we say that a person's constitution is out of balance, we actually mean that there is a disturbance in the innate dynamics among these biological forces. Like genetic traits, the prakriti itself is determined at conception (depending upon the parents' makeups) and remains the same throughout life. What changes, as Chapter 3 explains, is the condition of the doshas due to factors in our lifestyle and environment. This modified condition is known as the *vikriti*.

Our inherent capacities as well as our inherent weaknesses (both physiological and psychological), the natural tendencies of our thoughts and emotions, our physical features, and of course, our skin type, all owe their particular characteristics to the balance of our doshas. As Deepak Chopra has said, each of us lives in a world "colored . . . down to the smallest detail" by our leading dosha.

YOUR SKIN TYPE
AND GENERAL CHARACTERISTICS

It is, we hope, no surprise to you, now that you are familiar with their attributes, that Vata dosha produces *dry* skin, Pitta produces *sensitive* skin—"fiery" skin that is prone to redness and inflammation—and Kapha produces *oily* skin. These qualities refer to the natural tendency of the skin, but *normally the effect is slight.* For example, Pitta skin feels moister and warmer to the touch than Vata skin, which is cooler, rougher, and dryer; and Kapha skin is relatively smoother and more lubricated than either Pitta or Vata. In fact, *all skin types look clear and radiant when the doshas are in balance.* When the doshas are *not* balanced, however, the characteristic effect of each skin type tends to become more pronounced and may develop into a full-blown problem—for example, dry skin may lead to dry eczema; sensitive skin, to burning eczema; and oily skin, to wet eczema.

This is true not just for skin problems but for disorder and disease in general. *Depending upon our constitution, each of us is born with a propensity to develop specific types of health problems when we are imbalanced.* For example, someone with strong Pitta is not just naturally more likely to get inflamed skin conditions; he or she is also naturally more at risk for ulcers, heart disease, liver problems, and many stress-related disorders associated with the hot-blooded temperament of this fire type. By the same token, our mental and emotional traits also change in specific ways when we are imbalanced. Thus, ambition turns to aggression, and impatience to rage when Pitta is imbalanced. Kapha's natural serenity turns sullen, for example, and Vata's exuberance turns to overexcitability. What is our greatest strength when we are balanced often is our worst weakness when we are out of sorts. (In Chapter 3, you will learn about the factors that lead to imbalance.)

The overall traits and tendencies of each dosha and skin type appear in the charts on pages 51, 53, and 57. Note that no one will have *all* the characteristics of one type, however. All three doshas exist in everyone in some proportion, so everyone exhibits *some* characteristics of every dosha. Nevertheless, in most cases, each of us has *more* characteristics of one type. A few key traits (shown below) are fairly reliable markers of your skin type and leading dosha.

KEY CHARACTERISTICS

Dry skin (Vata):

Thin frame: thin, dry skin, fine pores; dark, scanty, frizzy, curly hair; restless.

Sensitive skin (Pitta):

Medium frame; sensitive, lustrous skin with oily T-zone and dry cheeks; soft, moderately thick straight hair with reddish or light tones; ambitious.

Oily skin (Kapha):

Large frame; prone to weight gain; thick, oily, soft skin with large pores; thick, wavy dark hair; calm.

You will have the chance to assess which category applies most accurately to you when you take the Skin Type Quiz at the end of the chapter. Before you go on to do this, however, we want to make an important distinction between your prakriti and your skin type as we define it. In all, Ayurveda classifies ten constitutional types. The additional seven are derived from combinations of the basic three: Vata-Pitta, Vata-Kapha, Pitta-Vata, Pitta-Kapha, Kapha-Pitta, Kapha-Vata, and Vata-Pitta-Kapha. In these double types, who make up the majority of the population, the two doshas are both present in high proportions; however, the *most* predominant one is always named first and is still considered the leading dosha. In so-called tridosha types, all three tie for the lead, although this constitution is rare. *In this book, we do not try to assess your actual body type, since our main concern here is the skin and its actual condition at this time.* However, if you happen to know your prakriti as the result of a professional analysis, keep in mind that your constitution and your skin type are related but not necessarily the same. In twenty-five years of practice, I have found that *every normal complexion has the distinct characteristics of the leading dosha only*—even in people with a two- or three-dosha prakriti. In other words, *Vata*-Pitta types tend to have dry skin characteristic of Vata; *Pitta*-Vata types have sensitive skin characteristic of Pitta; *Kapha*-Vata types have oily skin characteristic of Kapha; and so forth, as shown below.

PRAKRITI	SKIN TYPE
Vata	Dry
Vata-Pitta	Dry & slightly sensitive
Vata-Kapha	Dry with thicker texture
Pitta	Sensitive
Pitta-Vata	Sensitive & slightly dehydrated
Pitta-Kapha	Sensitive & slightly oily & thick
Kapha	Oily
Kapha-Pitta	Oily & slightly sensitive
Kapha-Vata	Oily T-zone with dry forehead & cheeks
Tridosha	Dry, sensitive, or oily

LIKE INCREASES LIKE:
THE PRINCIPLE OF BALANCE

Now that you have an idea of what determines your innate balance, the question is, how do you maintain it? The basic principle is very simple: In every aspect of life, stock up on the elements or attributes that your mind and body lack, and lower your intake of those that you naturally have in abundance. This is known in Ayurveda as *the rule of like increases like*.

Any dosha will become imbalanced when you continually eat foods, use substances, have thoughts and emotions, or otherwise engage in activities that, in Ayurvedic terms, are *qualitatively* "like" that dosha—that is, similar in attributes to the dosha. If your skin is naturally dry, for example, Vata energy—which is intrinsically cold, dispersing, quick moving, and irregular—is already strong in your makeup. If, then, you become frazzled by all the demands on your time—continually worrying about deadlines, grabbing cold leftovers and eating on the run, keeping long, erratic hours, and using stimulants like tea or coffee to stay apace—you are a classic candidate for Vata imbalance. In other words, you are overloading your naturally excitable constitution with more frenetic energy than it can handle and still maintain equilibrium.

Everyone's lifestyle should "nourish" all the doshas, so to speak, since all three are part of life. Generally speaking, however, dry skin types need to have more earth, water, and fire—more Kapha and Pitta influences—in their daily routine, and less space and air—less Vata—if they don't want their inner scales to tip. Sensitive skin types need more space, air, and earth—more Vata and Kapha influences—to stay cool. Oily skin types need more fire, air, and space—more Vata and Pitta—to lighten up. In the chapters to come, you will learn to recognize the specific activities and stimuli—both inner and outer—that are "like" and "unlike" your own nature so that you can achieve a balanced life.

Ayurveda describes beauty as the perfect balance of vastness like space, lightness like air, brilliance like fire, continuous flowing and renewing like water, and nourishing firmness like earth. In other words, beauty is the proper balance of all five elements in your constitution. The way to achieve this balance is to understand the "action and reaction" among the elemental attributes since we are linked by these energies to everyone and everything in life. In this sense, each human being is a microcosmic ecology linked by the senses and behavior to nature's macrocosmic system. We maintain inner equilibrium through a dynamic exchange of energies with our outer environment on all levels of our experience—body, breath, mind, and spirit. Below, we explain these strata of existence in terms of Ayurvedic "anatomy." With this complete picture of balance, you will enjoy looking at the "portraits" of Vata, Pitta, and Kapha and then taking the Skin Type Quiz to find out your own natural balance of doshas.

AYURVEDIC ANATOMY: THE FIVE STRATA OF EXISTENCE

Our bodily form, which is the subject of Western anatomy, is merely the "visible" aspect of the Self. According to Ayurveda, the so-called body also has a non-material form. This "invisible" anatomy is comprised not of flesh and bone, but of subtle networks of intelligence known as *koshas,* or sheaths. These koshas incorporate all aspects of human experience from subtlest to grossest, or invisible to visible. They are: *consciousness, ego, mind, breath,* and *body.*

Because these five strata unfold one from the other, starting in unmanifest consciousness, they are integrally connected. Everything that happens to us and everything that we do involves each aspect of existence *simultaneously*. Indeed, the five koshas—or "subtle bodies," as they are also called—are not separate parts, like layers of a cake, but constitute different energetic values within the continuum of consciousness, like visible colors and invisible rays within the spectrum of light. At the level of consciousness, there is no separation between the koshas; so there is no mind-body split in Ayurveda, as there is in classical Western science. Because of this essential unity, we cannot affect one layer of this anatomy without affecting the whole. To achieve a balanced life, therefore, we must balance every stratum. Hence, the many different kinds of balancing therapies you will find in this book.

FIVE STRATA OF EXISTENCE	EFFECT / FUNCTION
Consciousness *Anandamaya kosha*	Bliss (*samadhi*).
Ego *Viganamaya kosha*	Individuality, intellectual or elevated mind, feeling, intuition, desire, values.
Mind *Manomaya kosha*	Discrimination, decision making, control of behavior and attention.
Breath *Pranamaya kosha*	Vital energy of the body.
Body *Annamaya kosha*	Physical substance of the body.

OBSERVING THE DOSHAS: THE THREE FACES OF BEAUTY

When we determine someone's prakriti or skin type, what we are actually doing is detecting which dosha (or which pair of elements) is most dominant in the person's constitution. Ayurveda provides many ways to do this: by a technique of pulse diagnosis; by the Vedic system of astrology known as *Jyotish*, which is especially useful in assessing the mental aspects of the constitution; by specific examination of the lips, fingernails, tongue, eyes, or face, among others; or by direct observation of a person's general physical, mental, emotional, and behavioral characteristics. Typically, Ayurvedic physicians are trained in the subtle art of pulse diagnosis, which is the most effective means to assess and treat complex disease, as well as in the other forms of examination. Vedic astrologers also study their science for many years before they are skilled enough to apply it accurately. For the simple purpose of staying balanced and radiant, however, direct observation of the attributes is an easy, effective do-it-yourself method to pinpoint your dominant dosha, as you are about to see.

To give you an idea of how the attributes of each dosha might "look," we have sketched out "portraits" of the typical Vata, Pitta, and Kapha personality below. As you read these descriptions, please keep two thoughts in mind. First, the qualities of absolute beauty—poise, radiance, and vitality—are the same in everyone; yet they express themselves through our different natures. On the physical level, perfection has three forms—three *faces*. One has the features of Vata; another, Pitta; and the third, Kapha. Neither one is better than the other, just as a rose is no better than a tulip, and a tulip, no better than a daisy. They are simply different by design, and all perfect in their own right.

Second, the persons described below do not exist in the real world. True single-dosha constitutions are uncommon, and even the rare bird who has one will still exhibit some attributes of the other doshas. As you come to understand the many factors that affect the doshas, you will see that everyone occasionally comes under all their influences, no matter what his or her native makeup may be. A Vata type who starts to gain weight, for example, is experiencing the effects of too much Kapha. A Kapha type who has a flash of anger is experiencing the effects of too much Pitta. We have painted each portrait with a broad stroke to give you a clear picture of the dosha, but you may see parts of yourself in all three. This is only natural, since all three doshas are present in everyone at all times in some proportion.

A PORTRAIT OF VATA

Imagine the wind. Wherever it moves, it brings along cool, refreshing air that drives off dampness and dew. Sometimes it comes in wisps that barely rustle the treetops. Other times it gusts so strongly it pushes us along our way. It can make heavenly music on crystal chimes or send a hurricane howling. One thing is certain: It never sits still and never stays the same very long. Its nature is quick, subtle, changeable, and as close to ethereal as any matter gets.

When you understand the wind, you can appreciate Vata's beauty. Her skin is thin and fine-pored all over, and ever so slightly cool and dry. Her features are often refined and delicate, though sometimes irregular or elongated, and her cheekbones may be high because her bones tend to be prominent. Her hair is dark, like her eyes, and often curly, wispy with a coarse, flyaway texture. She may be short in stature or tall, but her figure is naturally slender. She looks serene in warm pastels and dazzling in jade or emerald and gold.

In love, as in all things, Vata is quick in passion and attachment. Her imagination is very creative and she tends to start things with great enthusiasm. Indeed, when she is happy, she is exuberant and bursting with energy that is bright and infectious, and she can get away with less sleep than most other people. But Vata spends herself (and her money) too easily. Her endurance is slight, and sometimes her excitability, her mood swings, and her lack of tenacity give her the appearance of being flighty. Her active mind grasps new ideas in the blink of an eye, but she may forget them just as fast.

This rarefied creature never does well in loud, noisy environments or cold, windy climates. Salad, raw vegetables, and cold leftovers are a poor diet for Vata, but a steaming bowl of cream of asparagus soup with sourdough bread fresh from the oven will satisfy for hours. Too much traveling or activity of any sort wear her thin nerves thinner, and an irregular routine can send her mind and body into a tailspin. When she is out of sorts, her unsettled attention becomes easily worried and anxious, her appetite diminishes, and her insides get constipated. Her complexion becomes excessively dry, with fine lines etched above the brow. There is nothing so unsettling to Vata's sensibilities as a harsh voice or crass language, and nothing so soothing and grounding as a soft touch, the sweet scent of jasmine or rose, and a summer morning when the sun is ascending and her feet are planted square on *terra firma*, her eyes look out to sea, and her ears catch the easy rhythm of the waves.

A PORTRAIT OF PITTA

Picture a blazing fire. Its heat is penetrating and stimulating, always dehydrating and sometimes scorching. Its flames are sharp, fierce, and difficult to temper. Sparks fly when it rages, and few things can stop its cruel path. Its nature is purposeful, fearless, and invincible. Like the sun itself, it generates intense energy, and the longer it burns, the redder it becomes until its last embers fade. In the meantime, its unctuous dance creates beautiful taffeta swirls of light, and its glow is utterly mesmerizing.

Once you've stared into her fiery gaze, you cannot forget Pitta's beauty. Her eyes are penetrating green, gray, or hazel, with perfect lashes and brows. Her

DOSHA: VATA

Skin type: Dry

Elements: Air and Space

Life force	Prana.
General attributes	Light, dry, mobile, rough, cold, thin, clear.
Frame	Thin, irregular, tall or short.
Weight	Hard to gain, easy to lose.
Shape of face	Thin, long, egg-shaped, small forehead.
Skin	Cold, dry, sallow, rough, small pores, thin, prone to premature wrinkles, dark circles under eyes, tans easily, delicate like baby.
Hair	Dark, dry, frizzy, thin, coarse.
Nails	Brittle, ridged, grayish, discolored.
Lips	Thin, dry, tendency to crack, bleed.
Eyes	Small, dry, itchy, brown, gray, scanty eyelashes, remain partially open at night.
Nose	Long, thin, crooked.
Teeth	Dry, rough, crooked, small.
Tongue	Dry, cracked, cold, rough.
Appetite	Irregular, eats small quantities frequently, likes warm food.
Stamina	Poor, exertive, weak.
Sweat	Scanty.
Elimination	Irregular, small quantity, tendency toward constipation.
Sleep	Scanty, interrupted, bothered by light noise.
Temperament	Fearful, indecisive, nervous, anxious, worried, low tolerance.
Good qualities	Accommodating, very active, creative, intelligent, natural teachers, musicians, spiritual leaders, artists, philosophers.
Sexuality	Cold, variable, the process of love is more important than the actual act of love.
Seats	Colon, skin, bladder, kidney, large intestine.
Prone to	Excessive dryness, psoriasis, dandruff, wrinkles, constipation, sharp pain, backache, arthritis, nervous disorders, insomnia.

DOSHA: PITTA

Skin type: Sensitive

Elements: Fire and Water

Life force	Tejas.
General attributes	Hot, sharp, slightly oily, sour, light, fluid.
Frame	Medium, proportionate.
Weight	Easy to gain, easy to lose.
Shape of face	Triangular with pointed chin.
Skin	Warm, soft, reddish, pink, lustrous, large pores in T-zone, red nose, prone to allergic reactions, burns easily.
Hair	Soft, straight, light or reddish, premature gray.
Nails	Soft, pink, well-formed.
Lips	Medium, soft, red, pink.
Eyes	Almond-shaped, sharp, penetrating, light brown, hazel, gray, green, unable to bear light.
Nose	Medium, sharp, pointed.
Teeth	Medium, white.
Tongue	Red, dark, pungent taste in mouth.
Appetite	Good, excessive, strong digestion, thirsty, likes cold food.
Stamina	Moderate strength.
Sweat	Profuse, sour smell.
Elimination	Loose, regular, large.
Sleep	Little, but sound.
Temperament	Angry, frustrated, jealous, hateful, aggressive, irritable, arrogant.
Good qualities	Adaptable, ambitious, sensitive, compassionate, sharp, intelligent leaders, good businesspeople, administrators, directors, pioneers.
Sexuality	Intense, see themselves as great lovers, but in reality often lack patience.
Seats	Liver, gall bladder, small intestine, heart, eyes.
Prone to	Acne rosacea, rashes, cold sores, allergic reactions, burning sensations, peptic ulcers, bleeding, liver disorders, hypertension.

straight hair is sandy brown, golden blond, copper, or blazing red. Her skin is warm and pink and soft, and her cheeks and nose are very probably dotted with freckles or beauty marks and are ever so slightly moist and dewy from the natural oils in her tissue. Her scarlet lips are evenly shaped, her face is sharply contoured, and her body is well proportioned. Cool greens and blues balance her intensity and pearls, moonstone, and silver bring out the sparkle in her eye.

In love, as in all things, Pitta is naturally hot. Her desires are strong, and her passion and will are equal to it. She knows how to get what she's after, and usually knows how to pace herself to reach the goal. When she is happy, no one is warmer or more delightful. Beware of getting burned, however. Under pressure, Pitta can be impatient, impulsive, irritable, angry, and even hostile. Then she is likely to use her searing intellect against you, and her criticisms can cut deep. She is also prone to jealous outbursts, and her need for perfection and control at these times can be overbearing. Even in her cooler moments, Pitta's actions may seem brash or hasty to shyer and more cautious types.

This firecracker is easily set off by hot weather, spicy food, and stimulants of any kind, but she can go sweaterless on cold days and have ice cream in any season without a shiver. When her energy's cooking, Pitta feels invincible. However, even this powerhouse needs some limits. Her motto in life should be "Chill out or burn out," or "Everything in moderation." Pushed to her extremes, she is the classic Type A personality. Her mood turns angry and argumentative; her behavior, intolerant and compulsive; and her body, prey to ulcers, hypertension, and heart disease. Even her stomach and breath turn sour. Her lovely rose complexion erupts into red rashes, burning eczema, acne rosacea, or other inflammations. Only a sunburn or an allergy could make the condition worse. Then Pitta has to cool down her system with a rest from activity, a change of scenery, a scent of sandalwood, meditation, and a moonlight stroll.

A PORTRAIT OF KAPHA

Think of the earth in all its varied forms. It is the massive planet we walk on; the dark, cool, sweet soil that bears the crops; the soft, wet, viscous clay smoothed by a sculptor's hand; the thick, sturdy slope of an old mountain; the meandering river valley; and the deep cradle of the ocean. It makes changes slowly, if it changes at all. But once an endeavor's begun, the earth advances steadily to the goal, no matter how long it takes. It has patiently nurtured many an acorn until it has become a great oak. Don't be fooled by this languid nature, however. You will get a fight if you pull up roots, try to change the course of rivers or mine mountains for gold. What the earth holds dear, it does not let go easily.

Once you've felt earth's sensuous embrace, you have understood Kapha's beauty. Her complexion is pale, soft, and ever so slightly cool and slick. Her generous face is square or round with cheeks that beg to be pinched. She has voluptuous lips and and big brown or blue "bedroom" eyes that boast the lushest lashes. With her abundant, wavy tresses, and full-bodied figure, she is the epitome of a Rubenesque woman. Rich reds and purples warm her cool skin, and garnet and ruby peak her energy and will.

In love, as in all things, Kapha is cautious, wise, honest, and steady. She rarely goes headlong into affairs—she is even a little shy—but prefers to take time for the rituals of courtship and deepening romance. There is much to be said for her civilized pace. Her graceful manner is charming, not to mention calming and undemanding, and her resourceful mind offers a wealth of insight to anyone smart enough to ask. Serene Kapha does not fall quickly, but her sweet heart does fall hard. Once she commits, her unswerving feelings and her stamina are a match for any lover. When her natural equanimity is upset, however, Kapha's attachments become possessive. She has great difficulty moving on after a disappointment, and she can be sullen and unresponsive once her mood shifts. If she doesn't take steps to get back on even ground, her state of grief is apt to develop into full-blown, can't-get-out-of-bed depression.

At such times, Kapha literally seems to collapse under her own weight. She is no longer earth itself, but Ms. Atlas carrying the world on her shoulders. Her placid demeanor, once so attractive, begins to look downright dull and lazy, and her already zaftig shape is prone to put on pounds. When she clings too tenaciously to her sadness, it literally seeps through the body in other forms: Mucus collects in her chest and sinus cavities, producing conditions ranging from colds and congestion to coughs and asthma; her normally active sebaceous glands "cry"

DOSHA: KAPHA

Skin type: Oily

Elements: Earth and Water

Life force	Ojas.
General attributes	Heavy, slow, cool, damp, oily, sweet, dense.
Frame	Broad, well built, evenly proportioned.
Weight	Easy to gain, hard to lose.
Shape of face	Round if more water, square if more earth.
Skin	Cool, fair, oily, large pores, blackheads, prone to cystic acne, scarring, deep wrinkles, tans easily.
Hair	Thick, oily, wavy, curly, black, shiny.
Nails	Clear, pale, square, white.
Lips	Thick, cold, pale.
Eyes	Sensual, big, black, blue, thick, long, oily lashes, watery eyes.
Nose	Straight, thick, good smell perception.
Teeth	Shiny, oily, strong, white.
Tongue	White, slimy, excessive salivation, sweet.
Appetite	Constant, likes spicy food, not so thirsty.
Stamina	Strong, good immune system.
Sweat	Moderate, sweet smell.
Elimination	Slow, regular, moderate.
Sleep	Needs more sleep, long, deep.
Temperament	Easily depressed, lazy, stubborn, attached, greedy, passive.
Good qualities	Calm, stable, dependable, nurturing, good memory, good providers, artists, dancers, doctors, accountants, teachers, good parents.
Sexuality	Warm, enduring, good lovers, romantic.
Seats	Stomach, pericardium, spleen, respiratory system, tongue, nose, hair, nails, muscles.
Prone to	Cystic acne, edema, swelling, sinus headaches, sore throats, respiratory problems, asthma, diabetes, cysts, tumors.

through the skin, causing excessive oiliness, cystic acne, or wet eczema, while the rest of the body retains fluids, causing a sagging, flabby appearance overall.

In these moments, a cold, rainy day, a sad, slow love song, and a sundae topped with thick whipped cream may seem a comfort for Kapha's gloomy mood, but they are the precise opposite of what this doleful, sweet soul needs. A vigorous massage with warming sesame oil, a pungent potpourri of eucalyptus, sage, and musk, a hike in sunlit woods followed by a hot-baked cinnamon apple will go a long way to ease her heart and mind. Add some lively music with a strong beat and the volume on high, and Kapha will be up and dancing again in no time.

THE SKIN TYPE QUIZ

Look at yourself in a magnifying mirror, then check the responses that best describe you, keeping in mind that not every part of an answer must apply. Check multiple responses to each statement if more than one is applicable; leave the question blank if no choice applies. When you are finished, score your answers as instructed.

1. My skin feels
   a. __ dry, thin, rough, cold   b. __ slightly oily, soft, hot   c. __ oily, moist, thick, cold

2. The pores on my face appear
   a. __ small and fine   b. __ large on T-zone, small and fine elsewhere   c. __ large and open

3. My skin tone* is best described as
   a. __ bluish   b. __ reddish   c. __ yellowish

Skin tone is not the same as skin color. Every type of skin, regardless of a person's race or ethnicity, exhibits a subtle but discernible undertone. Fashion professionals are looking at these undertones when they match clothing colors to your skin.

4. My complexion is marked by
   a. __ fine lines, prominent veins   b. __ broken capillaries, freckles, moles   c. __ blackheads, excessive oiliness

5. My skin appears discolored in spots such as

   a. __ dark pigmentation   b. __ reddish   c. __ white or brown
        on the cheeks   pigmentation all over   pigmentation

6. My skin is generally healthy but prone to

   a. __ dryness,   b. __ inflammation,   c. __ acne and pimples,
        dehydration, especially   rashes & sunburn,   especially in area of
        in cold weather   especially in hot   mouth, chin, or neck
                             weather

7. My face is

   a. __ small, thin, long,   b. __ average, triangular   c. __ large, round, or
        egg-shaped   with sharp contours   square with soft
                                             contours

8. My eyes are

   a. __ small, brown, or   b. __ sharp, penetrating,   c. __ wide, big, brown,
        black with scanty   green, gray, or hazel   or blue with thick
        short or long lashes   with average lashes   lashes and brows
        and thin brows   and brows

9. My lips are

   a. __ thin, dry, long,   b. __ average, soft, red   c. __ thick, large, even,
        irregular   firm

10. My nose is

   a. __ thin, small, long,   b. __ average, prone to   c. __ thick, big, firm, oily
        crooked, dry   broken capillaries

11. My hair is

   a. __ dry, thin, coarse,   b. __ moderate, fine,   c. __ thick, oily,
        curly, frizzy, wiry,   soft, golden or reddish,   abundant, wavy,
        scanty, dark   prematurely gray   dark or light
                             or balding

12. My fingernails are

   a. __ dry, small,   b. __ soft, medium,   c. __ thick, oily, smooth,
        crooked, brittle,   pink   white, strong
        rough, discolored

13. My perspiration is
   a. ___ scanty with no smell
   b. ___ profuse and hot with strong smell
   c. ___ moderate and cold with pleasant smell

14. My physical build is
   a. ___ small, thin, tall, or short with prominent bones
   b. ___ average with good muscles
   c. ___ large, well-developed, firm, prone to overweight

15. In physical activity, I tend to be
   a. ___ very active, quick to start and stop, with poor endurance
   b. ___ moderately active with average strength, but intolerant of heat
   c. ___ lethargic or slow to start, but strong with good endurance once I do

16. In mental activity, I tend to be
   a. ___ restless, erratic
      ___ hyperactive
      ___ undisciplined
      ___ very creative
      ___ unsteady
      ___ quick to learn, but with poor long-term recall
   b. ___ ambitious, motivated
      ___ purposeful
      ___ sharp
      ___ strong-minded
      ___ intelligent
      ___ disciplined with good memory overall
   c. ___ calm, steady
      ___ resourceful
      ___ dependable
      ___ cautious
      ___ insightful
      ___ slow to grasp, but with good long-term recall

17. My normal temperament is
   a. ___ changeable
      ___ unpredictable
      ___ anxious
      ___ insecure
      ___ quick in passion and attachment
      ___ impatient
      ___ unforgiving
   b. ___ brave
      ___ aggressive
      ___ invincible
      ___ irritable
      ___ jealous
      ___ cruel
      ___ quick to anger and cool off
   c. ___ calm
      ___ civilized
      ___ romantic
      ___ truthful
      ___ shy
      ___ sad
      ___ dull
      ___ overly attached

18. Under stress, I am prone to

a. __ anxiety attacks b. __ excessive anger c. __ depression

__ fear and/or worry __ frustration __ unresponsiveness

__ insomnia __ outbursts __ greed

__ trembling __ jealousy __ grief

__ hysteria __ hostility __ possessiveness

19. I am particularly sensitive to

a. __ dry cold and wind b. __ heat and sun c. __ cold and damp

20. I sometimes suffer from

a. __ low appetite, b. __ hot flashes c. __ sinus congestion

weight loss __ acid stomach __ coughs

__ kidney problems __ hemorrhoids __ colds

__ menstrual cramps __ inflammatory __ drowsiness

__ muscle spasms disease __ high cholesterol

__ arthritic pain __ ulcers __ asthma

__ constipation __ liver disease __ diabetes

__ low back pain __ hypertension __ weight gain

__ intestinal gas __ heartburn

21. I currently have one or more of the following skin problems

(An asterisk indicates that the condition is described on page 62.)

a. __ excessive dryness b. __ rashes, hives, or c. __ excessive oiliness

__ dry eczema* cold sores __ itching/wet

__ scaly skin __ burning eczema* eczema*

__ psoriasis* __ acne rosacea* __ cystic acne*

__ discoloration and __ burning of face, __ flabbiness

puffiness under eyes eyes, or feet due to water

__ dandruff __ whiteheads retention

__ cracks on palms __ excessive sweating __ deep scars and/or

or soles __ contact dermatitis* deep wrinkles,

__ wrinkles on forehead or dryness of scalp particularly laugh

when constipated, or hairline lines

stressed, or anxious __ premature wrinkles __ weight gain

or lines under eyes __ loss of skin tone

Acne rosacea: Dry, reddish pimples typically around the nose and cheeks, but may appear anywhere on face or back.

Burning eczema: Itching or burning dry, red patch around the joints or elsewhere on the body.

Contact dermatitis: Itchy, dry, red, scaly patches around eyes and brows.

Cystic acne: Pustular eruptions typically around the mouth, chin, and neck, or anywhere on face or back.

Dry eczema: Dry, red patch typically around the joints, but may be anywhere on body.

Psoriasis: Dry, flaky skin condition characterized by silvery scales.

Wet eczema: A sensitive, oozing red patch typically around eyes, brows, nose, or scalp, but may be anywhere on the body.

YOUR RESULTS

The Skin Type Quiz is actually designed to answer two questions: Is your basic skin type *dry, sensitive,* or *oily*? And, is your condition presently *balanced* (normal skin) or *imbalanced* (problem skin)? You need both answers to select the right personal skin care routine.

Your normal skin type: Score questions 1–21 by adding up the number of answers you checked in each column, *a, b,* and *c.* Write your totals (from 0 to 50) here; then circle your highest score:

a. ____   b. ____   c. ____

IF YOU CIRCLED COLUMN:	A.	B.	C.
Then your *leading dosha* is:	Vata ____	Pitta ____	Kapha ____
And your *normal skin type* is:	Dry ____	Sensitive ____	Oily ____

Your skin condition: Look at your answers to Question 21 only. For each column, mark "0" in the space provided if you checked no responses; mark "X" if you checked one response or more:

a. ____   b. ____   c. ____

Then your skin condition at this time is: Balanced _____ Imbalanced _____

(Normal skin) (Problem skin)

WHAT YOUR RESULTS MEAN

As we explained earlier, your skin type reflects the qualities of your leading dosha:

Dry skin is a characteristic of dominant Vata.

Sensitive skin is a characteristic of dominant Pitta.

Oily skin is a characteristic of dominant Kapha.

Sensitive skin is also known as *combination skin* because it often appears dry along the cheek area and oily in the T-zone that runs across the eyebrows and down the nose. This mix of qualities reflects that fact that Pitta, which causes sensitive skin, has certain properties that mimic the other doshas. For example, Pitta is hot, so it dries skin like Vata; and it is also slightly oily, so it stimulates sebaceous glands like Kapha. Because it can be dry and oily, sensitive skin is sometimes tricky to identify. If you are confused about it, check the appearance of your pores, which are a good indicator of skin type. Enlarged pores on the T-zone with small, fine pores elsewhere are a telltale sign of sensitive skin; small, fine pores overall are a sign of dry skin; and enlarged pores overall, a sign of oily skin.

Normal skin refers to the condition of the complexion when the doshas are balanced. Every normal complexion is naturally clear and glowing but will appear *slightly dry*, *sensitive*, or *oily* depending on your basic type. *Problem skin* includes any condition out of the normal.

While problem skin always indicates some imbalance of the doshas, you will see in the following chapter that flawless skin is not necessarily the equivalent of a clean bill of health. As proof of this fact, just think of someone you know who has high blood pressure, for example, and a perfectly lovely complexion. Because of a principle called end-product sensitivity, certain people may have any number of health conditions that do not produce obvious blemishes on the face but do take their toll on the body in other ways. Therefore, a normal, healthy complexion should never be misconstrued as a reason to forgo checkups or ignore medical advice.

By the same token, problem skin means that there is presently an imbalance in your constitution that needs to be corrected—*not* that you are doomed to have skin problems forever. If you do have a problem, in the following chapters, you will learn the source of problem skin and how to select the proper treatment to restore balance and health. Even if you do not have a complaint about your skin at the moment, you will find much important information in these chapters regarding health and beauty in general. Moreover, you will learn to recognize signs of imbalance so you can prevent future problems and also enhance your understanding of the many ways in which the complexion reflects your inner state of body and mind.

CHAPTER 3

IMBALANCE:
SKIN DISEASE, AGING,
AND THE MIND-BODY
CONNECTION

Her pure and eloquent blood
Spoke in her cheeks, and so distinctly wrought
That one could almost say her body thought.

JOHN DONNE

Skin is a vast subject. The largest human organ, skin has seven cell layers that together serve as the body's primary pathway for detoxification; a storehouse of fat, water, glucose, and salt; a main channel of absorption and secretion; a regulator of body temperature and water balance; and a major producer of endocrine hormones. Its 640,000 sensory receptors provide our source of touch, pain, pleasure, pressure, heat, and cold, as well as a direct link to the brain and every other part of the nervous system. Skin is also our first line of protection from the outside, and the *reflection* of everything that is going on *inside* of us.

This ultimate function of skin—as a physical indicator of the subtlest workings of mind and body—is the primary concern of this book. Until recently, it has also been the one function almost completely overlooked by Western science. Without this knowledge of how skin reflects the unique nature of an individual, skin care professionals, like the deer in the woods, are bound to search in vain for the perfect beauty product.

This chapter provides the scientific basis for a skin care program that works from the inside out. We will define "perfect" skin, what happens when it loses its perfection, and the causes of premature aging and skin disease. From the diverse points of view of modern and ancient mind-body medicine, we will see why skin is called the body's "second brain," how its appearance actually mirrors the state of our psychophysiology due to the hormone connection, what the specific symptoms of imbalance are according to Ayurveda, and why these symptoms are different for each person. The Skin Problem Quiz at the end of the chapter will help you determine if you have a skin problem right now, and what factors in your life may be contributing to it, so that you will be able to select the proper course of treatment to correct your condition.

AGING AND THE SKIN

Look at the face of any healthy infant. No matter how cute or homely the features may be, no matter whether the complexion is fair or dark, baby's skin is irresistible to touch. Soft, supple, smooth, dewy, bright, and blemish-free, its perfection is our natural inheritance at birth and the standard against which we measure encroaching old age.

Until adolescence, skin normally remains at its peak. The top horny layer is well hydrated. The intercellular cement that provides oil and moisture is in good condition. Cellular activity is fast and efficient. The elastin has plenty of "snap," and the network of connective collagen fibers is strong and plentiful. From the onset of puberty on, however, it is all downhill, biologically speaking. As you shall see, changes in body chemistry, along with the effects of environment, lifestyle, and stress, inevitably conspire to undo nature's original handiwork. As early as thirteen or fourteen, you can detect signs of deterioration by pursing your lips in a small, round kiss. If deep vertical lines appear around the mouth, you can reasonably conclude that poor health or nutrition, or a habit of smoking or drinking, has begun to dehydrate and prematurely age the complexion—although the first permanent wrinkles typically do not appear until age thirty.

By age forty to fifty, many skin functions will slow down significantly, and its structures will start to deteriorate. The epidermis thins, sweat glands decrease in size and in number, the dermis loses cellular content, blood flow slows, the immune response weakens, and collagen fibers lose their organization and therefore their strength, causing the skin to sag. If you are in this category, try a pinch test of the skin on top of your hand. You can measure the loss of elasticity, a common sign of aging, by how quickly or slowly the skin bounces back. Alas, it is true that men's skin does tend to "age" better than women's. Their dermis layer is thicker to start with and consequently retains its strength longer, and while sebaceous secretions taper off in women at the time of menopause—average age fifty—they typically keep lubricating the male complexion up to age seventy.

Although the ideal complexion is as short-lived as childhood, many imperfections we associate with aging skin are not due simply to the passage of time. In fact, there are two types of aging processes affecting the skin. The first is external. Like the heart, lungs, and brain, skin is a living organ, except it happens to be the only one completely exposed to the environment and to the constant destructive effects of sun, weather, pollution, and the chemicals in our toiletries and water. For this reason alone, skin demands special daily attention that the rest of the body does not usually require in order to prevent premature aging.

The second type of aging process happens within, but its effects are not out of sight. Psychophysiological imbalances due to lifestyle and stress alter the function and structure of individual skin cells and consequently alter the outward appearance of the tissue itself. The effects of this inner aging are more pervasive than the external process, yet the internal process is also more within your power to control. You cannot do much to influence the weather, but you can influence your neurophysiology by your eating habits, your breath, your thoughts and emotions, and even your sense of purpose in life, as you will see. Many so-called normal signs of aging are actually signs of premature aging and disease resulting from imbalance, and imbalance is something you can correct and prevent through Ayurveda.

THE THINKING BODY: STRESS, HORMONES, AND IMMUNITY

How does the body become imbalanced? What disrupts its normal functioning, and how does this effect show up on the skin? Mind-body medicine—both modern and ancient—answers that the fundamental cause of imbalance is *stress* and the hormonal changes it produces.

The term "stress" entered the Western medical lexicon about fifty years ago, when scientists began to study the body's self-regulating capacities and its balanced, integrated condition, which they called *homeostatis*. Hans Selye, a pioneer in the study of stress, suggested that it causes the failure of the body's homeostatic mechanisms, which in turn leads to physical or mental disease. "Stress" became the catchall term for any physical, chemical, or emotional factor that overloaded this homeostatic system, resulting in imbalance. Its characteristic psychophysiological effects, as you will see, make up the stress syndrome.

In Ayurvedic terms, stress is anything that overloads your innate balance of energies—that is, the natural "set point" of your doshas—with too many "like" energies. In other words, it is anything that disturbs nature's rhythms. Stress factors may be physical, psychological, or spiritual, according to Ayurveda. Physical stressors include any overload of the senses, wrong diet and exercise, poor habits, overwork, physical strain, lack of rest, excessive travel, incorrect breathing, excessive use of stimulants including alcohol, tobacco, caffeine, and hot spices, and chemical overloads from environmental pollutants, food additives and preservatives, synthetic soaps and lotions, and drugs. Psychological stressors include emotional crises, unsatisfying relationships, personal conflicts, and negative behaviors. Spiritual stressors include doubt, despair, and confusion—or lack of purpose in life—as well as lack of direct experience of consciousness, that is, the lack of inner harmony and peace of mind. The stress condition is called *santrasa*. Vasant Lad, Ayurvedic physician and founder of the first full-time Ayurvedic institute in the United States, defines this term as a continued discomfort of body and mind that upsets the balance of the five elements, the seven dhatus, or body tissues, and the three *malas*, or waste products produced by the healthy body.

Later in the chapter we will come back to this Ayurvedic description of imbalance, but now we want to look at the contemporary point of view. In Western terms, it is not too hard to imagine how foods or drugs or pollutants might alter the body chemistry and substance. But the tough questions are: How do external experiences produce physical outcomes in our bodies? And how do thoughts and feelings change the function and substance of our cells and show up on our skin? In other words, how does the mind make us old and sick, and can it also keep us young and well? In just the past two decades, modern scientists have found some answers to these questions in the relationship of our emotions, thoughts, and hormones—the biochemical messengers that govern most physiological functions, including the body's stress response. Their basic discovery is that no human experience, good or bad, happens "only in the mind."

HORMONES:
THE CHEMICAL LANGUAGE
OF THE BODYMIND

Stress—by which we mean psychological stress—is not a germ. It is not an invading organism that we can see under the microscope and fight with drugs. It is nothing more "real" than our perception. Psychological stress is like any other *thought*, and is literally created in the mind. Events occur in the world, and the brain decides which ones are stressful based upon our subjective interpretation. No event is intrinsically stressful, but it becomes so because of our reaction to it. Thus, two people go to see the same movie and one finds it frightening, the other, laughable. One person is devastated by divorce, another feels liberated. One individual wins the lottery and retires happily to Palm Springs. Another wins, has a heart attack from shock, and dies. Even the same individual does not necessarily react the same way twice to repeated events. We are so nervous, we fail our first driving test. With a little more knowledge and experience, we find the second test is a piece of cake.

Of course, not all stress is harmful. We may feel excitement about our upcoming wedding, for example, and also a little nervous. We can feel positive about a divorce and also feel uncertainty about our altered financial situation. Generally speaking, such heightened emotional states are *regulated arousal*—the pumped-up energy we create when we are motivated to get a job done or feel a sense of anticipation. This motivated state is sometimes called *eustress*, or "good" stress—it gives us a helpful "edge" but it also calms down naturally when the purpose is accomplished. However, when the state of arousal is *unregulated*—that is, when it is a conditioned reaction due to unconscious thoughts and beliefs—we call it stress, plain and simple.

This is the kind of psychological stress with which Ayurveda is concerned, and as we said, it arises from our subjective perceptions of reality. Whether our idea has objective validity is totally irrelevant to the body. Once the mind makes its decision that an event is stressful, the *thought itself* becomes our *molecular* reality in the form of biochemicals called *catecholamines*—the "stress" hormones that arouse key bodily systems to prepare for fight or flight. The brain, *mind*ful of a crisis or a threat, signals the hypothalamus, which activates the master gland, the pituitary, which in turn activates the adrenal glands, whose hormonal secretions provide the familiar adrenaline "rush" symptomatic of the stress response. It is through the combined and protracted effects of these stress hormones on the immune system that thoughts have the capacity to make us ill.

The discovery that stress affects immunity through chemical hormones is only several years old, but it has transformed Western understanding of the mind's influence on the body. Before 1974, the prevailing theory held that all brain-body communications were carried by the fixed "wiring" of the nervous system. The brain was command central, relaying all messages to and from the far-flung outposts of its network, neuron to neuron, while all other body systems either functioned autonomously or waited dumbly for orders to reach their nerve receptors.

Then scientists made a series of remarkable discoveries. First they found that nerve impulses are not only carried electrically along established neuronal pathways, but also chemically by "messenger molecules"—a special class of hormones called *neuropeptides*—that travel freely through the bloodstream to regulate cell functions. Next, they found that other nerve cells are *not* the only body cells that receive these chemical messages from the brain. And third—and most remarkable—they found that brain cells are *not* the only cells that produce and send neuropeptides.

Indeed, scientists have learned that the immune system—long believed to be functionally independent from the brain and central nervous system—along with the glandular system, has "brainy" cells, so to speak. These receive, send, and even rewrite amine-peptide nerve signals, just as neurons do, in order to monitor and modulate one another's activities. These three master systems, it turns out, "talk" to each other, and talk incessantly, in the same chemical language. Like the mobile-phone technology that liberated telecommunications from the tyranny of cables, these free-floating molecular messengers and their ubiquitous cellular receptors enable direct access (beyond the nervous system) between the brain and the body's trillions of cells individually. As physiologist Robert Keith Wallace declares, this discovery points to the conclusion that "our body is a thinking body in which information or intelligence constantly flows among all its innumerable parts."

As Western science is beginning to find, we can no longer regard body and mind as two separate entities with distinct functions and experiences. As Ayurveda has always taught, we must conceive of them as an undivided "consciousness" that acts and reacts as a single unit. In the end, whether the stress is physical, chemical, or psychological, it is all the same in its effects on the *bodymind*.

THE STRESS SYNDROME
AND IMMUNITY

"**W**herever a thought goes, a chemical goes with it," states Deepak Chopra. But how do a thought and a chemical become a pimple on the skin? The explanation in Western terms lies in the mechanics of the stress response itself and the impact of stress arousal on immunity.

The autonomic nervous system modulates the stress reaction. It has two branches, which together govern all involuntary physiological functions. The *sympathetic* nervous system induces the stress reaction, regulates the stress hormones, and affects related internal functions, such as the breath and heart rate. It also governs normal functions related to outward experience such as sensory response, including skin function and energy supply for action. The *parasympathetic* nervous system controls the internal life-maintenance functions, such as heart, lung, and digestive activity, as well as all glandular, or endocrine, activity. It also supplies energy to the sympathetic system during stress, and then triggers the restful response to restore equilibrium following stress arousal.

Under normal conditions, these two systems balance each other. Under stress, the sympathetic nervous system goes into high gear and activates the parasympathetic system to turn out adrenaline and other hormones. As a result, the senses go on alert; breathing patterns and digestive activities are interrupted; blood sugar levels jump to provide more energy; and cardiovascular functions go into overdrive to rush blood to the limbs in readiness for fight or flight. Indeed, the purpose of the stress response is to give us the mental and physical power necessary to survive attack. Once the threat passes, parasympathetic activity takes over again, the adrenaline stops pumping, the body and mind calm down, and no permanent harm is done. In and of itself, the stress reaction does not hurt the body—it is in fact necessary to life. When the arousal is *protracted*, however, as it is when we have the constant pressure of deadlines or personal crises, the sympathetic system continues to release stress hormones and the parasympathetic system continues to give up energy—and herein lies the problem.

Prolonged arousal affects every gland. The outer part of the adrenals starts to release cortisol, which is toxic to the body in large doses. The pituitary releases antidiuretic hormone, ADH; adreno-corticotrophic hormone, ACTH; and thyroid stimulating hormone, THS, which together affect water balance, sex gland function, metabolism and bone growth. Continued release of catecholamines into the bloodstream increases production of oxident molecules and leads to cellular breakdown, which is the first step in the aging process. At the same time, prolonged activation of the parasympathetic nervous system depletes the body's energy reserves. With its resources constantly rerouted to the emergency, other parasympathetic functions that are essential to life and health do not operate properly.

The cumulative effect of these hormonal changes is a variety of physical and psychological symptoms that make up the stress syndrome. Common digestive complaints include pain, abdominal cramps, diarrhea, constipation, indigestion, excess stomach acid, bowel spasms, ulcers, spastic colitis, and eating disorders. Cortisol secretions affect glucogen levels, leading to diabetes. Changes in estrogen and progesterone levels disturb the menstrual cycle and can cause infertility and PMS problems. Cardiovascular changes lead to high blood pressure, tension, heart palpitations, and heart disease. Other symptoms include muscle tension, teeth grinding, sweating, shoulder pain, asthma, allergies, headaches, and chronic fatigue. Psychological symptoms include poor concentration, nightmares, and stuttering.

While individually these stress symptoms are troublesome, they are not even the worst effect of prolonged arousal. The most damaging long-term consequence of this neurophysiological imbalance is the suppression of the body's immune functions due to the flood of stress hormones. Research indicates that the immune system, which is partly shut down by stress in the short run, does not adapt to it in the long run, but remains active at the lower level of functioning. Studies also show that once immune reactivity is weakened, we are more likely to become sick in response to new stress. In this sense, prolonged stress is the mother of all disease—including that of the skin.

STRESS, EMOTIONS, AND THE SECOND BRAIN

The three master control systems all maintain large headquarters in the skin. Via their abundant nerve links to the brain and their versatile chemical links to the neuroendocrine and immune systems, the skin cells are in constant communication with the entire body. This understanding is very new to Western science, which until quite recently was unaware that the skin had many metabolic functions beyond the capacity to produce the hormone we call vitamin D. Following the discovery of neuropeptides, however, researchers have begun to find numerous body organs—including skin—that participate with the brain in the regulation of the physiology. One of their most surprising discoveries in 1992 was that epidermal cells called *keratinocytes*, which form the outermost layer of skin tissue, produce the same messenger molecules that immune cells do, giving the skin a participating role in immune behavior in addition to its role in neural and neuroendocrine activity. As a result, the skin naturally reflects changes in every system of the body, and affects them as well. (An Ayurvedic skin massage, for example, actually boosts the body's immunity.) In fact, skin cells are so much like brain cells in their ability to accept and transmit communicator molecules and monitor bodymind activity that skin has earned the title of the body's "second brain."

As we saw, stress causes imbalance, and *all imbalance eventually shows up on the skin as accelerated aging.* However, stress has a direct impact on the skin as well. In times of crisis, skin tissue is the last to receive nutrients because the blood supply goes straight to the vital organs, including the heart, brain, and lungs. Dilated blood vessels result in broken capillaries, redness, and skin sensitivity. Increased perspiration on the palms, soles, and underarms exacerbate eczema conditions, causing oozing, itching, or inflammation. Chronic strain disrupts metabolism, slowing down cell turnover and the process of rejuvenation, producing clogged pores and the buildup of toxins under the skin, which gives it a dull, yellowish or whitish appearance. Stress also alters protein, potassium, phosphorus, and calcium levels in the body, which help to regulate body fluids and maintain acid-alkaline balance. Consequently, the skin becomes dehydrated and the pH levels change under stress. Other stress-related skin problems include hair loss, premature graying, increased pigmentation or loss of pigmentation (vitiligo), bruising, itching (Lichen chronicus simplex), and aggravated Lichen Planus, a condition characterized by tiny reddish or purplish elevations on the wrists, ankles, or other extremities.

Numerous skin problems are related directly to stress-induced hormonal changes. Too much androgen, a "male" sex hormone released under stress, causes increased sebaceous secretions, the loss of hair on the scalp, and the growth of hair on the face. Increased production of steroids also stimulates the sebaceous glands, leading to acne, oiliness, broken blood vessels, poor healing, thinning of the skin, and infection. Imbalanced insulin levels not only cause diabetes but produce numerous side effects, such as a flushed complexion, persistent itching, brownish or yellowish lesions, fungal, bacterial, and yeast infections, and poor healing. Thyroid imbalances affect metabolism and growth. High levels cause increased blood flow and flushed skin, excessive sweating, hair loss, and the separation of nails from the nail bed. Low levels produce dry, brittle hair, loss of eyebrows, and premature wrinkles. Finally, increased levels of neuropeptides released into the skin by sensory nerves due to stress can cause local inflammation, psoriasis, atopic dermatitis, eczema, hives, shingles, and other allergic reactions.

Of course, not all hormonal changes that affect the skin are induced by emotional stress. Many are a normal part of life. In females, the ovaries produce estrogen, the hormone that controls menstruation. Estrogen is necessary for smooth, hydrated skin, and also suppresses sebaceous gland activity. When estrogen levels naturally decrease prior to a woman's monthly period and at menopause, the consequent increase in sebaceous secretions often produces acne in sensitive and oily skin, especially if imbalances already exist.

Nevertheless, according to Boston dermatologist Robert Griesemer in *Emotions and Your Health*, emotional stress is the key factor in many types of skin disorders. He reports that 98 percent of severe itching conditions; 95 percent of warts; 94 percent of acne-like symptoms; 86 percent of pruritus, or itching; 68 percent of hives; 62 percent of psoriasis; and 56 to 70 percent of eczema conditions are stress-related or induced.

As the research documents, stress produces a broad range of effects on the skin and body. Through the language of hormones, the skin and immune system "know" exactly what we think and feel at every moment, and reflect our thoughts through their functioning. In this sense, Western science has confirmed Ayurveda's ancient teaching that most physical illness does have an emotional subtext. When we fail to "speak our mind" through the natural organs of expression, we create emotional stress. In the form of messenger molecules, our unspoken feelings inevitably find other outlets in the body, causing any number of ailments that express the feelings for us. Ulcers can announce that something is "eating at" us, coughs and congestion may indicate that there is something we need "to get off our chest," hypertension reveals that we are "steaming mad," and, as you will see, skin disorders often speak of a range of unexpressed sensations, each with its own "voice": wet eczema weeps for our sorrows; worry lines confess our fears; the burning rash seethes with our bridled rage.

The choice of language here is not merely metaphorical. Ted A. Grossbart, a psychology instructor at Harvard Medical School who treats people with skin disorders in private practice, notes that visible cutaneous symptoms often "reflect the [emotional] conflict symbolically." In *Mind/Body Medicine*, he cites the case of an unhappily married patient who developed a severe skin allergy on the finger where she wore her gold wedding band, although, he writes, "other fingers with gold rings were fine."

In my own practice, I have found that the destructive effects of "undigested" emotions, as I call them, are often apparent when a client is unable or unwilling to formulate and communicate feelings directly. One of the worst cases I have ever seen was an eight-year-old girl who suffered from mild autism and almost never spoke, although she was indeed capable of speaking a little and understood everything. In school, she had been forced by a misguided teacher to talk in front of her class. Afterwards, she ran home to her parents and broke out in a rash over every inch of skin from the top of her head to the soles of her feet. When they brought her to see me three weeks later, the rash remained unabated, since the unfortunate child still had no healthy outlet for her frustration and rage.

According to Ayurveda, none of these effects happens at random. Each of us translates the perception of threat or stress into a specific emotion—anger, depression, or fear—depending upon our constitution. The stress response mechanism is universal, but its effects will look different on individual skin types as a result of the specific hormones our emotions release. Any imbalance—indeed, any psychophysiological event—leaves its message on the skin quite clearly. However, most people do not know how to read its script. Ayurveda teaches us this remarkable language.

IMBALANCE AND IMMUNITY: THE AYURVEDIC VIEW

"He is the wise physician and philosopher who realises that in regarding the external appearance of his fellowman he is studying the external nervous system and not merely the skin and its appendages."

FREDERIC WOOD JONES

According to Ayurveda, ojas is the source of the body's immunity, and the subtle substance that gives healthy skin its natural glow. When ojas is depleted, the immune response is weak and the skin becomes lifeless. Low ojas and low immunity result from the same cause: an overload of physical, chemical, or psychological stress. Ayurveda describes this overloaded condition as an imbalance of the five elements (or three doshas), the seven dhatus, and the three waste products that together make up all the structures and substances of the bodymind. Therefore, a strong immune system and beautiful skin require strong ojas, a balanced constitution, and the proper transformation and function of the dhatus and waste products. The relationship, then, between the doshas, the dhatus, malas, and ojas is the key to understanding the causes of aging and disease in Ayurvedic terms.

THE DHATUS, MALAS, AND OJAS

The seven dhatus, or body tissues, are plasma, blood, muscle, fat, bone, bone marrow and nerve tissue, and reproductive tissue. As you read in Chapter 1, they develop in sequence; each dhatu is formed out of the "raw material" of the preceding tissue. The process starts with plasma, or *rasadhatu*—the first tissue, which is essentially the body's nutrients. Plasma forms blood, blood tissue gives rise to muscle tissue, and so forth. In this sense, all bodily tissue is rasadhatu *transformed.*

The process of dhatu transformation is the key to understanding the health of the skin. This is true for two reasons. First of all, each of the seven dhatus has a functional link to one of the seven layers of skin. As a result, whatever happens to any dhatu will have an effect on the skin. Conversely, any substances absorbed into the skin will also affect the individual body tissues. Second, the overall health of the skin depends directly upon the quality of rasadhatu, from which the skin derives its nutrients. In fact, the skin is known as the *rasasara*—"the cream that rises to the top." Just as cream contains in concentrated form all the components of the milk from which it comes, so too does skin contain the essence of rasadhatu. By analyzing the condition of the skin, a skilled Ayurvedic practitioner can detect deficiencies in rasadhatu—that is, in the body's nutrients—just as by analyzing the cream—the rasasara—we can see what is in the milk. In this sense, whatever we eat literally shows up on the skin.

Essentially, dhatu transformation is the conversion of nutrients into bodily substances in order to nourish, maintain, heal, and regenerate the cells. Rasadhatu gets its nutrients directly from the food we consume—both edible and emotional—and then provides nutrients in turn to the other dhatus. The complete assimilation of nutrients by all seven tissues takes about forty days, or five to six days for each dhatu. Of course, no dhatu can get the nutrients it needs if we eat the wrong substances, or do not digest them adequately because of low agni or stress. However, once a dhatu is not functioning properly because of an imbalance, the assimilation process itself will be disturbed. As a result, that tissue will not be able to get all its necessary nutrients—regardless of food quality—and neither will the tissues that form from it. Moreover, the improperly metabolized food becomes *ama*, or toxins, in the blood, which is a key factor in disease and particularly in skin disease. Of course, imbalance or dysfunction of a particular dhatu results in specific physical symptoms. Because the dhatus and skin layers are connected, an Ayurvedic practitioner can tell which dhatu is affected from the type of problems appearing on the skin. Indeed, the knowledge of the skin's relationship to the dhatus, and the *upadhatus*, or subsidiary tissues, is one of the most important diagnostic tools of the *vaidyas*, the Ayurvedic physicians.

For example, acne rosacea is a blood disease and therefore originates in the second tissue. Whether the stressor is the wrong food or an undigested emotion, it takes a couple of weeks from the time the stress occurs for this type of acne to break out. That is how long it takes for toxins from our undigested food or emotions to affect the blood tissue. Cystic acne, on the other hand, is a disease of the fat tissue—the fourth dhatu—and takes as long as three to four weeks to appear on the skin. Cancer, a disease of the bone marrow and reproductive tissues, or sixth and seventh dhatus, may exist in the body as long as seven years before we are able to diagnose its symptoms. As you will see, an experienced vaidya is able to diagnose and treat the imbalances that lead to cancer in much earlier and less critical stages of disease, based upon their knowledge of the dhatus and doshas. (Appendix A describes in detail the relationships between the skin layers and dhatus, as well as their functions and effects.)

Healthy dhatu transformation not only provides the raw material and nourishment for each tissue but also creates a waste by-product, or *mala*. In Ayurvedic medicine, the proper formation and elimination of waste is as important to balance as proper nutrition, and vaidyas can diagnose imbalance and disease from specific changes in the color, form, and quality of the feces, urine, sweat, and menstrual blood, for example. Although they are the "unretainable substances" of the body, these wastes are necessary for the functioning and nourishment of certain organs. When dhatu transformation is healthy, plasma produces mucus and menstrual discharge as waste; blood produces bile; muscle produces ear wax, tartar, and sebaceous secretions; fat produces sweat; bone produces beard, body hair, and nails;

and marrow and nerves produce head hair and breast milk. The seventh dhatu, the reproductive tissue, does not produce a waste product. As long as tissue formation and metabolism function properly, however, it does produce the super-fine essence of the skin and body—the ojas. Indeed, ojas is the vital essence of all tissue, and it must be properly distilled at each stage of tissue formation in order to create the super-fine essence that is the basis of our immunity as well as the source of absolute beauty.

A quick look at fetal development from the Western viewpoint brings into clear focus not only the process of dhatu transformation and the link between the dhatus and the skin layers, but also the notion that skin is our second brain and the mirror of our thoughts and emotions. As a fertilized egg develops into an embryo, it forms three cell layers known as the *ectoderm*, *mesoderm*, and *endoderm*—the outer, middle, and inner skins from which all other body tissues and organs subsequently grow. The cells of the outer skin differentiate to form the epidermis, central nervous system, sense organs, hair, nails, and so on. The middle cells form the dermis skin layer; muscle, blood, and lymph cells; reproductive tissue; kidneys; spleen; and connective tissue. The inner cells form the epithelial linings of the vascular system, lungs, thymus, thyroid, liver, pancreas, and the gastrointestinal, respiratory, and urinary tracts.

It seems self-evident from this developmental viewpoint that the skin and mind would mirror each other completely, since they both arise from the same embryonic tissue. In fact, the *upturned* surface of ectoderm becomes epidermis, the top layer of skin, and the *inturned* surface of ectoderm becomes the central nervous system, brain, and spinal cord. As Ashley Montagu observes: "The nervous system is, then, a buried part of the skin, or alternatively the skin may be regarded as an exposed portion of the nervous system. It would, therefore, improve our understanding of these matters if we were to think and speak of the skin as the external nervous system."

Indeed, Ayurveda recognizes a direct correlation between the quality of our emotional life and the quality of the dhatus. The fires of agni digest thoughts and experience as well as food and drink as they transform one dhatu into another out of these substances. In Western terms, we assimilate these mental "foods" into the body in the form of hormones. The Ayurvedic understanding is somewhat more encompassing. We would say that we absorb the subtle energies of all nutrients—physical and psychological—not only into the body, but also directly into consciousness through the operations of the three doshas. In this way, the qualities of our emotions are reflected in each of the dhatus, just as the qualities of our food are.

For example, rasadhatu is nourished by the experiences of joy and satisfaction, and likewise nourishes these qualities in consciousness. An imbalance in this dhatu will create, in addition to certain physical effects, *feelings* of lack, emptiness, and even malnourishment. The word "rasa" means emotion or essence; and as Lonsdorf, Butler, and Brown write in *A Woman's Best Medicine*, the first taste of food not only is transformed into rasadhatu "but also nourishes the first awakening of feeling, the first satisfaction of a desire, resulting in bliss." If rasadhatu is imbalanced, we literally may lose our "taste" for food and life. By the same token, strong and continuing feelings of dissatisfaction, depression, and apathy also disturb the functions of rasadhatu, and will show up as specific problems on the skin. Similarly, if there is a dysfunction of the bone tissue—the supportive tissue—we may feel weak and unable to "stand tall"; conversely, continual feelings of lack of emotional support can result in bone tissue disorders.

In the table opposite you can see the reciprocal affects of the dhatus and emotions. To fully understand the disease process in terms of Ayurveda, however, we also have to look at the role of the doshas, as well as the all-important factor of our own constitutional makeup.

THE DOSHAS AND DISEASE

In Ayurvedic terms, the process of aging and disease begins quite subtly in the body or mind when stress disturbs the natural balance of our doshas, and it develops into full-blown physical disease when the imbalanced doshas disrupt the functioning and transformation of the dhatus. When we are balanced, the subtle energies and substances of the doshas naturally are concentrated in the specific sites of the body that we described as the doshas' seats: Vata is localized in the colon and kidneys; Pitta in the liver and small intestines; and Kapha in the stomach and lungs. The first stage of imbalance occurs when a dosha starts to *accumulate* in its natural seat

IMBALANCE: THE MIND-BODY CONNECTION /81

DHATU	EMOTIONAL EFFECTS WHEN DHATU IS HEALTHY	EMOTIONAL EFFECTS WHEN DHATU IS IMBALANCED
Plasma (*Rasa*)	Joy, contentment, nourishment, satisfaction.	Depression, lack of energy, malnourishment, restlessness, eating disorders.
Blood (*Rakta*)	Stimulation, exhilaration, ambition.	Loss of zest for life, no excitement or ambition, anger, hate, jealousy.
Muscle (*Mamsa*)	Buffering, nurturing, forgiving, courageous, secure.	Helplessness, insecurity, lack of nourishment, increased passivity and attachment.
Fat (*Meda*)	Lubrication, love, commitment.	Loneliness, lack of love, not enough lubrication.
Bone (*Asthi*)	Support, courage, creative, active.	Indecisiveness, cannot stand still or tall, lack of support, courage, self-confidence, and creativity.
Bone marrow and nerves (*Majja*)	Fullness, self-assurance.	Loss of strength or confidence, feeling of "I'm getting old," holding on to the past.
Reproductive tissue (*Shukra*)	Vigor, romance, creative and procreative, purposeful.	No joy, no romance, life feels "dried up," ojas gets depleted.

because of stress and improper lifestyle. Typically, but not always, the first dosha to go out of balance is the one that is already most prominent in our constitution. If nothing is done to correct the imbalance, the dosha becomes *aggravated*. It starts to move first into surrounding areas and then deeper into the body—that is, deeper into the tissues. In the later stages of the disease process, the aggravated dosha again localizes, this time in tissue where it does not naturally belong, but where ama, or bodily toxins, has created a physical weakness.

It is only in the three final stages, when overt physical problems begin to appear in the tissue, that Western doctors detect what they consider symptoms of disease. Ayurveda, on the other hand, has catalogued specific symptoms of imbalance—both physical and psychological—for six distinct stages of disease development, which gives the vaidyas three more opportunities for disease prevention than their Western counterparts. Whether the symptom is as innocuous as a case of gas caused by accumulated Vata in the colon, or as pervasive as a case of psoriasis due to Vata moving into a deep tissue, a trained practitioner can diagnose not only which doshas and dhatus are involved, but also the origins and severity of the problem.

Again, Western science is discovering piece by piece what Ayurveda has taught for thousands of years. Medical researchers recently learned, for example, that people who gray prematurely are at greater risk of developing osteoporosis. This finding is good news for everyone, because it gives potential victims an early warning signal and a chance to take steps to avoid a crippling disease. Nonetheless, this is ancient news for Ayurvedic practitioners, who have long known that hair is a waste by-product of bone tissue formation, and that early graying is a sign of bone tissue imbalance which, left unchecked, typically progresses to bone disease.

The progression of imbalance is always through the same six steps, but the exact course of an aggravated dosha through the body tissue, and the resulting symptoms, differ depending upon our constitution and skin type, as well as upon the health of the dhatus, which is the source of our immunity. The symptoms differ because each of us metabolizes stress in a characteristic way according to our innate psychophysiological tendencies. Typically, dry skin types (Vata) get worried, fearful, and agitated under stress, inducing the post pituitary gland to release antidiuretic hormones that affect the kidneys and cause excessive dehydration. Sensitive skin types (Pitta) get angry, frustrated, or annoyed, triggering the release of adrenaline and resulting in the characteristic flushed, hot skin of someone ready for a fight. Oily skin types (Kapha) become depressed, negative, and inactive, inducing various hormones that stimulate excess sebaceous secretions and cause the body to retain water.

MENTAL TENDENCIES UNDER STRESS:	PHYSICAL SYMPTOMS UNDER STRESS:
VATA	
Anxious, nervous, fearful, agitated, distracted, indecisive, hysterical.	Excessive dryness, wrinkles on forehead, discoloration, dry eczema, psoriasis, dandruff, nails are brittle, bitten, or have longitudinal striations; cracked lips and feet, split ends on hair; constipation, gas, nerve pain, joint problems, fissures on palms and soles, epilepsy, kidney problems, bloating, trembling.
PITTA	
Overly aggressive, angry, frustrated, annoyed, rageful, critical.	Whiteheads, rash, allergic reactions, burning eyes and feet, acne rosacea, contact dermatitis, broken capillaries, burning eczema, redness, heavy sweating, horizontal ridges on nails; hyperacidity, peptic ulcers, bleeding tendencies, liver diseases, hypertension, inflammatory diseases.
KAPHA	
Negative, depressed, lethargic, possessive, unable to "let go,"	Excess oiliness; cystic acne and all cystic conditions including fibroid tumors; lack of skin tone, flabbiness, double chin, puffiness, edema, excessive sweat (due to excess water), oozing or itching eczema, swollen feet and ankles; weight gain; bumps or "parrot beak" at tips of nails; asthma, colds, cough, coronary heart disease, diabetes, urinary stones.

Again, not all stressors are emotional or psychological. In Ayurvedic terms, anything that imbalances the doshas is, in effect, a stressor. In the Skin Problem Quiz below and in the later chapters on specific balancing therapies, we describe the many physical, environmental, behavioral, and chemical factors that imbalance the doshas. According to the principle of like increases like, too much spicy food can imbalance Pitta just as easily as too much emotional stress. As you will see, any experience you have through the body, mind, and senses creates balance or imbalance depending upon your constitution.

Keep in mind, however, that your constitution itself never changes. The *condition* of the doshas changes day by day, but the natural balance of energies—the "set point" of your doshas—is unalterable. It is your nature at birth, and it is your nature for life. What we have in our power to change is not who we are, but *how we perceive.* Stress is the fundamental cause of aging and disease, and stress is nothing more than our perception. The easy flow of experience and peace of mind that we recognize as bliss results from living in accord with our nature, not in fighting against it. Change your *mind,* and balance and bliss will be yours. Just as emotional stress triggers a specific set of hormonal reactions, so does emotional contentment. Happy thoughts produce *endorphins,* the body's natural painkillers, that stimulate the immune system and promote health. Indeed, each constitutional type "expresses a different flavor of pure joy," as Chopra says, just as it expresses a different flavor of stress. When the doshas are balanced, Vata is naturally upbeat, exuberant, creative, flexible, buoyant, and alert; Pitta is energetic, joyful, affable, outgoing, and clear-minded; Kapha is loving, courageous, giving, forgiving, strong, steady, easygoing, and perceptive. In the final chapters of the book, you will learn how to find freedom from stress and the pure taste of joy by learning how to "change your mind" through meditation practices and the principles of action.

Whether or not you have a problem now, look at the Skin Problem Quiz and read the explanation that follows. Not only will it help you to recognize signs of imbalance should they arise later, but it will also enhance your understanding of the many ways we can "read" the state of mind and body on the skin.

THE SKIN PROBLEM QUIZ

This is the follow-up to the Skin Type Quiz for those who have a skin problem now. The purpose is to determine which imbalanced dosha (or doshas) is involved in your current skin problem. This is *not* a quiz to find out *if* you have an imbalance. As mentioned, it is possible to have an imbalance and not have a visible skin problem. If you have been diagnosed with other health problems, you can see the common physical disorders associated with each dosha in the table on page 83. Again, the quizzes and charts in this book help identify the causes of imbalance, but they are not tools for diagnosing disease and should not be used as a substitute for a professional medical examination.

Complete the following statements and score as instructed.

1. My skin condition is best described as

    (Check one or more responses if they apply.)

    a. __ dry, flaky scaly patches, and/or wrinkles

    b. __ burning red patches, and/or small reddish watery pimples and/or rashes

    c. __ large, whitish, pussy eruptions, sometimes with itching and/or deep scarring

2. The problem appears primarily on my

    (First look at choices a, b, & c. If one or two of these apply, check one or both. If all three apply, check d ONLY.)

    a. __ forehead

    b. __ T-zone and/or cheeks

    c. __ mouth, chin & neck area

    d. __ It appears all over my face and/or other places on my body

To score 1 & 2: Add up your responses in each column above and write the totals here. *Write "0" if you had no response in a column.*

   a. __   b. __   c. __   d. __

In the table below, find the set of numbers in the left column that matches your totals, then look in the right column to see what imbalance(s) you have.

IF YOU SCORED:				THEN YOU HAVE AN IMBALANCE OF:
a	*b*	*c*	*d*	
Group (I)				*One dosha*
2	0	0	0	Vata
0	2	0	0	Pitta
0	0	2	0	Kapha
Group (II)				*Two doshas*
2	1	0	0	Vata + Pitta
2	0	1	0	Vata + Kapha
1	2	0	0	Pitta + Vata
0	2	1	0	Pitta + Kapha
0	1	2	0	Kapha + Pitta
1	0	2	0	Kapha + Vata
Group (III)				*Two or possibly three doshas*
1	0	0	1	Vata + other(s) to be determined
0	1	0	1	Pitta + other(s) to be determined
0	0	1	1	Kapha + other(s) to be determined

Score not listed above. This may be due to a scoring error or to a multiple imbalance. Try redoing the questionnaire, taking more time to consider your answers. If your score stays the same, just follow the guidelines for Group (III).

For questions 3–6 below, check the responses that best describe you.

3. My problem tends to recur or get worse when

    a. __ I am very active or traveling   b. __ I have an allergic reaction   c. __ my period is due   d. __ I don't know

4. My problem is affected when the weather is

a. __ cold, dry,   b. __ hot   c. __ damp   d.__ I don't know
    or windy   and cold

5. My problem tends to recur or get worse under stress, especially when I feel

a. __ anxious   b. __ angry   c. __ sad   d.__ I don't know
__ worried,   __ jealous   __ depressed
    fearful   __ frustrated   __ overly
__ hysterical   attached

6. I tend to eat or drink a lot of

a. __ raw   b. __ spicy, oily   c. __ sweets   d.__ I don't know
    vegetables   food   __ fatty or
__ salads   __ shellfish   fried foods
__ cold food   __ sour fruit/   __ dairy, cheese
__ caffeine   juice   __ beef, liver,
                  __ soda   wheat germ

To score 3–6.

Add up your responses in each column above and write the totals here.

a. __   b. __   c. __   d.__

IF YOUR HIGHEST SCORE IS:	THEN YOUR BEHAVIOR MAY BE AGGRAVATING:
a	Vata
b	Pitta
c	Kapha
d	If you scored highest in this column, it is a "red flag" that you need to be more conscious of how your lifestyle affects your health and skin.

For statements 7 and 8 below, check as many responses as apply. Leave blank if none apply.

7. My problem appeared or worsened at the time of

    ___ puberty   ___ illness and/or its treatment

    ___ pregnancy   ___ other

    ___ menopause

8. I use(d) drugs and/or stimulants including

    ___ beta blockers and/or other   ___ tranquilizers
        blood-pressure or heart medication   ___ estrogen therapy

    ___ diuretics   ___ tobacco

    ___ antibiotics   ___ aspirin

    ___ prednisone or other drugs for   ___ beer, wine, or other alcohol
        hypersteroid conditions   ___ antihistamines

    ___ antidepressants   ___ coffee or caffeinated drinks

    ___ antihypertensives   ___ decongestants

    ___ cortisone   ___ other

Do not total your responses to 7 and 8. See discussion below.

. . . .

WHAT YOUR RESULTS MEAN

IMBALANCE AND ITS INDICATORS (QUESTIONS 1 & 2):

Group (I): If you scored in this group, then your problem is due to the imbalance of one dosha. Almost without exception, the imbalanced dosha will be the same as your leading dosha—that is, the one associated with your basic skin type. For example, a person with normally oily skin, characteristic of Kapha, is most likely to have skin problems because of Kapha imbalance.

Group (II): If you scored in this group, then your problem is due to the imbalance of two doshas—your leading dosha and the second dosha identified on the score chart.

Group (III): If you scored in this group, then your problem *may be* due to the imbalance of more than one dosha—probably your leading dosha plus *one* or *both* others. To pinpoint multiple imbalances often requires more experience and more data than you get from just filling in a questionnaire—as well as deeper knowledge of Ayurvedic principles. Keep in mind, however, that your basic skin type is the most important fact you need to know to use Ayurveda to improve your complexion and bring out your natural radiance, and even without knowing a specific imbalance, you can greatly improve your skin just by following the routines prescribed in Part III for your type.

In fact, no matter what imbalance you have, your dominant dosha and therefore your basic skin type and characteristics never change. A person born with dry skin, for example, may show signs of Kapha imbalance such as puffiness or weight gain in the face; but a closer look at the complexion always proves that its inherent Vata qualities—dry texture and fine pores overall—remain unchanged. In other words, if your leading dosha is Vata, you will most likely not have very oily skin or enlarged pores even with a Kapha imbalance; if your leading dosha is Kapha, you will probably not get very dry skin or fine pores even with a Vata imbalance.

Location, location, location—the physical key to detecting your imbalance. When you have a skin problem, one easy way to determine the source of imbalance is to observe *where* the problem appears on your complexion. In Ayurveda, every part of the face is associated with a body organ and every organ is associated with a dosha. The forehead is linked to the colon, the seat of Vata; the nose and cheeks are linked to the small intestine, the seat of Pitta; the mouth, chin, and neck are linked to the stomach and chest, the seats of Kapha. As you recall, the seat of a dosha is simply the place in the body where its energy is most concentrated, and consequently it is also the first organ affected when that dosha goes out of balance. Any change in the functioning of the organ produces a corresponding change on the related part of the face, thus creating an external signal of the internal imbalance. This makes single-dosha imbalances fairly easy to detect, since their symptoms tend to be confined to one area of the face—the area connected to the seat of the dosha. Thus, Vata imbalance typically shows up as dryness on the forehead;

Pitta imbalance, as inflamation on the nose and cheeks; and Kapha, as oiliness around the mouth, the chin, and the neck.

Indications of a two-dosha or three-dosha imbalance. If a skin condition is widespread rather than localized in one key area of the face, this is usually an indication that more than one dosha is imbalanced. For example, the classic symptom of Pitta imbalance is a rash or other reddish inflammation appearing on the nose and cheeks. A rash that spreads to the *forehead* usually indicates a secondary Vata imbalance; a rash around the *mouth* and *chin* indicates a secondary Kapha imbalance. Wherever it appears, redness is always a sign that the fires of Pitta are involved, whether we have hives from a food allergy or just turn red in the face due to hot weather or a hot temper. This is true for the symptoms of every dosha: extreme dryness anywhere indicates Vata imbalance, for example; oozing always indicates excess Kapha. However, when symptoms of one dosha appear on a part of the face associated with another dosha, then both doshas are probably imbalanced.

If you are unsure which dosha or doshas are involved in a problem, use the "face map" opposite to help identify the imbalances. It illustrates not only the typical Vata, Pitta, and Kapha areas, but also other telltale blemishes and lines whose appearance signals specific problems.

The emotional link. The site of your condition gives a clue to emotional as well as physical factors in your imbalance. Through the hormone connection, we know that our emotions affect the body in specific ways, causing characteristic changes. Fear, for example, stimulates production of antidiuretic hormones that affect kidney function, causing dehydration throughout the body and halting the secretion of digestive juices. Dry mouth, a classic symptom of fear, is an immediate and obvious sign of this effect. Less obvious is the constipation caused in the *colon* when fear or worry go unreleased. Therefore, the appearance of dry skin on the forehead not only indicates a colon problem and a Vata imbalance, but also points to deep feelings of fear, worry, or anxiety as a probable cause.

On the next page, you can see the relationship of the doshas, your physical symptoms, and their emotional causes.

THE FACE MAP KEY
PHYSICAL SIGNS OF IMBALANCE

1. *Horizontal lines on forehead:* worries; excess intake of liquids, sugar, fats.
2. *Right vertical line:* weak liver.
3. *Left vertical line:* weak spleen.
4. *Deep horizontal groove at root of nose:* prone to allergies; low sex drive.
5. *Crow's feet:* weak eyesight (squinting); weak liver.
6. *Bags under eyes:* water retention (kidney problem)
   or
   Purple discoloration: poor blood circulation.
7. *Tender area at center of cheek:* sinus congestion; digestive problems.
8. *Red nose tip:* overworked heart.
9. *Deep vertical line by side of mouth:* weakness in reproductive system.
10. *Prominent laugh lines:* pancreatic problems; diabetes.
11. *Vertical lines on top lip:* cigarette smoking; lack of sexual activity.
12. *Spots on lips:*
    Whitish discoloration: parasites in colon.
    Blue-purple discoloration: poor circulation due to constipation.
    Dark spots: colon problems.
13. *Double chin:* slow, weak thyroid.
14. *Ring around iris:* high cholesterol.
15. *Prominent temporal vein:* high blood pressure; anger; anxiety.
16. *Uneven jawline:* TMJ.
17. *Deep vertical groove on chin:* grief; frustration; strong sex drive.

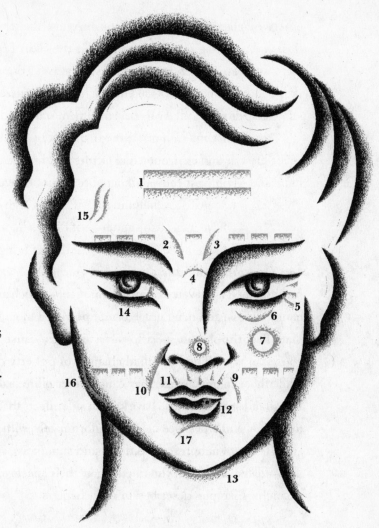

SYMPTOMS ON	LINKED TO	SEAT OF	EXPRESS
Forehead	Colon	Vata	Anxiety, fear, worry
Nose and cheeks	Small intestine	Pitta	Anger, jealousy, frustration
Mouth, chin, and neck	Chest, stomach	Kapha	Grief, depression, attachment

OTHER FACTORS IN IMBALANCE (QUESTIONS 3–8):

According to the principle of like increases like, a dosha becomes aggravated when you continually eat foods, use substances, have thoughts and emotions, or otherwise engage in behavior that—in Ayurvedic terms—is qualitatively *like* that dosha. The last group of questions in the Skin Problem Quiz is designed to increase awareness of how your lifestyle affects your doshas and, therefore, your skin condition. Remember, Ayurveda means *knowledge* of daily living—and self-knowledge is where all Ayurvedic treatment starts and ends.

Your habits and circumstances: In questions 3–6, each of the columns a, b, and c lists lifestyle and environmental factors that imbalance one specific dosha: Vata, Pitta, and Kapha respectively. Therefore reduce or avoid exposure to factors in:

Column a   if Vata is high

Column b   if Pitta is high

Column c   if Kapha is high

Growth, illness, and drug usage. Questions 7 and 8 list life events that may precipitate or aggravate skin problems. Factors such as puberty, menopause, pregnancy, illness, and drug usage create powerful biochemical changes in the body that affect the skin in specific ways, and may cause problems depending upon your skin type. The hormonal changes of puberty cause oiliness, for example, while those in menopause may cause either oiliness or dehydration and dryness. Drugs and stimulants also have known side effects that are particularly aggravating to certain skin types (see below). Unfortunately, with the exception of pregnancy and the use of nonprescribed drugs and stimulants, such events in life are largely unavoidable. However, you can mitigate their effects on your skin by using various balancing therapies described in later chapters.

DRUG OR STIMULANT	GENERAL SIDE EFFECTS	AGGRAVATES
Tobacco	Dehydration, wrinkles	Vata
Antihistamines, antidepressants, antihypertensives, anticancer drugs, antibiotics, decongestants, tranquilizers, diuretics, aspirin	Dehydration of skin, reduced blood volume	Vata

Beta blockers, other blood pressure or heart medications	Hair loss, dermatitis	Vata
Coffee, alcohol	Expansion of blood vessels causing acne rosacea, red nose, swollen capillaries	Pitta
Cortisone, estrogen therapy	Acne	Kapha
Prednisone	Edema, water retention, ruddy, flushed complexion, coarsening of facial features	Kapha

ALWAYS CONSULT YOUR PHYSICIAN

The Skin Problem Quiz is a useful tool to identify common skin conditions, basic imbalances, and tendencies in your health at this time; but it is not a replacement for a traditional constitutional analysis, which is best done in person by someone who is trained in the complex and subtle art of Ayurvedic diagnostics. Furthermore, neither this self-analysis nor one conducted directly by an Ayurvedic practitioner is a substitute for a thorough medical examination. If you have any health problem, whether a skin condition or otherwise, always consult a licensed physician before you begin any treatment.

The remainder of this book contains the many Ayurvedic therapies and treatments to help you achieve and maintain balance, beauty, and bliss. If you have had chronic skin problems and have been disappointed before by unfulfilled promises of relief, I urge you to open your mind to this very different approach to health and skin care, which I have seen help many.

About ten years ago, *Redbook* magazine sent me just such a client because they wanted to do an article about my work featuring a makeover. In her thirties at the time, this woman had suffered most of her life with acne rosacea, and though she had been treated by famous doctors and had tried everything from acutane and antibiotics to retin A, she had experienced no relief. By the time she came to me, her condition was one of the worst I had ever seen. Not only was her face covered with red pimples, but her forehead had deep vertical wrinkles from eczema, and the burning sensation on her skin was so severe she could not even apply water to her face without great pain. Moreover, she was outraged over her problem and at the world that had failed to help her.

Although I usually begin treatment of new clients with a facial to cleanse and nourish the skin, because of this woman's condition I began instead by suggesting adjustments in her diet, breathing exercises and meditation to calm her emotions, herbs for internal detoxification, and healing Ayurvedic oils to use externally at home on her infected skin. At this last suggestion, she suddenly spoke up, barely able to contain her hostility and cynicism. "Are you sure I should be using *oil* on my face? I haven't used it in thirty years."

I looked at her complexion and her suffering. "*This* is the result of those thirty years," I replied, "so why not try something different?"

Four days later the woman returned to my office. She had followed my advice, and for the first time in decades she actually saw and felt an improvement in her skin. Now that her worst pain was gone, I was able to apply a healing mask and soothing oils to her face. She then continued at home with the balancing therapies and daily applications of herbal oils, and within four weeks of treatment, the results on her skin were obvious.

This story has a happy ending for all concerned: My client got healthier, clearer skin, and relief from years of misery; *Redbook* got its feature; and I got six thousand calls for my products within two weeks—so that many other acne sufferers ultimately benefited from this success as well. My purpose in telling this story is not to boast, however, but to underscore two important points. First, for all the women and men, like my client, who have lived for years with a serious skin problem—and clearly there are many of you—I want to reassure you that there is a solution. Second, I want you to see how the knowledge of your Ayurvedic skin type and the principle of like increases like makes the difference between success and failure in the treatment of all complexion problems, as well as in the prevention of aging and the achievement of healthy, radiant skin.

From the moment this woman walked into my office, I could see from her general appearance, as well as from her reddened complexion and her angry disposition, that she was born with a fiery constitution and now had a severe Pitta imbalance. This is the information I had that the woman's famous doctors lacked: All the medication she had received—the antibiotics, the acutane, and the retin A—aggravate Pitta (as well as Vata). Inherently drying and stimulating, these drugs had exacerbated her already inflamed condition and had dehydrated her body to such a degree that her skin was raw for want of moisture and nourishment. Like most acne sufferers, she had believed that oil was the cause of her problem and conscientiously had avoided it all her life; yet in her case, soothing, cooling oil was exactly what was needed. In fact, as you shall see in Chapter 5, the *right* kind of essential oil, suited to your skin type and applied in appropriate amounts, is *good* for all skin—even if your complexion is oily to start. Ironically, had this client been born with more Kapha than Pitta in her constitution, and had she developed cystic acne—which results from the imbalance of the natural wateriness and oiliness in an earth constitution—the dehydrating, drying prescriptions from the doctors might have at least relieved her symptoms—though they would not have been a cure. Indeed, *no* topical intervention, medication, or diet alone—not even an Ayurvedic one—can bring total balance. Without techniques to reduce emotional stress, our "undigested" emotions continue creating imbalance and toxicity, which together are the root of all health problems.

Ultimately, the terms "skin problem" and "skin disease" have no practical meaning. From the point of view of Ayurvedic medicine, we may experience one of two conditions in life: a state of balance or a state of imbalance, each of which manifests on the skin in specific ways according to our innate makeup and our personal lifestyle. When we are in balance—when we are healthy in body and mind—we are naturally resistant to all disease, and the skin is naturally glowing. When we are out of balance, whether the symptoms appear on the surface of the skin or somewhere else in the body or mind, the cure is the same: To balance ourselves internally and externally on all levels of life—body, breath, mind, and soul.

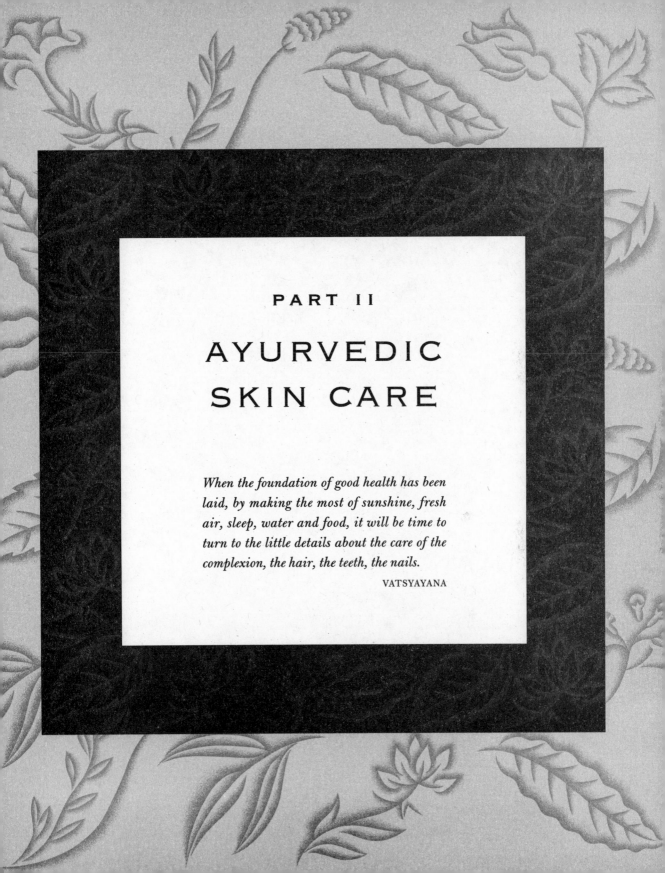

PART II

AYURVEDIC SKIN CARE

When the foundation of good health has been laid, by making the most of sunshine, fresh air, sleep, water and food, it will be time to turn to the little details about the care of the complexion, the hair, the teeth, the nails.

VATSYAYANA

CHAPTER 4

FOOD FOR
THE SKIN

Rejoice plants, bearing abundant flowers and fruit,
triumphing together over disease like victorious horses,
sprouting forth, bearing men safe beyond disease.
RIG VEDA

Hundreds of new beauty products come onto the market every month, and a burgeoning number of new "natural" bath and body shops have sprung up in suburban shopping malls and on city streets in just the past five years. Consumers have not only more skin care options than ever before, but also more confusion about what treatments are truly good for the body and skin. Misleading federal standards for labeling and advertising "natural" products do not protect the average buyer, who simply does not have the knowledge or experience to evaluate either the chemical contents in a product or the health ramifications of their use. If we are to believe the manufacturers, *every* product is good for everyone, and all of them are made with "exclusive" formulas rich in pure-sounding ingredients like "botanicals," "biominerals," "enzymes," and "extracts." While many of them do contain authentic natural ingredients, what the advertising headlines and large copy usually fail to mention are the many other synthetic and chemical ingredients—such as dyes, preservatives, emulsifiers, and fragrances—that also go into these products, and therefore into our skin. Despite FDA approval, many of the ingredients we use today have not been around long enough for us to know their long-term effects on health. Some may have questionable effects that government standards nonetheless consider to be at "acceptable" levels.

Ayurveda sets much higher standards than either Western medicine or government regulators, because its basic view of health, and its concept of "natural," is at once more encompassing and more refined. Moreover, having "tested" its theories for six thousand years directly on human "subjects," Ayurveda justifiably may claim to know the long-term effects of its treatments and products.

What difference does it make what we put on the skin? What is pure and natural by Ayurvedic standards? What ingredients go into Ayurvedic skin care products, and what is so special about them? You will find the answers to these questions below, and in Appendix B you will find simple instructions for making Ayurvedic formulations at home. If you prefer to buy the products, you can find my Bindi line in many health food stores around the country, or you can order both Bindi and Tej products as well as other Ayurvedic products and many of the ingredients for making them from suppliers listed in Appendix C.

SKIN EATS

If you were offered a meal made with cetyl alcohol, iodopropyl butylcarbamate, sodium dodecylbenzenesulfonate, disodium EDTA, BHT, red dye #17, and yellow dye #10, would you want to eat it? Millions of people do "eat" these chemicals every day (you may even be one of them), since they are standard ingredients in two of the most popular brand-name moisturizers and beauty soaps that Americans use on the skin—one of the body's major organs of absorption.

Although we get most of our nutrition orally, the skin ingests nutrients as well. In fact, unlike the food we chew and swallow, which is broken down in the stomach before it is absorbed, creams and lotions applied to the skin bypass the digestive process and *go full strength directly into the bloodstream*. Like all ingested substances, they become raw material for building new body tissue—or they become potent toxic waste. For this reason, I always say to my clients: Think of your beauty products not as cosmetics, but as *food. If you cannot eat it, do not use it on your skin*. This is the Ayurvedic standard for pure and natural.

Skin *eats,* and whatever you feed it, you feed to yourself. Therefore, if you are concerned about your health and well-being, it is as important to know the ingredients of what you apply externally as it is to know the ingredients of your diet and medications. Start reading the labels of your beauty products. If one contains a chemical or herb you do not know, try to find out what it is before you use it. Unfortunately, the word "Ayurveda" on a label, or any variation of it, does not guarantee the purity of the product any more than the word "natural." When we discuss the external skin care routine in the next chapter, we will give you some tips on how to read the fine print on beauty product labels.

My own commercial products and all the formulations in this book are completely free of synthetic substances. With one exception, they are also completely free of toxic chemicals and preservatives. The exception, which is unavoidable, is the moisturizing cream in my commercial lines. All moisturizing creams contain chemical preservatives because their hydrating ingredients are inherently unstable and would quickly break down without them. For your own use here, we will tell you how to use pure herbs and oils to give your skin all the moisture it needs without commercial creams.

These oils and all the skin care preparations in this book can be made easily at home using foods and natural food products exclusively. As in all genuine Ayurvedic preparations, their sole ingredients include plants and herbs or their essences and extracts; pure vegetable oils; pure water; clarified butter (ghee); honey; milk; and whole vegetables and fruits. Ingredients such as herbal essences and extracts that are not found in supermarkets are usually available in health food stores or pharmacies that carry naturopathic remedies, or through the mail, as mentioned above. You could safely eat any beauty product made of these authentic Ayurvedic ingredients, and indeed, a client in Switzerland once called to ask if she could use my herbal cleansing powder to spice her soup. She certainly could have, had she liked the flavor. However, as you will see, some active ingredients in Ayurvedic formulations are destroyed by cooking, and others, while good for the skin, are not very good to the taste.

Of course, throughout this book, you will be using only ingredients selected for your skin type according to their balancing properties. In general, dry skin types will use warming, grounding ingredients to balance Vata. Sensitive skin types will use cooling, soothing ingredients to balance Pitta. Oily skin types will use light, stimulating ingredients to balance Kapha. You will learn how Ayurveda determines the properties of ingredients below.

THE HEALING PROPERTIES
OF PLANTS

*"The universal all-pervading plants assail (diseases) as a
thief attacks a cowshed; they drive out whatever infirmity
of body there may be."*

RIG VEDA

Skin is alive, and life*less* chemicals cannot give life back to the skin. Synthetic
molecules lack intelligence, or what Deepak Chopra calls the self-contained
"know-how" of the building blocks in living organisms to preserve balance and
internal stability. To be "alive," beauty products and topical remedies for the skin
should be made purely of *plants*, or any of their parts or pure extracts, which are
balanced by nature and full of the intelligence—the vibratory energy—that constitutes life. By plants we mean all forms of vegetation, including trees, flowers, fruits,
vegetables, herbs, and spices. Technically speaking, plants are differentiated from
herbs by their woody stems above the ground. Plants also include spices, which
comprise all *pungent* plant substances such as cinnamon and cloves. Here, we use
these terms loosely with the understanding that plants include herbs and spices,
and that herbs and spices are plants. As you will see, these living substances contain all the purifying, nutritive, and balancing properties necessary to provide
external relief from complexion problems and to slow the aging process. They are
the perfect foods for the skin.

MEDICINAL HERBS IN AYURVEDA

Plants, like people, are composed of five elements and seven dhatus, or tissues.
According to *The Yoga of Herbs*, each of the plant's dhatus affects a corresponding
tissue of the human body: the juice of the plant, the watery liquid, works on plasma;
the sap, the thick milky liquid, works on blood; the soft part of the wood, on muscle;
the gum of the tree, on fat tissue; the bark, on bone; the leaves, on nerve tissue and
bone marrow; and the flowers and fruits, on the reproductive tissue. Seeds, which
contain in unmanifest form all parts of the tree, work on the body as a whole.

COMMON MEDICINAL PLANTS
AND HERBS

PLANT/HERB	PART USED	PLANT/HERB	PART USED
aloe vera	leaf sap	*horsetail*	aerial parts
ashwangandha	root, herbal oil	*jasmine*	flower, essential oil
basil	leaves, essential oil	*lavender*	flower, essential oil
bergamot	essential oil (from orange fruit)	*licorice*	root, extract
		neroli	essential oil (from orange flower)
burdock	leaves, root, seeds		
camphor	gum oil	*nutmeg*	kernel, essential oil
cayenne	fruit, essential oil	*patchouli*	essential oil
chamomile	flower, essential oil	*peppermint*	aerial parts, essential oil
cinnamon	bark, essential oil		
clove	essential oil, flower buds	*rose*	essential oil
comfrey	aerial parts*, root	*rosemary*	aerial parts, essential oil
coriander	seed, essential oil	*sage*	leaves, essential oil
cumin	seed, essential oil	*sandalwood*	bark, essential oil
eucalyptus	essential oil, leaves	*shatavari*	root, herbal oil
eyebright	aerial parts	*thyme*	aerial parts, essential oil
fennel	seed, essential oil		
fenugreek	seed, aerial parts	*valerian*	root, essential oil
ginger	root, essential oil	*vetiver*	essential oil
gotu kola	aerial parts, herbal oil	*ylang-ylang*	essential oil

The aerial parts include the leaves, stem, flowers and seedheads. They are used either fresh or dry, or both, depending upon the herb and remedy.

Also like people, plants have their own ojas—their own consciousness. The ojas of the plant is its *essence*, which we experience both as its *scent* and *taste*. It is the plant's soul—the fragrance of life. Like human ojas, it is the ultimate by-product of the ultimate tissue, the seventh dhatu, or the plant's reproductive organs, which are glands in the flower and fruit. Like ours, plant ojas is the subtle energy and substance of the plant's immune system. In fact, the plant's scent maintains its health and life by acting to kill bacteria, prevent infection, and even repel predators. As a result, the essence of the plant contains powerful healing properties; it is the active ingredient of medicinal plants. As you will see, when we ingest the plant's immune enhancer, it enhances our immunity as well.

In addition to their overall value as nutrients and immune enhancers, different plants and essences have different medicinal properties. Depending upon the part of the plant used and the form it is used in, plants can help to balance the doshas, nourish the body and skin, and heal specific diseases. For example, garlic and goldenseal are antiparasitical and are good for treating yeast infections; turmeric is astringent and will help stop external bleeding. Herbs such as valerian and manjista are emmenagogues—that is, they help regulate menstruation. In Appendix D, you will find a list of specific herbs and plants for each skin type, classified according to their many medicinal uses.

As you will see, Ayurveda uses plants and herbs internally as well as externally. In fact, we can ingest the essence of plants not only through the mouth and skin, but also in vaporized form through the nose. For example, some plant parts, like fresh-cut leaves, roots, and seeds, or dried leaves and powders, may be taken orally either in solid form or as juices or teas. Some are infused into liquids and then used in cooking (tarragon-infused olive oil, for example), or applied topically as a moisturizer in massage and skin care. Powdered and crushed herbs are used directly on the skin as cleansers and masks. Added to baths, massage oils, or incense, for example, liquid plant essences are the basis of aroma therapy. Depending upon the part of the plant used and the form it is used in, plants can help to balance the doshas, nourish the body and skin, and heal specific diseases. We describe each of these uses in detail in various chapters of the book.

Ayurveda determines which plants are good for a particular problem or person on the basis of its fragrance or flavor. According to Ayurveda, the fragrance of any substance derives from its taste, and the taste of anything derives from the dominant pair of elements in its own makeup. There are six basic tastes, or *rasas*, which are created by the pairing of different elements: sweet, sour, salty, pungent, bitter, and astringent. As the following table shows, each taste naturally has the same attributes as its dominant constituents:

TASTE	CONSTITUENT ELEMENTS	PREDOMINANT ATTRIBUTES
Sweet	Earth + Water	Cold, oily, heavy
Sour	Earth + Fire	Hot, heavy, oily
Salty	Water + Fire	Hot, oily, heavy
Pungent	Fire + Air	Hot, light, dry
Bitter	Air + Space	Cold, light, dry
Astringent	Air + Earth	Cold, medium

By the same token, each taste naturally has a different influence on the doshas. Disease diagnosis is based upon the three doshas; treatment is based upon the six tastes. Thus, salty, sour, and sweet tastes—which are predominantly hot, oily, and heavy—balance the cold, dry, light properties of Vata. Bitter, sweet, and astringent tastes—which are predominantly cold, heavy, and dry—balance the hot, light, slightly oily properties of Pitta. Pungent, bitter, and astringent tastes—which are predominantly hot, light, and dry—balance the cold, heavy, oily quality of Kapha.

	TASTE	PROPERTIES
Dry skin (Vata) *needs*:	Salty, sour, sweet.	Hot, oily, heavy.
Sensitive skin (Pitta) *needs*:	Bitter, sweet, astringent.	Cold, heavy.
Oily skin (Kapha) *needs*:	Pungent, bitter, astringent.	Hot, light, dry.

This system of identifying and classifying plants and herbs by their healing properties is one of the oldest in the world. Cultures as vastly different as the ancient Chinese, the ancient Greeks, Indonesians, and Tibetans have borrowed and learned from, and in turn contributed to, the Ayurvedic herbal pharmacology over the millenia. However, the ancient vaidyas studied and wrote about indigenous plants; and many of these were—and are—not found outside India. Consequently, as the knowledge of Ayurveda has spread, practitioners in other parts of the world of necessity have learned to apply its principles of classification and application to their native plant life. Even today, many medicinal herbs commonly used by Indian practitioners are not available to practitioners in the West. For your convenience, the formulations in this book use herbs that are available here, although sometimes the available herb is not necessarily the best one for the job according to Ayurveda.

In the chapters on nutritional therapy and aroma therapy, we will look at the Ayurvedic theory of tastes in greater depth. For now, however, we are interested in how we can use plants *externally* to improve the health and appearance of the skin.

MEDICINAL HERBS
IN MODERN MEDICINE

Contemporary Western medicine also recognizes the healing potential of botanical substances. Long before the discovery of modern chemistry, plants and herbs *were* the world's medicines. They were widely used even in Western countries as recently as the 1930s, and are still commonly sold in European pharmacies. In fact, a high percentage of pharmaceuticals prescribed today are pure plant compounds or synthetic forms of drugs that are found in plants. The chemical heart stimulant digitalis, for example, came into use after an herbalist first obtained it naturally from a plant called foxglove.

Nevertheless, there is a crucial difference between using herbs as medicine and taking a synthesized version of the active ingredient, or even taking the natural chemical that has been removed from the herb itself. According to Ayurveda, the effectiveness of a medicine does not depend wholly upon the molecular structure of its active ingredient; rather, the active ingredient works by virtue of the checks and balances provided by other elements present in the plant. Thus, the right chemical "fit" is necessary for a drug to work, but it is not sufficient for creating health and balance. The ultimate efficacy of herbal medicine lies in the balance of life forces—"the delicate web of intelligence"—that structures and maintains the body of the plant itself. A molecule of digitalis, for example, is chemically the same whether it is found in foxglove or synthesized in a laboratory. However, one is alive and one is not. Separated from the source of its intelligence, the digitalis molecule is just a collection of dumb atoms—it makes a fine chemical, but lacks the savoir faire to meet the infinitely changing demands of life and its precise rules of etiquette. It is literally out of the loop of the body's complex feedback system with the brain, and like a gregarious person with an unfortunate lack of right connections and social graces, constantly insinuates itself in situations where it is not really wanted. Hence, the adverse side effects common to man-made medications—for example, the cold tablet that dries up the cold symptoms but also constricts blood vessels, and consequently causes dizziness or even high blood pressure, especially in hypertensive individuals.

In *Quantum Healing*, Deepak Chopra traces the movement in the bloodstream of a biochemical—one of the body's "living" drugs—to demonstrate the almost unimaginable organizing power necessary for a single molecule to find among the body's *trillions* of cells the single cell receptor site where it is needed. He writes: "When a blood cell rushes to a wound site and begins to form a clot, it has not traveled there at random. It actually knows where to go and what to do when it gets there … The molecules themselves actually seem able to pick and choose among various sites—it is uncanny to follow their tracks under an electron microscope as they make a beeline to where they are needed."

This purposeful and precise movement of a single cell is all the more astonishing when you consider that this intricate choreography is just one minute part of the dance. The body "can release hundreds of different chemicals at a time and orchestrate each one with regard to the whole," Chopra notes. It is this "perfect timing" and "superb coordination of a dozen related processes" that is absent when we replace the missing clotting factor, for example, with a prescription drug.

As individuals have become increasingly reluctant in recent years to take man-made drugs, because of their unwanted side effects, and as the market has grown for natural remedies (The *New England Journal of Medicine* reports that Americans spent over thirteen billion dollars on so-called alternative treatments in one year), pharmaceutical companies have started to forage the world's rain forests looking for "new" sources of medicinal herbs. At the same time, some contemporary scientists have begun to study the effects of so-called folk remedies with renewed vigor. In a joint study by the Fox Chase Cancer Center in Philadelphia and the University of Madras, India, researchers investigated the effects of the *phyllanthus amarus* plant on hepatitis B surface antigen. They found that in 59 percent of hepatitis B carriers, the surface antigen disappeared and did not reappear during the nine-month follow-up. The same result occurred in only 4 percent of the control group. The authors of the study note that Ayurvedic physicians prescribed the same herb two thousand years ago for treating jaundice, a classic hepatitis symptom.

ESSENTIAL OILS
AND THE SKIN

"Both the plants that hear this prayer, and those which
are removed far off, all coming together, give vigor to this
infirm body."

RIG VEDA

What we term the "essence" is the substance that contains the active ingredient of the medicinal plant—it is the so-called taste of life. In order for the skin to "eat" this healing food, however, a plant first must be in a form that the skin can absorb externally through its hair follicles and glands. Ayurvedic skin care formulations use plants in a variety of ways to nourish and rejuvenate the skin, including fresh-cut parts, juices, teas, dry crushed leaves or powders, and liquid extracts. However, pure liquid essences—or essential oils—are the most concentrated and refined extracted form of the plant, and therefore the most effective for this purpose. They are 70 to 80 percent more concentrated than herbal powders. Yet their molecular density is so fine, they penetrate the skin to the cellular level and produce effects 60 to 75 percent stronger than the herb taken whole, helping to nourish the tissue and stimulate new cell growth. Moreover, as the source of the plant's own immunity, they have natural antibacterial, antiseptic, antifungal, and preservative properties, which help to heal wounds and infections on the skin. As a result, essential oils are widely used in Ayurvedic skin and body care.

Pure essences, which are extracted by distillation in alcohol, are not only the most potent plant extraction but also the most costly to produce. One ounce of pure rose essence, for example, requires 180 pounds or more of flower petals. One pound of rose essential oil costs about six thousand dollars. Because of the cost and effort, it is impractical to distill essences in your home. Rather, nowadays you can buy small bottles of essential oils—which are pure essences diluted in a base liquid—in most health food stores and natural perfumeries. Unfortunately, the quality of these products varies. Their labels do not indicate the percent of dilution or whether or not they contain synthetic extracts, so you cannot judge either the potency or the purity of what you are buying. A useful gauge, though not an absolute one, is the price of the essential oil. The more pure essence the manufacturer puts in the dilution, the more you are likely to pay.

In my own factory in New York, we use only naturally distilled pure essential oils purchased directly from the distiller, many from India, and we mix them in a ghee base, which is more penetrating and soothing than vegetable oil. To guarantee the purity and quality of my products, I personally prepare the formulations each week to replenish our stock. In Ayurvedic tradition, I also say specific *mantras*, or sounds, as I mix the preparations, in order to enhance the ojas of the plant or oil and thereby to enhance our ojas when we ingest its essence through the skin.

There are an estimated half million plant species known worldwide, but most have been untested for essential oils, and only a few hundred so far have yielded amounts sufficient to be produced commercially at a reasonable cost. The most common essential oils for each skin type appear below (a more complete list appears in Appendix E). The oils for dry skin are primarily sweet, soothing, and warming; the ones for sensitive skin are sweet, soothing, and cooling; and the ones for oily skin are pungent, stimulating, and warming. As you will see, you need only a small assortment of essential oils in your medicine cabinet to prepare most of the Ayurvedic skin care formulations in this book.

ESSENTIAL OILS

For dry skin (Vata):	Nutmeg, cardamom, ginger, saffron, champa, jasmine, geranium, red rose, red sandalwood, lemon, neroli, vanilla.
For sensitive skin (Pitta):	Rosewhite, sandalwood, vetiver, coriander, cumin, mint, ylang-ylang, camphor.
For oily skin (Kapha):	Patchouli, eucalyptus, camphor, clove, lavender, bergamot.

Note that throughout the book we usually refer to essential oils simply as "rose" or "rose oil," for example, rather than "rose essential oil."

THE BASIC HERBAL FORMULATIONS

As we said, nearly all the Ayurvedic formulations in this book, including skin and body care preparations, massage oils, and aromatic oils, are available through Bindi, Tej, and other suppliers of Ayurvedic products listed in Appendix C. If you are the do-it-yourself type, however, you can easily make them at home once you have the basic ingredients, including the basic essential oils for your skin type, the base liquids (vegetable oil, ghee, water, or milk), and the necessary herbs or herbal extracts. By making your own formulations, you not only save money, you also control the purity of the ingredients, as we explain in the following chapter.

In addition to essential oils, you will come across a few other types of herbal formulations in this book that you can use in your Ayurvedic skin care routine. These are moisturizing oils, massage oils, teas and decoctions, infusions, herbal extracts, herbal oils, pastes, and herbal baths, all of which are made with herbs or essences, or both. Their uses and descriptions appear below. If you want to make your own, you will find the basic recipes for these formulations, as well as directions for storing them, in Appendix B.

BASE OILS

As mentioned, liquid essences are too potent in their pure form to use directly on the skin, and consequently they are usually sold in diluted form. Since it is impossible to know how diluted they are, however, I recommend that you always dilute essential oils in a base oil (or a carrier oil, as it is also known) as a precaution before applying it to your skin. For this purpose, use a base of vegetable oil or clarified butter, also known as ghee (see box below). These base liquids have their own nutritive and balancing properties that can complement and enhance the effect of the herbs if used properly.

BASE OILS

For dry skin (Vata):   Black sesame, sesame, avocado, olive, almond, walnut, peanut, castor, ghee.

For sensitive skin (Pitta):   Almond, coconut, sunflower, apricot kernel, olive, ghee.

For oily skin (Kapha):   Canola, corn, safflower, mustard, grapeseed, almond, apricot kernel.

Vegetable oils are themselves plant extracts. Unlike "volatile" essential oils, however, vegetable oils are "fixed"—that is, they do not evaporate. However, their subtle nutritive qualities are easily damaged or destroyed by improper extraction. Therefore, we recommend that you use organic vegetable oils that have been made without the use of heat, solvents, or harmful chemicals or preservatives. These oils are available in any good health food store and will be labeled "pure cold-pressed" or "pure expeller-pressed." We recommend organic products because the toxic contents of processed oils or oils made from plants that have been chemically sprayed or fertilized go directly into your bloodstream when applied to the skin. Ghee is also an excellent base oil because of its penetrating and soothing qualities. Ghee is usually available in health food stores, and of course in

Indian markets, or you can make it easily at home (see the recipe in Appendix B). It requires no refrigeration and has an unlimited shelf life—in fact, the older the ghee, the more medicinal value it has.

MOISTURIZING AND MASSAGE OILS

Moisturizing and massage oils are made from selected essential oils added to a base oil in order to balance the doshas and nourish the skin. The proportion of essential oil to base oil will differ, depending upon whether you are using the formulation for massaging the entire body or for the face alone. Face oils contain a higher concentration of essential oils because the complexion, which is always exposed to the elements, needs more repair and protection than other areas of skin. At the same time, the larger skin surface of the body needs a lower concentration of essential oils to absorb the same overall amount of nutrients from them.

DECOCTIONS

Less potent and penetrating than essential oils, decoctions are essentially strong teas. They are made by boiling fresh or dried herbs—typically the woodier parts of the plant—to extract their essence. Decoctions may be taken as a tea (mild decoctions) or used in a bath (strong decoctions). They also are used in the preparation of herbal oils (see below).

HERBAL EXTRACTS AND OILS

Herbal extracts are another form of concentrated plant essence, though not as strong as pure liquid essences. They are taken orally for medicinal purposes— a typical dosage is 3–4 drops on the tongue—or used externally in the form of herbal oils.

Extracts are made by soaking herbs in a solvent such as alcohol, stirring the mixture daily for 3–10 days, depending upon the herb, and then straining the liquid. This process is used to extract the essence of herbs that do not easily produce a liquid essence by distillation. Like pure essences, extracts are sold widely in health food stores.

Herbal oils are very concentrated medicinal oils, which are used in undiluted form in massage therapies or as the base for essential oils to make them more potent. Herbal oils are made in two ways: by adding herbal extracts to a base oil, as we said, or by adding a strong decoction of a medicinal herb to a base oil and then boiling off the extra water.

Traditionally, Ayurvedic practitioners make medicinal herbal oils with no regard to their scent, and in most cases, their scent is quite unpleasant. When I came to the United States, I realized Americans would not use an ill-smelling beauty product, no matter how beneficial it might be for their skin. I began adding essential oils to my remedies, which had never been done in India. Adding essential oils to medicinal oils not only improves the aroma of the herbal oil, helping to balance the mind as well as the body, but also enhances the healing properties of the herbal oil.

HERBAL EXTRACTS AND OILS

For dry skin (Vata): Shatavari, ashwangandha, basil, bala, vacha, colows, comfrey, gotu kola, or ginger.

For sensitive skin (Pitta): Neem, shatavari, amalaki, licorice, fennel, cardamom, mint, gotu kola, bhringraj, manjista, saffron, burdock.

For oily skin (Kapha): Sage, neem, rosemary, triphal

HERBAL INFUSIONS

Typically, infusions are made from fresh or dried leaves, flowers, or other aerial parts of the plant that are unsuited for boiling. They are made using water or oil as the base liquid.

HERBAL PASTES

Pastes made from herbal powders (fine-ground herbs such as those found in seasonings) are used as cleansing scrubs or as face and body masks for deeper exfoliation. They are made by adding a small amount of the selected liquid (usually oil, water, or milk) to an herbal powder until the mixture achieves a pasty consistency.

HERBAL BATHS

Essential oils or dried herbs wrapped in cheesecloth added to bathwater create a balancing aromatic effect, depending upon the ingredients.

In the chapters to come you will learn how to use these "edible" formulations to cleanse, nourish, and moisturize the face and body, and to relieve complexion problems. Again, if you do not want to make them, you can buy most of them in a good natural foods store or directly through the mail—and if you prefer to buy your beauty products over the counter, we will tell you what ingredients to look for or avoid in the following chapter. Either way you do it, it is time to say good-bye to your harsh synthetic scrubs and lifeless lotions, and to make some room on the bathroom shelf for a gourmet feast for your skin.

CHAPTER 5

CLEANSE, NOURISH, AND MOISTURIZE: THE EXTERNAL SKIN CARE ROUTINE

Get thee a skin of exquisite texture, of a soft and
delicate bloom, and a complexion pure and clear.
YOGA SUTRA

The first rule of Ayurvedic skin care is: Cleanse, nourish, and moisturize. No matter what type of skin you have, this three-step routine for the external care of the skin is essential to counteract the daily effects of environment, stress, and the skin's natural process of cell degeneration. It is the very minimum you must do to maintain a normal, healthy, youthful complexion. And even if you do nothing else for an existing skin problem, keeping this regimen every day can help to improve your condition, sometimes in a short period of time.

In this chapter, you will learn the basic requirements for good skin, plus the daily cleansing, nourishing, and moisturizing routine that is right for your complexion type, and special rejuvenating treatments for every part of the body from head to toe. This total beauty program—which features exfoliating face masks, Ayurvedic hair care, rejuvenating body baths, and seasonal skin care tips, among many others—is ideal for everyone, no matter what your skin condition is at this time. If you do have a problem, however, you will also find specific head-to-toe treatments in the second half of the chapter for the most common complaints, including acne, eczema, psoriasis, premature wrinkles, age spots, dandruff, cellulite, cracked feet, and more. In both cases, you will learn the specific recipes for the various skin care products and remedies, as well as how these formulations differ from popular commercial beauty treatments. If you prefer to buy your skin care products rather than make them, the information in this chapter will make you a more knowledgeable consumer, especially when it comes to distinguishing the true natural products from the many pretenders on the market today. In addition, all the products described below can be purchased ready-made from either of my companies, Bindi or Tej.

Once again, these treatments are *external*. While they are vital to achieving a flawless complexion and an all over healthy appearance, alone they are not sufficient to eliminate all skin problems completely, or to achieve absolute beauty. They alleviate only the manifestations of disease and aging, not the cause itself. For that you must also cleanse and nourish the body and mind *internally*, where all disorders begin. Indeed, I cannot overemphasize the importance of the internal aspects of care, especially diet and the all-important stress-reducing techniques of meditation, massage, and breathing. Over the years, I have cured numerous cases of skin disease and have helped many clients take years off their appearance by rejuvenating the skin. But we could not have achieved these results if the individuals themselves had not been willing to take at least a few extra steps in addition to the external beauty routine.

Nevertheless, I never push my clients to do anything. I simply let the the Ayurvedic methods speak for themselves. Once you have enjoyed the unique sensation of the herbs and oils penetrating deeply into the tissues, and have seen for yourself the improvements in your complexion just from the daily routine, curiosity and the desire for the best naturally will lead you to investigate further into Ayurveda's beauty secrets. Start here with your daily external skin care routine; the rest of your journey to absolute beauty is inevitable.

THE BASIC REQUIREMENTS FOR GOOD SKIN

All normal skin is rosy, lustrous, unblemished, smooth, evenly colored, soft, firm, and elastic. Dull, sallow, ruddy, pale, blemished, dry, discolored, sagging, puffy, and wrinkled skin are indications of imbalance and premature aging due to stress, poor habits, exposure to the environment, and of course, improper skin care. All these symptoms result from a decline in skin functions such as new cell growth, elastin, and collagen production, blood circulation, secretion of ground substances, immune activity, and enzyme activity, which give the complexion its color and glow. Therefore, in order to keep the skin looking young and radiant, your beauty products and treatments must at least provide exfoliation to remove dead skin cells; epidermal stimulation for new cell growth; antioxidant properties for cellular rejuvenation and repair; improved capillary blood flow; immunostimulation; and penetrating moisture and nutrients to replenish all seven layers of skin tissue. At the same time, they should not harm your health in any other way, since anything that imbalances the body or mind at some point also causes premature aging or other damage to the skin. The three-step process of cleansing, nourishing, and moisturizing the skin—using only herbs and oils appropriate for your skin type—fulfills all these basic requirements.

PROPER CLEANSING

The secret to proper cleansing is to get rid of any dead cells and toxins not thrown off naturally by the skin; remove all makeup, dirt, grime, and chemical pollutants; clear out clogged pores; and eliminate infection-causing bacteria without also stripping away the skin's natural oils and water. Most cleansing products, including soaps, shampoos, and scrubs, dry the skin and alter its pH balance, causing it to become more alkaline. In fact, advertisers have led us to believe that the best cleansers are the ones that make the most suds and leave a "squeaky clean" sensation. In principle, this result sounds good, especially if you have oily skin. But in fact the body always seeks to normalize its processes—that is, to balance and heal—so when we remove the skin's natural oil content, it responds by producing more oil to compensate for the loss. If we dry the skin too much, it will produce excessive amounts of oil that may exacerbate an existing acne condition. Of course, if you have dry skin already, you do not want to make it drier, and if you have sensitive skin, you do not want to irritate it with harsh chemicals and soaps.

Ayurveda uses herbal powders to clean and exfoliate the skin on a daily basis. The herbs act as a gentle scrub to clear away the dirt, toxins, pollutants, and dead cells, but they do not strip away the necessary moisture in the skin. At the same time, they are balancing, nutritive, healing, and totally free of toxic ingredients.

For deeper exfoliation, we recommend herbal masks at least once or twice a week. (A professional chemical peel, which gives the deepest exfoliation, should not be done more than once a year.) The skin sheds cells at the rate of a million per hour, so it is important to do a mild exfoliation daily. At that rate, it takes the skin about one month to completely regenerate itself, and the herbal mask "peels" away the old cells so new ones can grow. The dead cells are also the top protective layer of the skin, however, and if we remove too much too soon, we leave the new cell growth unprotected. If the cells are not growing in at the rate we remove them, skin problems will develop.

For this reason, alpha hydroxy, retin A, and glycolic acid, which are popular chemical peels, are much too drastic in their effect to be used on a frequent basis. They are particularly harsh on sensitive skin. In my practice I have seen all too many new clients—like the woman from *Redbook*—whose skin has become red, raw, and painfully dry in the hands of well-meaning but poorly informed aestheticians and dermatologists who have mistakenly tried to clear Pitta complexions with chemical peels. Glycolic acid has the advantage over the other two popular peels of being a food product—it is found naturally in some fruit juices, for example. However, beauty products often contain glycolic acid in a synthesized or extracted form (the active ingredient is isolated and removed from the natural source), so its natural "intelligence" is absent. While you can use pure apple or lemon juice mixed with a few drops of water occasionally as a natural peel with good results, glycolic acid, even in an unadulterated form, is—like alpha hydroxy and retin A—too strong to be used on a daily basis for dry or sensitive skin. A few drops a day of apple or papaya juice and water are okay for very oily skin *except* when there is also skin sensitivity.

PROPER NOURISHING AND MOISTURIZING

When the soap bubbles have burst and the skin starts to squeak, the first thing we look for after using harsh cleansers is a soothing lotion or cream to replace the natural fluids we have so thoroughly washed away. Unfortunately, these products rarely do the job as well as nature. The first obstacle to their success is often the skin itself. Many products are simply too dense on the molecular level to permeate the tissue adequately, so they never penetrate to the cellular level where their nutrients can be used. What happens instead is that the unabsorbed substances leave a film on the surface of the skin—hair conditioners create the same problem in their attempt to undo the damage of harsh shampoos. This film clogs pores, collects dirt and grime, and generally requires astringents to be removed, thus perpetuating the dry skin problem that the lotion or cream was meant to relieve. If the product does penetrate the skin surface, in many cases its nourishing and moisturizing benefits are outweighed by the toxic effects of its chemical contents on the body.

Skin eats, as we have said, but we want to nourish it with what is pure, natural, and balancing to body and mind. In Ayurveda, we feed the skin only pure essential oils, which are naturally hydrating and rich in nutrients, and also fine enough to penetrate the skin completely to rejuvenate the cells. Gently massaged into the skin, essential oils help to improve circulation and strengthen the connective tissue, thereby reducing wrinkles. Their aroma also helps to balance the doshas, and the essence itself provides protection from infection. For those of you with overactive sebaceous glands, the idea of using oils to remedy oily skin may seem contradictory, if not plain crazy. However, I urge you to give it a try. Essential oils penetrate all seven layers of the skin within a few minutes to supply nutrients and restore the body's subtle intelligence. They disappear directly into the cells, where they are needed; they do not leave any greasy residue on the surface of the skin.

We apply the nourishing face oils (essential oils diluted in a base of vegetable oil or ghee) mixed with water (for young or normal skin) or with *liposomes* (for mature or scarred skin). Made of lipids, the same substance that composes the cell wall, liposomes are penetrating agents that have the ability to carry other materials through the cellular membrane and deposit them in the cell itself. In other words, they facilitate the absorption of the essential oils into the skin. Liposomes also carry additional moisture into the cells, which helps the rejuvenation process as well. While I recommend that you use liposomes, I do not recommend that you try to make them, since the process is complex. Most commercial liposomes are made with egg protein; Bindi and Tej liposomes are made with bean extract and oil. Either kind will do the job.

In truth, essential oils mixed with liposomes supply sufficient moisture to the skin so that a moisturizing cream is not necessary. This, however, is a very difficult concept to convey to Westerners, who long ago were sold on the idea that creams and lotions are the answer to dry skin and wrinkles. For those of you who feel, as do many of my clients, that you want a moisturizer in addition to the hydrating essential oils, we have provided recipes in this chapter. If you prefer to use a commercial cream, avoid those containing fragrances, colors, and mineral oil.

CHOOSING OVER-THE-COUNTER PRODUCTS

Ayurvedic preparations for cleansing and nourishing the skin primarily contain essential oils, ghee or vegetable oil, and herbs. They are simple to make and naturally have a long shelf life, so you do not have to make them frequently. Nevertheless, many of you will not have the time or interest to prepare all of your own beauty products. If you choose to buy rather than make the prescribed formulations, keep in mind these basic principles when you are purchasing the product:

- It should contain pure plant ingredients exclusively—no synthetic substances, no mineral oils, no chemical additives, no dyes, no preservatives, no chemical fragrances, no known carcinogens.
- It should heal and soothe the skin, not irritate and dry it.
- It should contain ingredients appropriate for your skin type.

On pages 122–125 you will find a list of chemical ingredients commonly found in commercial skin care products. Many of them appear on the FDA's own list of known potential carcinogens and toxic substances. (A number of these toxic ingredients were found on a label of a product that is packaged and sold as an "Ayurvedic" formulation; its ingredients were, in order: water, sunflower oil, *fifteen different chemicals*, herbal extracts, and fragrances.) You will also find as a comparison the list of actual ingredients in three commercial beauty products, including two brand name products and my own moisturizer. Whenever you buy a beauty product, read the label. If you cannot eat what is in the bottle, do not use it on your skin.

Whether you make them or buy them, Ayurvedic herbal preparations and oils recommended in this book ultimately are simpler to use and much less expensive than any other over-the-counter soaps and lotions—and they are also the best for you. They are simpler because the twice-a-day cleansing and nourishing routine is all the skin care you need on a daily basis. Indeed, the Ayurvedic cleansing and nourishing formulas completely eliminate the need for toners, scrubs, eye creams, neck creams, day creams, night creams, and antiseptics, because these two products alone do all these jobs—and do them better. Toners, which are nothing but alcohol and water or witch hazel, are only necessary when you use soaps that upset the skin's pH balance in the first place. All moisturizing creams are nothing but mixtures of oil and water held together with chemical emulsifiers. As you will see, you can nourish and moisturize your skin using a few drops of essential oil and water, mixed fresh each time you need them directly in the palm of your hand. Consequently, you can get the hydration and lubrication you need without the added chemicals and without the heavy oils that clog pores. Moreover, the essential oils are naturally antiseptic; and the herbal cleansers provide the proper level of exfoliation and nourish the skin simultaneously. As for the collagen complex that is often added at a high price to many skin care products, this "miracle" ingredient, which is a derivative of animal tissue, has no effect on the skin's own collagen, according to medical experts.

When you cleanse, nourish, and moisturize the skin properly with herbs and oils—in fact, when you do any of the Ayurvedic treatments in this book— you will see and feel an immediate improvement in your skin because the ingredients are alive, nutritive, healing, and soothing all at once. At my clinic, new clients coming for their first facial from me are often amazed at the difference in their experience. They are used to scheduling their treatment for a time when they can run home and hide for the rest of the day because their skin looks so red and blotchy afterwards. I tell them to schedule an exciting date right after their Tej facial, because that is when their skin will look the most flawless and radiant. Indeed, the idea that you have to squeeze every pimple to properly clean the skin, or massage the face for a long time is simply not true. Squeezing only breaks fine capillaries unnecessarily, and too much massage irritates the skin, especially if the lotions contain chemicals. The right skin care product or treatment should leave the complexion soothed, healthy, and glowing—not irritated. This is my standard, based upon a quarter century of treating skin, and I recommend it to you as well.

THE CHEMICALS
ON THE LABEL

CHEMICAL DYES

FD&C colors, D&C colors, HC colors (peroxide dyes).

Potential harmful ingredients: FD&C colors are U.S-certified artificial colors made from coal tars that are permissable for use in foods (F), drugs (D), and cosmetics (C). Almost all of these have been proven to be carcinogenic in studies on animals, although six coal tar colors are on the FDA's permanent "safe" list. FD&C Yellow #5 causes allergic reactions in people sensitive to aspirin and can also cause induce asthma attacks. Other possible effects are dizziness, headaches, and confusion. Some hair coloring dyes have been known to cause skin rash, eczema and bronchial asthma. Many do-it-yourself home hair colors contain dichlorobenzidene, a carcinogen that is easily absorbed through the skin and may also cause anemia, jaundice, central nervous system problems, kidney, and liver damage.

FRAGRANCES

"Fragrances" means synthetic fragrances. Otherwise, the label would indicate "essential oils."

Potential harmful ingredients: The word "fragrance" can indicate the presence of up to 4,000 different unlisted ingredients. Complaints to the FDA have included headaches, dizziness, rashes, skin discoloration, violent coughing and vomiting, and allergic skin irritation.

DETERGENTS

NDELA (nitrosodiethanolamine) formed with TEA, DEA, MEA, sodium lauryl sulfate.

Potential harmful ingredients: On FDA list of suspected carcinogens. Also may cause dermatitis, flu-like and asthmatic conditions, severe eye damage, and severe upper digestive tract damage if ingested.

PETROLEUM PRODUCTS

Paraffin, mineral oil.

Potential harmful ingredients: On FDA list of suspected carcinogens. Paraffin may cause adverse reactions in people sensitive to petrochemicals.

BLEACHING AGENT

Hydroquinone.

Potential harmful ingredients: On FDA list of suspected carcinogens.

DRYING AGENTS

Phenols.

Potential harmful ingredients: On FDA list of suspected carcinogens. May also cause skin eruptions and peeling, swelling, pimples, hives, burning, numbness, cold sweats.

SURFACTANTS

PEG-8 (polyethylene glycol).

Potential harmful ingredients: On FDA list of suspected carcinogens.

HUMECTANTS

(Help skin absorb & retain moisture.)
Propylene glycol, glycerin, sorbitol, butylene glycol.

Potential harmful ingredients: Except for PEG-8, most glycols, natural or synthetic, are considered safe. Natural glycols are typically labeled natural glycerin or vegetable glycerin; unspecified glycols are listed as glycerol or glycerin, and petrochemicals are listed as glycol, glyceryl, ethylene glycol, PEG or polyethelene glycol, and propylene glycol. FDA studies indicate that polyethylene glycol may be highly allergenic.

EMOLLIENTS

(Soften and soothe skin tissue.)

Mineral oil, lanolin, silicones, such as dimethicone; fatty acids such as stearic and isostearic acid; fatty alcohols such as cetyl alcohol, stearyl alcohol, myristyl alcohol; esters such as isopropyl myristate, spermaceti, octyl palmitate, butyl stearate, isopropyl isostearate; tryglycerides, such as vegetable oils.

Potential harmful ingredients: Mineral oil is asuspected carcinogen and also interferes with vitamin absorption in the body. Prohibited for use as a food coating in Germany. It is more harmful when ingested or rubbed on skin than when inhaled.

EMULSIFIERS

(Prevent oil and water, the main ingredients in moisture cream, from separating.)

Glyceryl stearates, carbomer 934, ethers like steareth 2, laureth 4, beeswax, sorbitan stearate, cetearyl alcohol, polysorbate 60 & 80.

Potential harmful ingredients: Polysorbate 80 is a known carcinogen.

PRESERVATIVES

(Extend shelf life of product.)

Parabens such as methyl, propyl, butyl, quaternium-15, imidazolidinyl urea.

Potential harmful ingredients: All preservatives are potential allergens.

ANTIOXIDANTS

(Prevent spoilage.)

BHA (butylated hydroxyanisole), BHT (butylated hydroxytoluene), tocopherol (vitamin E).

Potential harmful ingredients: Animal studies show that BHT is a suspected human carcinogen and that BHT and BHA may cause metabolic stress, damage to liver, baldness, and fetal abnormalities. Physicians report that BHA and BHT may cause adverse reactions in people with sensitivities to petrochemical derivatives.

Lanolin (from sheep's wool), ceteareth 20 (cetyl alcohol + stearyl alcohol, made from sperm whale oil).

Potential harmful ingredients: Lanolin clogs pores and can cause blackheads.

Alcohol, formaldehyde, propyl alcohol, toluene 2, 4 diamine, EDTA (ethylenediamine tetra acetic acid).

Potential harmful ingredients: All of these ingredients are suspected carcinogens. EDTA may also cause numbness and tingling in fingers, lightheadedness, dizziness, vertigo, sneezing, nasal congestion, headache, skin irritation.

SAMPLE LABELS: A COMPARISON OF THREE PRODUCTS

A popular "gentle" cleanser: Mineral oil, water, beeswax, petrolatum, stearyl alcohol, fatty acid, lanolin alcohol, ceteareth–20, magnesium aluminum silicate [coloring agent], sodium dehydroacetate, methylparaben, propylparaben, butylparaben; FD&C Blue #1, FD&C Yellow #5.

A popular "exceptional" moisturizing beauty lotion: Water, mineral oil, propylene glycol, collagen complex, propylene glycol stearate, lanolin alcohol, tea stearate, triethanolamine, kaolin [clay], lecithin [emollient made from egg yolk], petrolatum, talc, magnesium aluminum silicate, cellulose gum, fragrances, trisodium EDTA, quaternium-15, methyl paraben, propylparaben, FD&C Yellow #5 and FD&C Red #4.

Pratima's moisture cream (depending upon body type): Cocoa butter or almond butter; olive, sesame, sunflower, or safflower oil; orange juice or strong herbal decoctions; essential oils or rosewater.

THE DAILY FACE CARE ROUTINE

The herbs and oils used in the daily skin care routine, as well as the methods of application, differ according to complexion type, but the basic guidelines for the procedure are the same for everyone. They are:

- Do the three-step routine twice a day, once in the morning and once before bedtime. (If you have oily skin, do the full routine three times a day in the summertime.) The complete routine takes five to ten minutes.
- Before you begin, remove face and eye makeup. Use ghee or pure vegetable oil (*not* essential oil) on a cotton ball and wipe gently. Sesame oil is good for dry skin; sunflower oil or safflower oil is good for sensitive and oily skin.
- If possible—and if applicable—do the skin care routine after exercising but before meditating.
- Cleanse the face before or during a bath or shower, and apply the nourishing oil and moisturizer afterward.
- Always massage oil on wet skin.
- Always massage the face and neck using a gentle stroke in an upwards and outwards direction only (see illustration on opposite page). Do not scrub, or rub up and down, to avoid stretching the skin.
- Always massage around eyes in a circular direction from the outside corner of eye to the inside using the ring finger to ensure the gentlest touch (see illustration on page 129).
- After you complete the final step, wait at least 2–3 minutes before applying makeup.
- For deeper exfoliation and added nourishment, do a weekly fruit face mask (see page 131).

The specific recipes and instructions for cleansing, nourishing, and moisturizing your skin appear in the following pages.

DRY SKIN (VATA)

To cleanse.

- Mix 1 tsp almond meal + ½ tsp dry milk + 1 pinch sugar. Store in a spice jar.
- In your palm, make paste using ¼ tsp cleanser + warm water.
- Apply paste all over face and neck and gently massage into the skin for about one minute. *Do not scrub.* Rinse well with warm (not hot) water. *Do not dry.* *Special instructions:* If your skin is very dry, do *not* follow the cleansing routine above. Instead, wash with a mixture of 1 Tbsp heavy cream (the dairy product) + 2 drops lemon juice.

To nourish:

- Mix 1 oz sesame oil + 10 drops geranium oil + 5 drops *each* neroli and lemon oil. Store in dark glass bottle with a dropper.
- In the palm of your hand, mix 3 drops of nourishing oil (above) + 6 drops water. (Replace 6 drops water with 6 drops Tej or Bindi liposome liquid, if you have it.)
- While your skin is still wet, gently massage mixture all over face and neck for about 1 minute, or until skin absorbs all the essential oils.

To moisturize:

- Melt 1½ oz cocoa butter in double boiler. Add 4 oz avocado oil. Remove from heat. Using a dropper, add 1 oz orange tea *one drop at a time* while stirring the mixture. When it is cool, add 3–4 drops *each* geranium and rose oil.
- Gently apply moisturizing cream over surface of face and neck. Do not massage it into the skin. Apply as needed during the day.

SENSITIVE SKIN (PITTA)

To cleanse:

- Mix 1 tsp almond meal + ½ tsp ground orange peel + ½ tsp dry milk. Store in a spice jar.
- In your palm, make a paste using ¼ tsp of the mixture + rosewater.
- Apply paste all over face and neck and gently massage into the skin for about one minute. *Do not scrub.* Rinse well with cool (not cold) water. *Do not dry.*
- *Special instructions:* If your skin is very sensitive, use this cleanser only once

a day at bedtime. In the morning, wash only with plain heavy cream (the dairy product). Rinse thoroughly with cool water, then continue with the normal nourishing and moisturizing steps below.

To nourish:

- Mix 1 oz almond oil + 10 drops *each* rose and sandalwood oil. Store in dark glass bottle with a dropper.
- In the palm of your hand, mix 2–3 drops of nourishing oil (above) + 4–6 drops water. (Replace 4–6 drops water with 6 drops Tej or Bindi liposome liquid, if you have it.)

- While your skin is still wet, gently massage mixture all over face and neck for about one minute.

To moisturize:

- Melt 1 oz cocoa butter in double boiler. Add 3 oz sunflower oil. Remove from heat. Using a dropper, add 2 oz rose tea *one drop at a time* while stirring the mixture. When it is cool, add 5–6 drops sandalwood oil.
- Gently apply moisturizer over surface of face and neck. Do not massage it into the skin. Apply extra moisturizing cream as needed throughout the day.

OILY SKIN (KAPHA)

To cleanse:

- Mix 1 tsp barley meal + 1 tsp lemon peel + ½ tsp dry milk. Store in a spice jar.
- In your palm, make a paste using ¼ tsp of the mixture + warm water.
- Apply paste all over face and neck and gently massage into the skin for about one minute. *Do not scrub.* Rinse well with warm (not hot) water. *Do not dry. Special instructions:* Do *not* use soaps or astringents containing alcohol to reduce oils. If you have whiteheads, wash first with a paste made of ¼ tsp neem powder + 1 pinch sugar + water. Then follow steps as above.

To nourish:

- Mix 1 oz sunflower oil + 10 drops lavender oil + 5 drops *each* bergamot and clary sage oil. Store in dark glass bottle with a dropper.
- In the palm of your hand, mix 2 drops of nourishing oil (above) + 4 drops water. (Replace 4 drops water with 4 drops Tej or Bindi liposome liquid, if you have it.)
- While your skin is still wet, gently massage mixture all over face and neck for about one minute.

To moisturize:

- Melt 1 oz cocoa butter in double boiler. Add 3 oz almond, safflower, or canola oil. Remove from heat. Using a dropper, add 2 oz rosemary or basil tea *one drop at a time* while stirring the mixture. When it is cool, add 1 drop camphor oil + 2 drops bergamot oil + 3 drops lavender oil.

• Gently apply a very small amount of moisturizer over face and neck. Do not massage it into the skin. Use once in the morning and evening only. If your skin is very oily, use Bindi or Tej Soothing Lotion before the moisture cream.

THE WEEKLY FRUIT FACE MASK

To make an exfoliating mask, use:

For dry skin:	Banana or avocado pulp.
For sensitive skin:	Banana or pineapple pulp.
For oily skin:	Strawberry or papaya pulp.

Do your normal cleansing routine only. Then apply pulp to face and neck, and lie down for 10–15 minutes with legs raised to increase blood supply to face. Rinse with water, then finish with your regular nourishing and moisturizing routine.

CLEANSE, NOURISH, AND MOISTURIZE FROM HEAD TO TOE

The "cleanse, nourish, and moisturize" rule applies to every part of the body, not just the face. However, the Ayurvedic cleansers and oils for the face are not adequate for washing and nourishing the rest of the body. The face cleanser is designed for the delicate areas of the complexion, while most other areas of the skin can tolerate a stronger scrub. On the other hand, the face oil is too concentrated (too high in essential oils) to use in the large quantity necessary to cover the total skin surface of the body. Consequently, the body oils contain less essential oil.

Below are cleansing and nourishing treatments for the body as a whole, as well as for its various parts, since the quality of the skin, and life's demands upon it, differ from head to toe.

DO'S AND DON'TS
OF EXTERNAL SKIN CARE

DO'S

DO wash your face twice a day, morning and evening, with a soft herbal cleanser and lukewarm milk or water.

DO nourish and moisturize your skin daily with the appropriate essential oil.

DO avoid excessive exposure to sun, saltwater, wind, cold, and snow.

DO facial exercises (see pages 259–260) once daily.

DON'TS

DON'T use heavy night creams. They clog pores and cause puffiness.

DON'T use soaps or harsh detergents on the face.

DON'T use chemical makeup removers, heavy eye creams, or oils. Instead, use cotton dipped in plain vegetable oil to remove eye makeup.

DON'T use harsh scrubs like loofahs, chemical powders, or pumice stones.

DON'T use tissues to remove makeup. Wood pulp in paper is harmful to skin.

DON'T use chemical astringents or products containing alcohol.

DON'T use very hot then very cold water to wash. It breaks capillaries.

DON'T wear face makeup to sleep no matter how tired you are.

BODY BEAUTIFUL

To cleanse (all skin types):

- To make the body cleanser, mix equal parts chickpea flour + dry milk powder in a plastic spice jar.
- Sprinkle 2 tsp of body cleanser into palm of hand, add some water to make a thin paste, and scrub lightly over wet skin in shower or bath. Rinse off and pat dry.

To nourish and moisturize:

- To make body oil, mix 1 oz almond oil + 10 drops essential oil appropriate for your constitution. Massage on wet skin following shower.

NATURAL UNDERARM DEODORANT

Mix 1 oz sandalwood powder + 1 oz arrowroot powder + 10 drops lavender oil. Apply as needed.

HAIR CARE

Hair is ninety-seven percent protein and three percent moisture. The average person has one hundred thousand hairs on her scalp, which grow at the rate of .37 millimeters per day and shed at the rate of fifty to one hundred per day. The factors that affect hair growth and hair loss include age, health, diet, hormones, seasonal changes, and climate. Trauma, stress, and anxiety also affect growth, because they reduce the blood and oxygen supply to the scalp, which is essential to healthy hair. Weekly scalp massage is, therefore, one of the most beneficial treatments for maintaining beautiful hair and alleviating or reducing the symptoms of common hair problems, such as male pattern baldness, alopecia (hair loss), alopecia arebets (bald patches), dandruff, psoriasis, and premature graying. Scalp massage not only increases the oxygen supply to the brain, it also improves circulation of the life-giving sap, cerebrospinal fluid, which stimulates brain development, relaxes the nerves and muscles, reduces fatigue, and loosens the scalp. For dry and sensitive skin, the added essential oils also penetrate the roots to strengthen the hair and alleviate dryness.

Ayurvedic hair treatments, including hair cleansers, use only herbal powders, herbal essences, and oils. As you will see, the "shampoos"—both wet and dry—do not produce any lather when you use them because, unlike brand-name shampoos, they contain no sodium or aluminum lauryl phosphate or other sudsing agents that destroy hair protein and strip the natural oils. Herbal cleansers wash away the dirt without washing away the natural moisture, so you do not need conditioner to undo the damage of a "squeaky clean" shampoo. The dry shampoo is a good alternative when you are sick or do not have time to wash and dry. It is also good for dull or oily hair. For weekly conditioning treatments, use the scalp massage and hair "mask."

DO'S AND DON'TS
OF HAIR CARE

DO'S

DO massage scalp lightly in morning or evening with a few drops of lavender or rosemary oil, then brush hair 50 times in a downward direction from scalp to ends.

DO wear a hat or other head protection in sunlight.

DO keep all hair implements clean.

DO avoid using chemical sprays, coloring agents, perms, and drying shampoos.

DO avoid tight barrettes, clips, or hair ties, and never wear them to sleep.

DON'TS

DON'T use very hot or very cold water to wash hair.

DON'T use detergent shampoos or blow dry every day, especially if your scalp is dry.

DON'T use metal brushes or combs.

WET SHAMPOO

- Make a strong herbal decoction using:

  For dry skin: Lavender or geranium.

  For sensitive skin: Chamomile.

  For oily skin: Sage, bay leaf, or rosemary.

- Make a shampoo using 8 parts herbal decoction (above) +1 part liquid olive soap, or use Bindi or Tej Herbal Hair Wash. (These natural shampoos create very little lather.)

- Shampoo as usual. If desired, rinse with color enhancer (see below). Then massage scalp with 2 drops lavender or rosemary essential oil before drying. For added shine, follow shampoo with a hair rinse made of the juice of ½ lemon in 1 cup water.

DRY SHAMPOO

- Mix equal parts ground cornmeal + ground almonds + orris root.
- Massage a small handful on dry scalp, then brush off.

WEEKLY CONDITIONING HAIR MASK

- Make a paste with 1 tsp each triphala, neem, sandalwood, and licorice powders + 10 tsp water.
- Apply paste to dry scalp, leave on for ½ hour, and rinse thoroughly with warm lemon water.

NATURAL HAIR COLOR ENHANCERS

After shampooing, rinse with:

For black hair:	Decoction of sage, rosemary, or black walnut.
For red or copper hair:	Henna paste or a pinch of saffron diluted in water.
For blond hair:	Decoction of chamomile flowers.
For dark brown hair:	Decoction of walnut.
For golden-toned hair:	Decoction of mullein

THE WEEKLY SCALP MASSAGE

For dry and sensitive skin: Massage 1 tsp warm sesame oil or gotu kola herbal oil (available from Bindi or Tej) into scalp for 10 minutes. Wrap head in hot towel and leave on for 5–10 minutes. Do once a week for sensitive skin and twice a week for dry skin or dandruff. The best time for this treatment is first thing in the morning or before bed. Massage increases circulation to the scalp. Therefore, do not do this massage right after a meal, when the blood supply is needed for proper digestion.

For oily skin: With head down, dry brush hair 50 times from roots to end to spread natural oils.

EYE CARE

Clear, vibrant eyes are a sign of good health. However, the area directly under the eyes does not have sebaceous glands to lubricate the skin. As a result, it is the most delicate part of the complexion and the first to show signs of aging. Stress, anxiety, worry, insomnia, water retention, sun, dim or bright light, overwork, and excessive alcohol all strain the eyes and lead to crow's-feet and fine lines. Three simple methods to relieve and avoid eyestrain, prevent wrinkles, and give more sparkle to the eyes are:

• *Blinking.* Blinking helps to prevent strain, improve vision, and lubricate the eyes. Normally, the eyes blink three or four times per second. Notice how a baby does it so naturally and delicately. When you are reading, get into the habit of blinking your eyes at least two times per line. Or, rotate the eyeball in a complete circle, first clockwise, then counterclockwise.

• *Palming.* Every hour, close your eyes and cover them gently with your palms to cut out all external light. Relax a couple of minutes as you experience this "blackout." If you still "see" light in your mind's eye, take a few more minutes until you experience darkness. This technique helps relieve bloodshot eyes.

• *Massage.* Oil massage is particularly beneficial, since the eyelids (along with the scrotum) are ten times more permeable to lipids (oils) than any other part of the body. To prevent lines, gently massage eyes at bedtime using warm ghee, almond or olive oil + a few drops rose or sandalwood essential oil.

TIPS FOR BEAUTIFUL EYES

- To add sparkle and tone to surrounding tissues, make an eye wash of fennel or eyebright tea. When the tea has cooled, pour it into an eye cup or shallow bowl. Bathe each eye for 30 seconds by blinking in the wash.
- For relief of bloodshot eyes, dip cotton pads in rosewater, lie down, and place pads on closed eyes for 10–15 minutes.
- For thicker lashes and brows, put a touch of castor or olive oil on them nightly.
- For sound sleep and relief from eyestrain, massage scalp and feet nightly with castor oil. Use brahmi or coconut oil instead if weather is hot or your feet burn.

- To refresh the eyes, spray a light mist of water on closed lids as needed during the day. Splash eyes with cool water once an hour when reading or working for extended periods of time.
- To protect eyes, wear sunglasses outdoors (lenses should be smokey or green, not purple or blue), and do not read in the sun.

HAND CARE

Soft, beautiful hands and healthy, well-groomed nails reflect not just your personality but also your age. Like the face, the hands need extra care because they are constantly exposed to the environment—and also constantly in use. To keep your hands soft and young-looking:

•Carry body oil in your cosmetic bag and massage some into your hands whenever you wash them throughout the day.

•At bedtime, make a mixture of 1 Tbsp almond oil + 1 tsp buttermilk, and massage it into hands. Put on cotton gloves, and go to sleep. In the morning, remove gloves and rinse hands. This will also help to strengthen the nails.

FOOT CARE

Feet are our faithful servants. Yet, unless we suddenly have trouble walking as the result of an injury or a foot problem, we rarely give them much thought in modern life. At one time in history, to anoint the feet was an act of honor or a preparation for worship. It is a custom worth reviving. Indeed, regular foot care is important, not only because it reduces the chance of developing problems that could limit mobility, but also because the feet are connected by the body's subtle energy channels to all body organs, including the brain. The techniques of reflexology—the massage of energy points on the foot—are based upon this understanding. Consequently, when the feet are tired, so is the mind and body. On the other hand, beautiful, healthy feet are literally the foundation of physical grace and stamina.

TIPS FOR BEAUTIFUL FEET
- Keep feet clean, but do not suppress foot perspiration; doing so can cause damage to internal organs.

- Do not wear tight shoes or very high heels.
- Take a relaxing foot bath for 5–10 minutes in warm water + 2–3 drops witch hazel + 1–2 drops each comfrey, lavender, sage, and rosemary oil.
- Do a weekly foot "mask" with herbal paste of almond meal + lentil flour. Leave on 10 minutes; rinse with warm water; then massage with olive, sesame, or avocado oil. This foot massage is also good for eliminating insomnia.
- Put raw green peas in the bottom of your socks and walk around for 5–10 minutes to give yourself a mini-massage.
- Roll each foot over a tennis ball or rolling pin while sitting, to strengthen foot muscles and increase flexibility in joints.
- Take an early morning barefoot walk on the dewy grass to relax.
- Detoxify the whole body before bedtime with a 5–10 minute foot soak in warm water + 2 tsp rock salt followed by a foot massage with olive, sesame, or avocado oil.

BODY BATHS

In many cultures throughout history, the bath has been—and is—an important ritual for purification of body and mind. In the Vedic tradition, we honor the temple statues with a "five nectar" bath of milk, honey, ghee, yogurt, and banana, which are the five perfect foods, according to Ayurveda. Since it is the best we have for ourselves, it is the least we can offer the deities.

Baths *are* holy—that is, they purify us and make us whole. Indeed, there is nothing so rejuvenating to the spirit and relaxing to the body than a leisurely soak in the tub. Consequently, Ayurveda considers the bath to be a vital part of daily life. Of course, the pace of modern life leaves most people with barely enough time to take a quick shower once a day. Nevertheless, taking half an hour to yourself once a week for a cleansing and nourishing body bath is really the least you can do for your health and peace of mind. It is also one of the fastest and easiest ways to increase your Tej factor.

THE TEJ THIRTY-MINUTE HOME SPA TREATMENT AND BATH

Step 1. Make a body "mask" using a thin paste of ground almonds + wheat flour

+ lentil powder + water or Bindi Herbal Body Cleanser. Rub paste all over body and let dry. Then brush off with dry towel to exfoliate. Do not use brush or loofa.

Step 2. Massage body with sesame oil + 1 pinch turmeric powder to stimulate circulation, strengthen agni, and improve ojas.

Step 3. Take a warm bath for 15–20 minutes, adding to the bath water:

For dry skin:	1 tsp honey + 7–8 drops rose or Bindi essential oil.
For sensitive skin:	1 handful dry milk powder.
For oily skin:	5–6 drops lavender, rosemary, or lemon oil.

Step 4. After bath, pat dry and use body oil to nourish and moisturize skin.

PANCHA AMRIT SNANA—THE FIVE NECTAR BATH

Give yourself a "divine" rejuvenating bath with the "five nectars." Combine the pulp of one banana with 2 Tbsp milk + 1 tsp ghee + 1 tsp yogurt + 1 tsp honey. Massage mixture onto body and then soak for 10–20 minutes in a warm bath to nourish, soften, soothe, and revitalize the skin.

BATHING WITH THE FIVE ELEMENTS INSIDE AND OUT

We purify the doshas using the five elements: Space purifies (or balances) air, fire, water, and earth; air purifies fire, water, and earth; fire purifies water and earth; and water purifies earth. In this sense, we can use the five elements to "bathe" or cleanse the body and senses inside and out:

Space baths

Deep meditation bathes the mind and body in bliss to purify all levels of life.

Air baths

Internal:	Deep breathing.
External:	Sunrise or moonlight walk.

Water baths

Internal:	Drinking water and herbal drinks.
External:	Hydrotherapy (tub baths).

Fire baths

Internal: Eating hot spices to increase digestive fire.
External: Sunbathing, saunas, and steam baths.

Earth baths

Internal: Cleansing mud drinks.
External: Mud baths.

ALL-WEATHER SKIN CARE

Along with stress and diet, environmental factors are some of the major causes of premature aging. Seasonal and climatic changes not only imbalance the subtle internal energies of the body and mind, as you will read in Chapter 6, but also have an effect externally on the skin. Sun, wind, and cold can be particularly damaging unless you take action to counter their effects.

Sun and skin: The sun has always been a prime offender against the skin, but in recent years its harmful effects have increased as the protective ozone layer has weakened. Solar radiation is a factor in cataracts, cell degeneration, and the breakdown of collagen and elastin, which leads to premature aging and wrinkles. It is the leading cause of skin cancer, and also increases free radical generation in the skin because it decreases production of antioxident enzymes. Drugs such as blood pressure pills, diuretics, and thorazine; and foods and herbs such as artificial sweeteners, carbonated drinks, lemon, lime, carrot, fennel, parsley, and bergamot are photosynthesizers. When ingested and absorbed into the body they will cause hyperpigmentation if the skin is exposed to too much sun. Of course, summer heat aggravates Pitta in general, leading to increased skin sensitivity and sebaceous secretions.

In general, sensitive skin types tend to turn red rather than tan in the sun, and also develop freckles. Dry and oily skin types both tan easily. Nevertheless, nothing ages the skin more quickly than a sunburn, so individuals of every skin type should take precautions (as suggested below) to protect themselves from the sun.

HOT WEATHER TIPS FOR BEAUTIFUL SKIN

- Warm weather is Pitta season, so adjust your lifestyle and diet to avoid "overheating" the metabolism. Chlorinated pools, saltwater, and sun dehydrate and age the skin, so the body needs extra moisture inside and out all summer long.

- Use a sunscreen with an SPF of 15 to 30, depending upon your skin. If you don't burn easily but do tan easily, use lower SPF. If you burn more easily than you tan, use higher SPF. *Try not to use sunscreen unless you are in the sun*, since it tends to increase skin sensitivity in general.

- Sesame and neem oil both are excellent natural sunscreens, according to Ayurveda.

- Avoid sunbathing. If you want to be in the sun, stay active (do errands, take a walk), but it is better to stay out of the direct sun as much as possible. However, if you do sunbathe, olive and coconut oils are natural tanning agents, since they absorb ultraviolet light and provide even coloring. A daily dose of B-complex vitamins also helps produce an even tan.

- After sunbathing, take a cool bath, adding a few drops rose, sandalwood, or vetiver essence to the water. Afterwards, pat the skin dry and when it is still moist, massage all over with your body oil or Pitta body oil from Tej.

- To treat sunburned skin, rub on *either* crushed cucumber, cucumber juice, aloe vera juice, an ice infusion of elderflower and chamomile, or cold buttermilk.

Wind and skin: Overexposure to dry, cold wind also damages the skin. In addition to their dehydrating effects, cold and wind put pressure on the epidermis, the horny protective top layer of skin, which is composed of dead cells. When the cold and pressure are extreme, the dead cells desquamate—they burst—leaving the living, young cells of the dermis layer exposed prematurely to the elements. As a result, the unprotected cells become inflamed, causing a tingling sensation on the skin, a problem that is particularly common among the elderly in cold season. Too much peeling with acids produces the same result. The best protection against this effect is not face creams, but essential oils.

COLD WEATHER TIPS FOR BEAUTIFUL SKIN

- Cold weather and cool winds aggravate Vata and cause dry skin conditions; so adjust your lifestyle and diet to add warmth, lubrication, and moisture.
- Take fewer baths and showers in cold season to avoid dryness. When you do bathe, add a few drops of essential oil to the water to provide lubrication. Massage with oil before and after showering or bathing.
- Keep the temperature of your home as low as you *comfortably* can and use a humidifier or keep a bowl of water on your radiators to add moisture to the air.
- Take an occasional steam bath, but avoid saunas, which are dehydrating.

COSMETICS FOR YOUR SKIN TYPE

The practice of enhancing the appearance with cosmetics is almost as old as civilization itself and is not likely to stop any time soon, no matter how radiant we become. Unfortunately, it is probably more difficult to find cosmetics made with pure natural ingredients than it is to find pure natural skin care products. Foundations, whether liquid, cream, or cake, are typically composed of water (if liquid or cream); mineral oils, fatty esters such as butyloleate or isopropylmyristate; fillers such as titanium dioxide and talc; emulsifiers such as alkali, borax, or cellulose; dispersing agents such as lanolin and lauryl sulfate; preservatives such as BHA and BHT; and pigments such as iron oxide derivatives. Many of the dyes used in blushers, lipsticks, and eye makeup are carcinogenic.

On the next page, you will find simple recipes for natural cosmetics. However, if you prefer to wear store-bought products, read the labels and look for the kind that is best for your skin type: Oil-based foundations, which lubricate the skin and give a dewy appearance, are good for dry and mature skin; water-based foundations (the color is suspended in a base that is part oil and part water) give a translucent look and are better for sensitive, or combination, skin; oil-free foundation (the color is suspended in water only) is best for oily skin—although individuals with clogged pores or acne problems are better off to avoid wearing foundation too often. Powdered blushers, which are mostly talcum and have little oil, are better for oily skin; cream blushers are high in oil and are good for mature skin; liquid and gel blushers are water-based and are good for sensitive skin—as long as other ingredients contained in it are hypoallergenic and do not irritate the skin.

NATURAL COSMETICS
TO MAKE AT HOME

FOUNDATION

Mix 1 oz red jamaica flowers or chopped raw beets + 4 oz almond oil in a saucepan and gently heat on a low flame until desired color is achieved. Strain the oil and set aside. Melt ¼ oz beeswax or cocoa butter in pan. Slowly add strained oil into melted wax and beat with a wooden spoon until it cools and has the consistency of a creamy liquid. Store in a jar.

LIP COLOR AND BLUSHER

Mix 1 tsp henna or alkanet root + 4 oz almond oil in a jar. Keep at room temperature for 10 days, stirring once a day, and then strain the oil. Melt ¼ oz beeswax in a pan. Slowly add strained oil into melted wax and beat with a wooden spoon until it cools. Optional: Add 2–3 drops banana, strawberry, or rose essential oil to the mixture as a scent when using as a lip color.

EYE COLOR

Fill an oil lamp with castor oil, light the lamp, and then cover with a copper colander or other perforated copper vessel. Let the oil burn for about 6 hours or until a black ash collects in the copper vessel. Mix a small amount of the ash with a dab of butter and rub around the eyelids. This not only adds color but also is soothing to the eyes. Note that this eye color is neither smudge proof nor waterproof, however, and smears easily.

TREATING THE MOST COMMON SKIN PROBLEMS FROM HEAD TO TOE

When our internal chemistry is normal, so is the skin. When the chemistry changes because of improper diet, improper breathing, or stress, toxins form in the body and the skin throws them out as acne, boils, eczema, psoriasis, rashes, or other kinds of eruptions. The type of toxic manifestation will depend upon each person's constitution. In general, skin disorders fall into six categories:

- Disorders of the sebaceous glands include seborrhea, which is a Vata or Pitta imbalance, and blackheads, whiteheads, cystic acne, sebaceous cysts and excessive oiliness, which are Kapha imbalances.
- Disorders of the sweat glands include excessive sweat, prickly heat, and red rash, which are Pitta problems; lack of sweat, which is a Vata problem; and abnormal foul-smelling perspiration, which is an imbalance of all three doshas.
- Pigmentation problems include blackish discolorations (Vata); brown or reddish discolorations (Pitta); and whitish pigmentation (Kapha).
- Skin allergies include psoriasis (Vata), dermatitis (Pitta), and eczema (Vata, Pitta, or Kapha).
- Infections (fungal, bacterial, or viral) are due to low immunity (depleted ojas) and occur in all skin types.
- Changes in tissue growth, which include dandruff (Vata or Pitta); moles, acne rosacea, and birthmarks (all Pitta imbalances); and cysts and tumors, which are Kapha imbalances.

In the following pages, you will find the external remedies for the most common skin problems from head to toe. (Several of them are the contribution of my colleague, Dr. Kirit Pandya, one of India's foremost Ayurvedic physicians specializing in skin disease.) All these problems are affected by (and in many cases even caused by) our behavior. As we have said, "undigested" emotions, tension, and stress, which create hormonal imbalance and weaken immunity, are leading factors (along with undigested food) of bodily toxins—and therefore of skin disease. For example,

DO'S AND DON'TS
OF PROBLEM SKIN CARE

DO'S

DO cleanse, nourish, and moisturize externally every day with herbs and oils.

DO apply herbal remedies to the problem areas, including lavender essence diluted in almond oil to avoid infection and scars.

DO have professional pancha karma treatment to cleanse the body internally.

DO eat a healing diet of foods to pacify the imbalanced dosha(s) (Chapter 7).

DO drink aloe vera juice mixed with water on an empty stomach every morning to detoxify the blood.

DO take ½ cup of warm milk + 1 tsp of ghee at bedtime to avoid constipation.

DO drink lots of water daily.

DO address personal issues to avoid having "undigested" emotions that create toxins.

DO practice meditation (Chapter 13) and self-massage (Chapter 8) daily to reduce stress.

DO exercise ½ hour daily, but DON'T overdo it.

DO live a balanced lifestyle for your constitution.

DON'TS

DON'T squeeze or open deep pimples.

DON'T eat too much refined, canned, processed, or fried food, sugar, chocolate, seafood, or red meat.

psoriasis is exacerbated by worries and anxiety; acne rosacea, by anger and frustration; eczema, by a variety of stresses, depending upon the person's constitution; and cystic acne, by depression, "holding on" to upsets, and emotional attachments in general. Other significant factors in skin disorders include the overconsumption of refined, devitalized, canned, and processed foods, sugar, sweets, chocolate, fatty and fried foods, salt, seafood, and red meat; low digestive fire and poor elimination (constipation); a lifestyle and habits that imbalance the doshas; hormonal changes; hereditary factors; lack of exercise and rest; and improper external cleansing and skin care.

Nevertheless, an *occasional* bout of dry skin or rash or pimples—although it is a frustrating experience at any age—is also a normal part of the flux of life. As weather changes, as hormone levels rise and fall in the course of a month, and as our diet and activity vary from day to day, the subtle balance of the doshas naturally shifts as well, creating swings in our mood and even changes in our physical appearance. Ayurveda gives us the basis to recognize these subtle imbalances and the tools to correct them before they become something more. Indeed, *imbalance does not necessarily mean disease*—we can be quite healthy and still experience symptoms of imbalance. However, imbalance can *lead* to disease if it is not corrected in its early stages.

One of Ayurveda's fundamental concepts is that health *and* disease are holistic—that is, whatever happens in one part of the body happens to the whole body. There are no isolated health problems. Any physical or psychological symptom of imbalance indicates an imbalance in the bodymind unit. Keep this holistic picture in mind as you look over the common skin problems below. Right now you might not have any significant complaints. Your complexion may not appear especially dehydrated, for example, but you may have dry, cracked lips. Although this condition certainly is not life-shattering—it is something with which you easily can live—it does indicate that Vata dosha is high. If you take the appropriate steps to eliminate this minor symptom—which is not hard to do—you will automatically take care of *all* Vata-related symptoms.

On the other hand, if you ignore this early sign and continue to do things that aggravate Vata, eventually you will see the *progression* of symptoms as the dosha starts to accumulate and circulate to vulnerable tissues and organs. In the case of dry lips, for example, symptoms may progress to include dry cracked feet, brittle nails, dry scalp, split ends, dandruff, and premature wrinkles. If toxins continue to build up in the body because of imbalance, this could lead to cellular breakdown and Vata disorders such as psoriasis or dry eczema—which in severe form can be personally shattering. You can see the potential progression of symptoms for each type of imbalance below. For each dosha, symptoms tend to appear in groups, because the bodymind responds to imbalance as a whole:

Vata: Slightly dry skin, dry lips, excessively dry skin, cracked feet, brittle nails, dry scalp, split ends, dandruff, psoriasis, dry eczema, wrinkles on forehead, skin discoloration, dark circles under eyes.

Pitta: Slightly sensitive skin, broken capillaries, whiteheads, acne rosacea, burning sensations, burning feet, burning eyes, nosebleeds or slight bleeding of pimples, dermatitis (eyebrows are dry and flaky), burning scalp, burning eczema all over body, hives, allergic reactions, wrinkles around eyes, moles, changes in pigmentation.

Kapha: Slightly oily skin, blackheads, enlarged pores, excessively oily skin and hair, white itchy dandruff-like patches on scalp (due to too much oil, which is the opposite of dandruff), cystic acne, eruptions with itching and oozing, swollen ankles and feet, weight gain, puffiness under eyes, cellulite, cysts, and tumors.

If you have a skin problem now, you will see an improvement just from doing the basic daily skin care regimen and using the remedies prescribed below. Once the symptoms have become widespread, however, it indicates that toxins have accumulated in the body and the doshas have begun to move deeper into the tissue. At this point, no external remedy alone—whether chemical or "alive"—is sufficient to eliminate the internal causes of toxicity. The visible symptoms may diminish, but as soon as you stop the treatment, they quickly reappear, usually in full force. In order to restore balance once disease or premature aging is present, you must cleanse and nourish yourself *within*, beginning with a professional purification treatment known as *pancha karma*, or the five actions (described in Chapter 6), or with a home detoxification treatment (described in Chapter 8). If you do not detoxify internally first, then any pure substances you feed the skin and body will simply go to waste. It is equivalent to adding clean water to dirt—all you end up with is mud.

Yet even a detoxification treatment is only a temporary remedy. If you change nothing else about your lifestyle, new toxins will soon appear. A client once came to me with eczema all over his body. He wanted a cure in a bottle or a pill, and except for trying my skin care products, he refused to try anything else, not even a simple meditation. It was very unfortunate. He hated his wife, and she hated him, and they stayed together for the sake of their child. There is no physical remedy that can cure this kind of disease, because the toxins that devastate his skin come directly from his thoughts.

Ayurveda's teachings and techniques are all-encompassing for precisely this reason: We cannot ignore one level of life without a cost to life as a whole. This understanding is the source of hope, not gloom. Ayurveda gives us the responsibility of having a balanced lifestyle, but also the freedom to take care of all psychophysiological symptoms of disease with a single course of treatment. For example, you do not need one medicine for acne, another for depression, and another for water retention. These are not different problems, but three symptoms of the same imbalance—in this case, aggravated Kapha. Therefore, all of them can be corrected by the same means—a Kapha-pacifying lifestyle.

In the second half of the book, we will show you how to create a natural, fulfilling lifestyle that balances the doshas on all levels of life—body, mind, and soul—and ultimately eliminates aging and skin disease at their root. Begin here, however, with the appropriate external remedies, and enjoy the immediate healing and soothing effects of these herbal treatments.

DANDRUFF

Imbalance: Vata.

Aggravated by: Dry scalp, harsh shampoos, improper rinsing, poor nourishment due to poor metabolism, anxiety, worry, poor circulation, hot spicy foods, drugs, medications.

Treatment:

- After regular shampoo, rinse hair with herbal tea made of burdock, arnica, or horsetail.
- Mix 1 egg yolk + ½ tsp lemon juice + 2–3 drops camphor. Apply to scalp;

leave on for 10 minutes; rinse with lukewarm water.

- Do hot oil scalp massage 2–3 times a week (see 135).
- *Supplements:* 400 units vitamin E + 15–20 mg zinc daily.

PREMATURE GRAYING

Imbalance: Pitta & Vata.

Aggravated by: Worry, anger, frustration, anxiety, sudden shock, too much mental work, stress, menopause, thyroid problem, lack of copper, zinc, folic acid, pantothenic acid, and paba.

Treatment:

- Make an herbal infusion of 2 tsp each sage + walnut leaves. Apply 1 tsp of the liquid to hair root where it is gray every night.
- *Supplements:* Take recommended daily dosage of horsetail, nettle, alfalfa, and fenugreek herbal tablets; biotin; vitamin E; B-complex; lecithin; kelp and silicone. Eat a diet rich in protein, nuts, and minerals.

HAIR LOSS

Imbalance: Vata, Pitta, or Kapha.

Aggravated by: Stress, poor nutrition, hormone imbalance, cigarettes, drugs, alcohol, overuse of hairdryer or shampoo, excessive sun.

Treatment:

- To encourage new hair growth, make paste using a pinch of crushed black pepper or ½ tsp fenugreek powder + ¾ cup coconut milk. Rub briskly on scalp, cover with plastic cap, leave on 30 minutes. Wash with gentle shampoo. Do this treatment nightly, or whenever you shampoo.
- Brush hair nightly with mixture of 2–3 drops each lavender + rosemary in coconut oil.
- Regular headstands help prevent hair loss according to Ayurveda, but consult physician before you try them.
- Do scalp massage using cooling, soothing oils such as brahmi, bhringraj, triphala, or gotu kola to improve circulation.
- Make an herbal oil using 1 part dry hibiscus flower decoction + 4 parts

coconut oil. Massage on scalp to stimulate hair growth.

- Overactive sebaceous glands due to hormone imbalance may also cause hair loss if the oil clogs pores. In this case, wash hair more frequently and do a light oil massage only, using 2–3 drops lavender or rosemary essential oil.
- Use Tej Hair Tonic, which includes an herbal tablet and a hair oil, to restore hair growth.

LINES ON FOREHEAD

Imbalance: Vata & Pitta. (The habit of frowning will produce lines even when there is no imbalance.)

Aggravated by: Anxiety, worry, excessive dehydration, too much sugar or protein, habitual frowning, alcohol-based astringents, excessive use of lemon, tomato, or cucumber juice.

Treatment:

- Mix 3 drops Bindi or Tej Vata essential oil in water and use as a daily mist to hydrate.
- Make a hydrating massage oil using a base of apricot kernel, avocado, sesame or almond oil + 2 drops each of sandalwood and geranium + 1 drop each of lemon and cardamom. With your fingers, massage the oil on forehead using a horizontal stroke.
- Do daily facial exercise: Alternately stretch and tighten forehead muscles; hold and release 3 times.
- Twice weekly, make a firming herbal mask using a paste of 1 tsp cornstarch or potato starch + 2 tsp aloe vera juice or egg white. Apply mask and lie down for 30–40 minutes. Cleanse, nourish, and moisturize as usual.
- Weekly, do an exfoliating enzyme mask: Apply pineapple or papaya pulp to face and lie down for 10 minutes. Cleanse, nourish, and moisturize as usual.

PREMATURE WRINKLES

Imbalance: Vata & Pitta.

Aggravated by: Dryness (less oil), dehydration (less water), stress, sun, wind, extreme temperatures, excessive exercise, travel, alcohol, coffee, tobacco, sweets, spicy foods, hot or cold water, sudden weight loss, water pills, hormone medication, diabetes, lack of purpose, lack of loving relationships, hereditary factors.

Treatment:

- Twice weekly, do a facial mask using a paste of 1 Tbsp sandalwood powder + 1 drop camphor oil + 3–4 drops lotus oil + 2 tsp water. With ring finger, gently massage a few drops of face oil made with sandalwood + rose oil directly under eyes for protection, then apply paste to rest of face. Cover eyes with wet cotton pads dipped in rosewater and lie down for 10–15 minutes. Cleanse, nourish, and moisturize as usual.
- Make a decoction of 1 Tbsp dry geranium in ½ cup water, then apply to face using a cotton ball.
- Do natural face-lift and face exercises (see Chapter 8).
- *Supplements:* Take recommended daily dosage of vitamin E and primrose oil capsules. Drink 6–8 glasses of water daily.

CROW'S-FEET, DRY EYES, AND EYE STRAIN

Imbalance: Pitta.

Aggravated by: Age, stress, worry, insomnia, alcohol, dehydration, squinting.

Treatment:

- Avoid chemical makeup removers and heavy eye creams. Use cotton dipped in plain vegetable oil to remove eye makeup.
- Wear sunglasses in daylight and avoid reading in the dark. Never look directly at sun.
- Twice daily, pinch the skin between your eyebrows, starting at the bridge of the nose and moving outward. Repeat 3–4 times.
- Do blinking, palming, massage, and eye bath treatments described on page 136.

DARK CIRCLES UNDER EYES

Imbalance: Brown circles: Vata; green-gray circles: Pitta.

Aggravated by: Anemia, ill health, lack of sleep, poor circulation, anxiety; hormonal imbalance, menstrual disorders; too many fried, frozen, and canned foods, beans, peanuts, salads.

Treatment:

- Lie down on slant board with feet raised for 5–10 minutes.
- Soak cotton pads in cold milk, rosewater, fig juice, or crushed mint juice, and place over closed eyes for 5–10 minutes.
- Apply crushed mint leaves around eyes for 5–10 minutes.
- Before bed, gently massage around eyes with saffron or almond oil
- Do daily blinking and palming exercises (see page 136).
- *Supplements:* Take 2–4 gms ashwangandha, shatavari, or ginseng herbal tablets or powder before lunch and dinner.

PUFFY EYES

Imbalance: Kapha.

Aggravated by: Hypertension, liver and kidney problems, poor elimination, low digestive fire, water retention, lack of sleep, hormonal changes.

Treatment:

- Make eyepads using either black tea bags soaked in warm water, cotton puffs dipped in witch hazel or celery juice, or gauze squares stuffed with 1 tsp grated raw potato. Place on closed eyes for 20 minutes.
- With your ring finger, press gently underneath the eye one point at a time from the inside corner to the outside corner to help drain the lymphatic fluids.
- *Supplements:* Take 1 tsp triphala every night; take 1,000 mg vitamin C, and eat black raisins and figs every day.

DRY, LINED, OR CRACKED LIPS

Imbalance: Vata.

Aggravated by: Smoking, drugs, cold, dryness, dehydration, age, excessive talking, licking lips. (Vertical lines above lip indicate unfulfilled sexual desire.)

Treatment:

- As often as you like, apply vitamin E oil, unsalted butter, or ghee directly to lips. Or use a mixture of 5 drops *each* rose and sandalwood oil in 1 oz avocado oil.
- Melt in a double boiler 9 tsp lanolin + 1 tsp castor oil. Remove from heat, add

3–5 drops rose oil, and let it solidify. Apply over lipstick to seal, moisturize, and add gloss.

- Massage lips nightly with 1 oz sesame oil + 2–3 drops glycerine.
- Take a mouthful of water and slosh around the inside of the lip area for 1 minute.

HEAT RASH AND PRICKLY HEAT

Imbalance: Pitta.

Aggravated by: Sun.

Treatment:

- For heat rash, make a paste using 1 tsp sandalwood powder + a pinch of camphor + buttermilk. Apply all over body before you bathe.
- For prickly heat, combine equal parts peaflower, sandalwood, and coriander powders + 1 pinch nutmeg + 2–3 drops vetiver essential oil. Make paste with rosewater, apply to body, leave on for a few minutes, then rinse or soak in a cool bath. Pat dry and dust all over body with sandalwood powder. If you do not have time for paste treatment, just apply sandalwood powder after shower.
- *Supplements:* To relieve itching, soak 1 tsp of cumin seed + 1 tsp coriander seeds in water overnight. In the morning, strain and drink the liquid.

SUDDEN RASH OR HIVES

Imbalance: Pitta.

Aggravated by: Allergies.

Treatment:

- Combine juice from macerated dried sweet basil leaves with any bland vegetable oil; apply to rash. Or just apply ice.
- Apply sandalwood oil to relieve itching.

WARTS AND OTHER SKIN GROWTHS

Imbalance: Vata, Pitta & Kapha.

Aggravated by: Warts are a common viral infection. However, toxicity and poor elimination are factors in all types of skin growths.

Treatment:

- Apply a few drops of castor oil nightly to warts and peel off skin.
- Soak cotton in fresh pineapple or lemon juice and apply as a bandage to dissolve warts.

FRECKLES

Imbalance: Pitta.

Aggravated by: Sun.

Treatment:

- To bleach, rub on cottonseed oil or crushed pumpkin seed kernels + olive oil.
- Mix 1 tsp yogurt + 2–3 drops honey to make a natural bleach. Apply, leave on for ½ hour, then rinse.

WHITE SPOTS AND LEUCODERMA

Imbalance: This condition is usually hereditary.

Aggravated by: Small white spots are caused by stress, excess salt. Leucoderma, or large white patches, is hereditary.

Treatment:

- Internal cleansing.
- Sunlight.
- Apply a few drops *each* neem & bakuchi herbal oils (available from Tej) directly to spots to reduce external symptoms.

AGE SPOTS

Imbalance: Vata or Pitta.

Aggravated by: Old age, cold weather.

Treatment:

- Do internal cleansing treatment at change of season.
- Massage with Vata-pacifying oils or Tej Saffron Oil.

BLACKHEADS

Imbalance: Kapha.

Aggravated by: Excessive oil secretions.

Treatment:

- To loosen, add a pinch of Epsom salts to a cup water. Dip cotton ball in mixture and wash face.
- Apply a mask of ground fresh parsley on oily area. Lie down for 10–15 minutes, then cleanse, nourish, and moisturize as usual.

PSORIASIS

Description: Silvery flakes mostly on scalp but may appear on any part of body. Characterized by chronic and excessive dryness and irritation.

Imbalance: Vata & Pitta.

Aggravated by: Liver dysfunction, anxiety, stress, ungroundedness.

Treatment:

- Bathe with horsetail herbs wrapped in cheesecloth in your tub.
- After bath, apply neem oil + ghee or karanj oil twice a day.
- Do daily self-massage (see Chapter 8).
- Do yoga or other nonaerobic exercise until you are sweating mildly (good for detoxification, stress reduction).
- *Supplements:* Take recommended daily dosage of primrose oil, cod liver oil, lecithin, vitamin E, and zinc.

DRY PATCHES

Mix 10 drops sandalwood oil + 1 oz castor oil and apply to dry area.

ECZEMA

Description: There are three types of eczema: Dry patches (dry eczema) or moist, burning, inflamed red patches (burning eczema) typically appear around the joints but may appear anywhere on the body. Pussy, oozing patches (wet eczema) *or* dry itchy patches typically appear around eyes, brows, nose, or scalp, but also may be anywhere on the body.

Imbalance:
    Vata (dry).
    Pitta (moist, inflamed, red, burning).
    Kapha (pussy if moist, itchy if dry).

Aggravated by: Improper diet, blood toxicity, constipation, stress, undigested emotions, excessive sun, saltwater.

Treatment:

- Mix equal parts neem + brahmi + basil herbal oils (all available from Tej), and apply.
- Mix ½ tsp camphor + 2 tsp zinc oxide + 7–8 tsp corn or potato starch, and apply.
- Apply a light compress of rose petals + nettle + comfrey in water.
- Cut an aloe leaf and apply sap directly to skin.
- Add a few drops of chamomile, geranium, juniper, or lavender essential oil to coconut oil, and apply.
- Take baths with comfrey and nettle decoctions.
- Massage feet and scalp at night with brahmi oil.
- *Supplements:* Daily take 1–2 tsp cod liver oil or primrose oil, 800 units vitamin E, 30 mgs zinc, recommended daily dosage of vitamin B-complex and lecithin. Take ½ tsp triphala at bedtime as laxative. Avoid salt, sugar, fats, onions, garlic, radishes; eat carrots and musk melon.

NATURAL CORTISONE TREATMENT

Take a soothing bath with licorice tea added to the water. Helps heal eczema and psoriasis.

ACNE VULGARIS (CYSTIC ACNE)

Description: Very oily skin, large pores, blackheads, large pussy pimples, deep scars.

Imbalance: Pitta or Kapha.

Aggravated by: Overconsumption of sweets, fats, oils, red meat, seafood, coffee, alcohol, tobacco; emotional stress and attachments (inability to "let go"), inactivity, feelings of possessiveness, depression, and purposelessness.

Treatment:

- Exfoliate skin with Tej or Bindi herbal powders.
- Take steam baths using rosemary or eucalyptus essential oils.
- Take weekly detoxifying bath using Epsom salts or ginger + rock salt.
- Exercise vigorously for ½ hour 3 times a week to achieve sweating.
- Apply Tej Soothing Lotion on pimples, and do a pimple "mask" once or twice a week. To make, mix ½ tsp crushed cumin seed + 1 tsp coriander + a few drops water, and apply paste over pimples. Leave on for 20–30 minutes, then rinse.
- Do soothing weekly facial mask using red sandalwood + neem + lodhra powders.
- For stubborn, large cysts, apply piece of warm onion 2–3 times a day to break it, but do not squeeze.
- *Supplements:* Take daily 10,000 units beta carotene, 1,000 mg vitamin C, and 15–20 mg zinc. Increase dietary fiber, including bran, fresh fruit, and produce. Drink fresh carrot, beet, or apple juice. Drink detoxifying herbal teas of burdock, goldenseal, echinacea, neem, or turmeric. Drink ½ glass *warm* water every hour. In morning, take aloe vera gel: 1 tsp for Pitta; 2 Tbsp for Kapha.

ACNE ROSACEA ("RED" ACNE)

Description: Red rash on nose and cheeks that may be very sensitive or burning, excessive oiliness on T-zone, broken capillaries, and thickened skin on nose.

Imbalance: Pitta.

Aggravated by: Anger, frustration, disappointment, anxiety, overambition, stress and pressure, unsatisfying or acrimonious relationships, overactivity, hot weather, sun, hot spices, sour fruits, fermented foods, tomatoes, seafood, canned or preserved foods, soda, pastry, chocolate, pizza, french fries, sweets; antibiotics and harsh chemical peels.

Treatment:

- Do not use astringents, toners, or any harsh substances on skin. Use only gentle, soothing treatments.
- Wash with milk + white sandalwood, manjista, and neem herbal powders.
- Soak towel in cool herbal tea of comfrey or nettle, and apply wet compress to face.

- Grind fresh cilantro + mint. Add water to make paste, and apply to face as a mask. Lie down for 10–15 minutes, then rinse off. If condition is very dry, red, or burning, apply soothing lotion of ghee + juice from crushed dried neem leaves, or use neem herbal oil. Additional internal and external remedies are available from Tej.
- *Supplements:* Take daily 10,000 units beta carotene, 1,000 mg vitamin C, 15–20 mg zinc, 400 units vitamin E, and daily recommended dosage of B-complex. Drink detoxifying herbal teas of burdock, goldenseal, echinacea, neem, or tumeric. Drink a glass of water (at room temperature) every hour. In morning, drink aloe vera juice.

CUTS AND WOUNDS

To treat a bleeding or open cut and prevent infection, sprinkle pure turmeric powder on the wound. To prevent infection and scarring, rub 2–3 drops lavender oil on wound.

CELLULITE

Imbalance: Kapha or Vata.
Aggravated by: Low agni, poor digestion, accumulation of water and fat.
Treatment:
- Kapha pacifying diet.
- Take ½ tsp trikatu powder after lunch and dinner + 4–6 Medohara pills in the morning (both are available directly from Tej).
- Drink ½ cup aloe vera juice early in morning for detoxification.
- Stomach and thigh massage with any warming oils such as bergamot or rosemary body oil or Tej Kapha or Medodhara Oil.
- Physical exercise.

DRY, CRACKED FEET

Imbalance: Vata or Pitta.
Aggravated by: Cold weather, dehydration.

Treatment:

- Massage feet at night with brahmi oil for Vata, castor oil for Pitta, or with cashew nut oil for all types. Cover feet with socks and wear to bed.

CORNS AND CALLOUSES:

Soak cotton in fresh pineapple or lemon juice and apply as a bandage to dissolve corns.

ATHLETE'S FOOT, FOOT ODOR, AND BURNING FEET

Imbalance: Pitta.

Aggravated by: Heat, exercise.

Treatment:

- Make a deodorant foot powder using 1 oz orris root powder + 2 oz arrowroot powder + 20 drops lavender or rose oil. Sprinkle in shoes or directly on feet.
- For burning feet, massage with sandalwood oil, unripe mango juice, or mango leaf.

The proper external skin care routine is the first necessary step to a beautiful complexion, and the sooner you start to do these treatments regularly, the more quickly you will have some relief from problems and premature aging. The ultimate cure is not in these herbal remedies, however, but deep inside the body and mind, where Ayurveda takes you by means of the many other therapies in this book—from diet and massage to breathing and meditation. I hope you will try some of these additional balancing treatments, but try them at your own pace.

One of my first American clients, who had suffered from terrible acne from the age of thirteen, came to me for many years before she was willing to try what she called my "strange" ideas—even though the herbal remedies alone had given her a degree of relief she had been unable to find for twenty years. Now, almost twenty years later, she is one of my most outspoken advocates for the benefits of Ayurvedic mind-body therapy—and she points to her own smooth and youthful complexion as her proof.

PART III

BODY PURIFICATION: BALANCING THROUGH THE SENSES

DEHA SHUDHI

With beauty before me, I walk.
With beauty behind me, I walk.
With beauty above me, I walk.
With beauty below me, I walk.
From the East beauty has been restored.
From the South beauty has been restored.
From the West beauty has been restored.
From the North beauty has been restored.
From the zenith in the sky beauty has been restored.
From the nadir of the earth beauty has been restored.
From all around me beauty has been restored.

A NAVAJO PRAYER

CHAPTER 6

AN INTRODUCTION
TO INTERNAL SKIN CARE

Adopt the pace of nature: her secret is patience.
RALPH WALDO EMERSON

The basic cleansing, nourishing, and moisturizing procedure takes only a few minutes of time each day, which is not a lot to give in exchange for keeping a youthful appearance. However, this regimen alone is not enough to prevent cellular breakdown due to stress and disease, or to maintain overall balance and promote absolute beauty. Nor is it enough in most cases to correct an existing imbalance—although it definitely will bring a degree of relief from the external symptoms. We start on the road to absolute beauty with what we can see and touch. It is the necessary first step, but it takes several steps more to achieve lasting radiance.

As my first attempt to treat acne taught me twenty-five years ago, Ayurvedic skin care treatments produce immediate visible results. However, I also learned from early experience that such improvements are rarely permanent, because external treatment does not change the reason for disorder and disease, which always lies deep within. As I became more knowledgeable about Ayurveda, I discovered that a change in diet produces a much more enduring effect than external remedies alone; yet even nutritional treatments did not clear up most problems completely. Until I introduced methods to reduce the effects of stress, balance the mind and emotions, harmonize the lifestyle, and change the cell memory that creates aging and disease, my clients' skin problems eventually returned to some degree.

As the mother of all medicine and the original holistic system, Ayurveda has always understood this truth: To eliminate any skin problem, restore balance, and stay healthy and vibrant, we need to cleanse and nourish ourselves internally as well as externally on all four levels of life. Ayurvedic therapies accomplish this by using the five senses, the breath, the mind and consciousness—or the techniques of body, breath, mind, and soul purification, as they are called. Below, we will introduce you to these therapeutic options and also show you how to choose among them to create a personal holistic beauty regimen that is suited to your pace, your taste, and your level of commitment. In addition, we will tell you how to achieve maximum benefit from these therapies by integrating them into your routine in accord with the cycles of nature, as well as by performing them with patience and focus. As you will see, more is not necessarily better, and according to the most fundamental Ayurvedic principle, what is good for someone else is not necessarily what is good for you. In Mark Twain's words, "We can't reach old age by another man's road. My habits protect my life, but they would assassinate you."

BODY PURIFICATION: BALANCING THROUGH THE SENSES

"The perception is the knowledge one has by the senses coming in contact with the soul."

JAIMINI'S MIMANSA SUTRA

Ayurveda offers various forms of balancing therapy using each of the five senses. Nutritional therapy involves taste; massage therapy uses touch; aroma therapy, smell; sound therapy, hearing; and color and gem therapy, sight. Together these compose the methods of *deha shudhi,* or body purification, described in Chapters 7 through 11.

The five senses exist at the boundary of body and mind, and connect human consciousness with the physical world to provide our experience of life. We "digest" the world through the senses in much the same way that we digest food—

that is, the force of agni transforms all the energy and intelligence that is outside us—taste, smell, sound, color, form, texture—into the energy and intelligence that is us—thoughts, emotions, nutrients, cells, tissue. In this way, all sensory experience creates balance or imbalance in the bodymind. Consequently, whether it uses herbal oils, tastes, colors, sounds, or touch, we always select treatment, and each component of a treatment, according to its energetic influence on the doshas.

As you may recall, each of the five senses arises from one of the five elements. As a result, the senses that correspond to the most dominant elements of your constitution naturally will be your most developed—that is, your most sensitive modes of perception.

SKIN TYPE	CONSTITUENTS	DOMINANT SENSES
Dry (Vata)	Space + Air	Sound + Touch
Sensitive (Pitta)	Fire + Water	Sight + Taste
Oily (Kapha)	Water + Earth	Taste + Smell

Because of this correspondence, every person has an affinity to specific sense therapies. Dry skin types have sensitive hearing and touch, so gain the greatest benefit from sound and massage therapies. Sensitive skin types have refined vision, and respond most readily to color and gem therapies and to visual experience in general. Oily skin types have well-developed gustation and olfaction, so nutrition and aroma therapy are important for them.

In fact, these sense-related preferences are evident in every aspect of life. In the romantic arena, for example, Vata people are attracted to softness and touch. A woman with dominant Vata will not be interested in a man who sounds harsh when he talks or who has rough skin, no matter how handsome he is. Similarly, Pitta types are drawn to good looks and beautiful surroundings, while Kapha types tend to be attracted to people who smell good, cook well, and love to eat.

An innate preference for one kind of therapy does not preclude getting value from the others, however. When you are out of balance, the most effective way to restore it is to use the variety of purification techniques at your disposal.

BEYOND THE SENSES
TO THE SOUL: BREATH, MIND,
AND SOUL PURIFICATION

In terms of Ayurvedic "anatomy," the senses are an aspect of *annamaya kosha*, or the gross physical body, while breath, mind, ego, and consciousness are aspects of our "subtle" bodies. Like the visible sheath, these unseen strata are governed by the doshas and influenced by our lifestyle and environment. They are linked to each other and to the physical body by many thousands of fine energy channels that run through every part of our gross anatomy. The five sense therapies—the techniques of deha shudhi—work on the physical body, but also affect the subtle bodies through these channels. However, other Ayurvedic techniques balance these unseen layers of intelligence directly: Breathing exercises affect the pranic body— the vital sheath—and help integrate body and mind; meditation enlivens the deepest level of consciousness and harmonizes all layers of life at once; mindfulness or concentration techniques work on the mind and ego to change the patterns of perception that produce stress and to develop clarity of purpose and focus in action. Together these make up the techniques of breath, mind, and soul purification—or *prana*, *manas*, and *atma shudhi*—which you will find in Part IV.

As in all creation, the subtlest layers of the body have the greatest potential energy, so when we effect a change in the breath, mind, or consciousness, we produce a more holistic and powerful result than when we effect a change on the gross physical level. An ordinary piece of wood illustrates this idea. If we change it physically by carving it into a bat, it is capable of the kind of force that sends a baseball out of the stadium at a hundred miles per hour. If we change it molecularly by setting it on fire, it can power a steam engine. If we change it atomically, we can power a nuclear energy plant.

As we have said, the field of consciousness underlying human experience and the unified quantum field underlying the subatomic universe are one and the same, according to Ayurveda. When we experience this absolute level of existence through this ancient science, we develop the capacity to achieve perfect balance, because we are working from the field of infinite intelligence, which is the source of natural law. The legendary miracles performed by "yogis" of the East and West are merely demonstrations of our human power to master matter and heal the body when we live life from this fully developed state of awareness. This is the ultimate vision of life's possibilities contained in the knowledge and techniques of Ayurveda.

HOW TO CREATE YOUR IDEAL BEAUTY REGIMEN

There are three basic concepts to keep in mind as you read the book and create your personal beauty routine from the many balancing therapies. They are:

- Your leading dosha, your skin type, and its condition at this time.
- The attributes of the doshas, including their physical and mental traits.
- The law that "like" attributes produce imbalance and opposite attributes produce balance.

Your general physical appearance and temperament tell the story of the innate balance of your doshas—your prakriti. The current appearance of your skin tells the story of the condition of your doshas *now*—your vikriti. In Chapter 2, you answered questions about these aspects of yourself to determine: one, your *leading dosha* and *skin type*; and two, whether your *constitution* is *balanced* or *imbalanced* at this time, based upon the condition of your complexion. These are the first pieces of information you need to know in order to select the proper skin care regimen. If you do not know your skin type or its condition now, go back and complete the Skin Type Quiz.

The next set of information you need to know is the *basic attributes,* or energetic qualities, of the doshas. As we said, the characteristics of your skin are determined by the attributes of your leading dosha:

- *Vata*, which is made of space and air, is *cold, light, clear, thin, rough,* and *dry*; dominant Vata creates *dry skin*.

- *Pitta*, which is fire and water, is *hot*, *light*, and *slightly oily*; dominant Pitta creates *sensitive skin*—that is, skin prone to inflammation. (Because heat always dehydrates and oiliness creates a greasy quality, Pitta skin is also known as combination skin—it is slightly oily in the T-zone across the eyebrows and down the nose, and slightly dry around the forehead, cheeks, and chin.)

- *Kapha*, which is water and earth, is *thick*, *cold*, *heavy*, *soft*, and *oily*; dominant Kapha creates *oily skin*.

When the doshas are in balance and the skin condition is normal, these differences are subtle. When the doshas are imbalanced, these qualities become more prominent. For example, dry skin develops conditions such as dry eczema when Vata is aggravated, and oily skin develops wet eczema or cystic (oozing) acne when Kapha is aggravated.

Finally, you need to remember the basic *principle of balance*, which states: As the cause, so the effect. In other words, like qualities increase like, and thus create imbalance. To correct an imbalance, we *diagnose* the condition in terms of the *doshas*, and *treat* it in terms of the *attributes*. Ask yourself: What doshas are dominant in my condition now? And what attributes balance these doshas? Then choose the therapy, food, or activity whose qualities are unlike your dominant dosha(s) and avoid the ones that are like your dominant dosha(s). *To maintain balance, you need the opposite qualities of what your makeup naturally is.*

How will you know the attributes of all these things? We will tell you directly as we describe the various treatments and techniques. Or we will simply give you lists of what to favor and what to avoid according to your skin type. For easy reference, many of these lists appear together in the appendixes. In the meantime, if you need a general reminder, the table at the end of this chapter provides a summary at a glance of the elements and attributes that balance your doshas.

These guidelines apply whether or not you have skin problems. Below, you will find instructions specific to the condition of your skin at this time—balanced (normal skin) or imbalanced (problem skin).

FOR NORMAL SKIN

If you have a normal complexion at this time, there are three progressive approaches you can take towards absolute beauty. Clearly, the more you do, the more radiant you become.

The first option is the basic maintenance program—that is, the daily cleansing, nourishing, and moisturizing routine using herbal preparations prescribed for your skin type. As we said, this is necessary simply to counter the day-to-day effects of the environment. However, because of the nutritive and rejuvenative properties of the Ayurvedic herbs and oils, this daily routine also visibly improves the vitality of your complexion. However, it does not change the toxic thoughts or behaviors that are at the root of aging and disease.

The second approach adds an overall health maintenance and prevention program to the daily beauty routine. This might include modest lifestyle changes in accord with Ayurvedic principles, such as modifications in your diet (Chapter 7) and daily schedule, and doing self-massage (Chapter 8) to keep the doshas balanced, reduce stress, improve immunity, and increase your sense of well-being.

The third choice is the total approach to achieving radiance and bliss. It encompasses the first two steps and adds successively more Ayurvedic strategies to your daily routine to create a holistic lifestyle in total harmony with nature and with your own purpose in life. This includes techniques to balance all levels of life—body, breath, mind, and soul. For example, you might incorporate daily breathing (Chapter 12) and meditation (Chapter 13) practices, as well as a seasonal pancha karma treatment (Chapter 8) to achieve optimum health and beauty.

Of course, this kind of Ayurvedic lifestyle is not something to implement overnight. It is a gradual and continual process of recreating and refining your habits of mind and behavior—your way of being in the world—so that you grow clearer in your purpose, more effective in your activity, more fulfilled in yourself, and whole in spirit. Later, we will give a few tips on how to begin this life journey in the most comfortable and practical way.

FOR PROBLEM SKIN

Once you know that you have an imbalance and what your imbalance is, there are five basic steps to correct problem skin:

- Begin the appropriate external skin care routine.
- Do a program of internal purification and light fasting.
- Follow up the detoxification program by modifying your normal diet and taking medicinal herbs to balance the doshas.
- Meditate daily to reduce stress and balance the emotions.
- Introduce other balancing treatments as desired.

The Skin Problem Quiz in Chapter 3 is designed to help you determine which dosha or doshas are imbalanced now, based upon your skin condition. Again, you need to know this information before you can select the proper balancing therapies. If you are not sure what your imbalance is, redo the questionnaire on page 85. In addition, review the follow-up, which explains how to detect the physical and emotional factors in your imbalance as well as how to choose therapies when more than one dosha is imbalanced. If you still have questions about your imbalance and cannot get to a professional Ayurvedic practitioner for a diagnosis, then simply aim towards balancing your leading dosha. Whenever you are in doubt, this is the best course to follow.

The first step of treatment is always your daily skin care routine. As we have said, proper cleansing and nourishing brings some relief from your outward symptoms, and also helps prevent aging due to environmental causes. Nevertheless, no external skin care routine can rid the body of the toxins that cause disease or balance the aggravated doshas. For that, you must cleanse and nourish within.

Thus, the second step to healthy skin is to detoxify *internally*. The ideal method for this is a professional pancha karma treatment, as we explain on page 178. However, if that is not an option, you will find instructions for a home detoxification and fasting program in Chapter 8 under the Tej Home Spa Treatment. Once you have completed this internal purification program, start to adjust your diet to restore balance to the doshas. Do not strain. Modify your diet as much as you comfortably can.

If you do not meditate already, begin a daily meditation practice. As we have said, emotional stress is a major factor in most skin disease—anxiety, fear, and worry imbalance Vata; anger, jealousy, and frustration, Pitta; and grief, depression, and attachment, Kapha—so address the underlying emotional issues in your skin problem. Other techniques for reducing stress include massage and breathing exercise.

Whatever you choose to do, try to do it on a regular basis for maximum results. Therefore, we recommend that you do not attempt to do too much at once. You will get more benefit from consistently doing one or two new practices—such as the daily cleansing routine plus some dietary changes—than by trying many different balancing techniques for just a few days at a time. Follow the pace of nature, as Emerson advised. There is very little we can do to alter the rate at which a flower blossoms. Let your own beauty unfold in the same way, enjoying each phase of change as it comes.

These two principles—enjoying the present moment and following the lead of nature—are essential to balance and bliss. As we describe below, they are represented in Ayurveda by the practice of *sadhana* and by living in accord with the cycles of nature.

SADHANAS: BALANCING THROUGH THE ACTIVITIES OF DAILY LIFE

Ayurveda reveals more than the secret to flawless skin; it shows the way to achieve a sublime life. Yet what makes Ayurveda the ultimate holistic knowledge is that it teaches us how to achieve this perfection through even the most mundane activities. We do this by the performance of sadhanas. When we order our daily activities, including purification therapies, into a consistent routine in order to have a balanced lifestyle, we refer to the activities themselves as sadhanas.

The Sanskrit word is often translated as daily "discipline" or "practice." Although this meaning conforms to the letter of the language, so to speak, it misses the spirit. A sadhana does take self-discipline and sometimes practice, and even helps build character, to use the military phrase. However, its intent is fulfilled not by our mere completion of the duty, but by our total surrender to and enjoyment of the act itself. This capacity to yield the heart and mind completely to the task at hand is the essence of grace and beauty. It is a way to flow with time so that we are always in the present, which is the experience of dynamic balance.

Life as a whole is already a "discipline" and a "practice"—an ever-repeating cycle of simple, routine acts like making the bed, brushing teeth, getting dressed, commuting to work, checking the e-mail, taking meetings, paying bills, talking to friends, cooking dinner, washing dishes, and tucking the kids into bed. This is not just the human condition. All of nature evolves through routine and regularity, as we discuss below. We call nature's discipline the tides, the planetary orbits, the rhythm of the days, the seasonal cycle, the inexorable growth from seed to sapling to tree to fruit. What makes the human condition different from the rest of the natural world, however, is our conscious capacity to choose what we do—and this act of will is what makes the difference between slavish routines and sadhanas.

I prefer to define sadhana as any activity performed with a spiritual mind, or what others call "mindfulness," or concentration. Mindfulness is the act of giving full and focused attention—full mind—to what you are doing *in the moment you are doing it*. In childhood, we do this effortlessly. With no concept of past or future, we naturally live in the present. But as we mature, our cognitive faculties, with full support from our overstressed emotions, start to intrude on nature with endless "good" reasons why the mind should be somewhere other than where we are. By adulthood, awareness is so preoccupied—so *thought* full—that it takes conscious effort, or at least a conscious decision, to be *mind* full at every moment.

Sadhana is the intentional *exercise* of consciousness, the continual act of choosing where to put our attention. Because it involves choice, sadhana also involves *desire*—the wish to carry out the task; in this case, the task of creating inner wholeness. In other words, sadhana is any action performed with all our heart, soul, and might. In the deepest sense, it is an act of devotion. By the attention and intention of a sadhana, we elevate even the most trivial task to a sublime status, and in so doing, invite the highest support for our action. This is the purpose of sadhana: to bring success to everything we do by means of the power of our consciousness. If you have ever observed an Olympic gold medalist in her winning performance, then you have witnessed a person acting mindfully. If so, you do not need the teachings of Ayurveda to know that any action performed with such one-pointed awareness and surrender produces the most powerful results.

Ayurveda explains this phenomenon by the fact that *whatever we put our attention on, grows*. Attention is energy, and energy has force. The more focused attention we give to a thought or activity, the more powerful that thought or action becomes, and the greater the sphere of influence of its effect. Thus, when we do an Ayurvedic therapy *as* a sadhana—that is, when we perform it with a spiritual mind—we bring balance to a deeper level of life and produce more holistic effects. *Any* action performed as a sadhana not only produces a more profound result but also a more *sattvic*—that is, a more evolutionary—influence in ourselves and our environment.

Successful action brings achievement and the fulfillment of our desires, and this is the basis of bliss. When you understand that performing sadhana—and performing every action as a sadhana—is the secret to attaining the highest pleasure, then every action becomes an opportunity to achieve absolute beauty, and the "discipline" itself becomes a labor of love. When you give yourself over completely to what you are doing, you win no matter what the ultimate outcome is, because you have already enjoyed the process one hundred percent. In this "flow," you are no longer bound by a desire for accomplishment to any particular result, since you experience a sense of fulfillment at every point along the way. Paradoxically, by freeing the attention in this manner from the *thought* of success, we create the very condition of single-mindedness necessary to achieve it. In this way, performing sadhana transforms the day-in, day-out routine of life from an experience of limitation to a means of liberation.

Indeed, the principle of performing sadhanas is inseparable from the principle of routine, and as you read about specific therapies, you will find that the season and time of day are often considerations in Ayurvedic treatments. The purpose of such routine is not for the sake of keeping schedules, however, but as a necessary means to balance our internal "environment" with our external one. Such inner and outer harmony is the foundation of absolute beauty.

A TIME AND A SEASON
FOR EVERYTHING: CREATING
A NATURAL LIFESTYLE

*"No one can say that a life with childhood, manhood and old age
is not a beautiful arrangement; the day has its morning, noon and
sunset, and the year has its seasons, and it is good that it is so.
There is no good or bad in life except what is good according to its
own season. And if we take the biological view of life and try to live
according to the seasons, no one but a conceited fool or an impossible
idealist can deny that human life can be lived like a poem."*

LIN YUTANG

According to Ayurveda, a change of seasons affects the human psychophysiology
just as it stimulates change in the life of a tree. Solar and lunar cycles also subtly
influence human life, as they do the lives of owls and bears. Made up of the very
same elements and forces as all things in nature, we exist in the give-and-take of a
cosmic ecological system to which our inner nature must continuously adapt.
While the rest of the animal kingdom instinctually live in harmony with these
forces, human beings, by virtue of free will, can—and often do—choose a way of
life that is counter to our own natures. So-called primitive cultures and traditional
agrarian societies live in attunement with natural patterns to a degree that technological societies do not. Modern life, dictated by clock-time and Daytimers—yet
virtually time-less thanks to electric lights, climate-controlled buildings, and
supersonic travel—leaves us disconnected from our biological rhythms. When we
live out of step with these innate rhythms for extended periods of time, the doshas
become imbalanced and illness of some kind inevitably ensues.

Ayurvedic routine resynchronizes our activities with energy patterns throughout nature. Just as the doshas govern different life processes in the body, they govern
different times of the day and year; and their energetic influence increases in the
environment during those times. Kapha, Pitta, and Vata (in that order) alternately
dominate in six four-hour intervals beginning at sunrise and in three "seasonal"

intervals. A Kapha-Pitta-Vata cycle also governs the arc of individual life from child-hood to old age. By adjusting our lifestyle to move in time with these "master cycles," our inner balance is not overthrown.

THE DAILY CYCLES

Sunrise marks the beginning of the morning Kapha period. In the table below, we have indicated 6 A.M. as the start of the day, but this time will vary depending upon where you live and the season of the year.

Kapha	6 A.M. to 10 A.M.	6 P.M. to 10 P.M.
Pitta	10 A.M. to 2 P.M.	10 P.M. to 2 A.M.
Vata	2 P.M. to 6 P.M.	2 A.M. to 6 A.M.

Each period of the day reflects the energies of the dominant dosha. Inertia is high during Kapha periods, which is why we tend to feel lethargic when we sleep into the late morning hours and tend to slow down in midevening. High noon is high-energy time, thanks to Pitta's fueling effect. Consequently, digestion is strongest at lunch hour, and so is our physical stamina. Vata governs the nervous system, so the mind is more alert in the afternoon, while physical energy typically slumps around 3 or 4 P.M.

In general, the best time to go to sleep is before 10 P.M. when Kapha's settled energy still prevails. If you stay up late, you are likely to get a second wind when Pitta peaks at midnight. At night, Pitta's energy goes to work on digesting supper and rejuvenating body tissue. Vata brings on the active dream phase known as REM sleep. At the peak of the morning Vata cycle, about 4 A.M., the plant and animal kingdoms start to reawaken. In fact, you can just about set your clock to the calls of certain birds who have distinct morning songs that they perform at the same hour every day. This rarefied atmosphere at midcycle of the early-morning Vata period has a powerful enlivening effect throughout nature, and if you are inclined to get up with the early birds (which is likely if you have dominant Vata), it is an excellent time to meditate and begin your own day. If not, try to awaken at least a half hour before sunrise, when Kapha's groggy nature takes precedence again. You will actually feel more refreshed to awaken while Vata energy is still lively.

Obviously, you need to be most concerned about these energetic effects during the cycles that correspond to your dominant dosha. Everyone has more difficulty getting up during morning Kapha hours, for example, but people with Kapha body types have the greatest problem. One of the worst things such individuals can do at that time is give in to their natural tendency to oversleep, since doing so only aggravates the dosha even more. Invigorating exercise, a stimulating massage, or even a cup of coffee, if that is what it takes, is a better morning choice for Kaphas than sleeping in. On the other hand, a late afternoon catnap will help balance Vata types during the peak of the enervating Vata cycle. Pittas face a different kind of challenge altogether. When Pitta is high at noon, their tempers tend to flare. Thus, they are wise to avoid scheduling difficult discussions or negotiations during lunchtime hours.

You will find many more such suggestions throughout the book to help you design a balanced beauty regimen, and in the appendixes, you will find recommended sadhanas (Appendix F) and daily schedules (Appendix G) for each type. As you grow more familiar with Ayurvedic principles in general, you also may think of many other ways of your own to coordinate your daily routine with nature's to achieve maximum balance, energy, and effectiveness.

SEASONAL CYCLES AND PANCHA KARMA THERAPY

The doshas also have a yearly cycle, which we experience as the seasons. There are three seasons in the Ayurvedic calendar, which coincide with changes in climatic conditions rather than with lunar cycles. Consequently, the starting date of each season varies according to geographic location—hot season in New York City is cold season in Sydney, Australia, for example.

In North America, the approximate seasonal cycles are:

Kapha The cold, damp months of late winter and spring (mid-February through May).

Pitta The hot, humid months of summer (June through September).

Vata The cool, dry, windy months of fall and early winter (October through mid-February).

As we said, these dates are approximate—February weather is quite different in Tampa than it is Toronto—so use your judgment to determine when the cycle has shifted to a new phase. Another way to identify the change of seasons is in terms of the elements themselves: The moist, cold qualities of Kapha's earth bring winter's deep freeze and the liquefying nature of its water brings the spring thaw and mud. Warmed by Pitta's fire, melted snows feed the rushing rivers and water the lush gardens of June. Eventually, fire evaporates water, leaving us the muggy, then scorching, days of July, August, and early September. Vata's rushing air brings rough winds in autumn; space brings the cool, clear, cloudless days we think of fondly as football weather.

Naturally, each of these climatic changes brings out similar types of changes in each of us. According to the principle of like increases like, each dosha accumulates in the body during its own season, just as it does during its daily cycles. Thus, as Pitta accumulates in nature in the summer months, everyone becomes a little fiery—that is, more irritable and hot-tempered, and Pitta types get especially volatile. They also experience more effects on their skin, such as sunburn, freckling, allergies, rashes, and melanoma. Similarly, everyone gets slightly dehydrated in the drying Vata season—and more so once the indoor heating comes on. Vata types, however, develop cracked lips and feet, dandruff, and other dry skin problems in wintertime, unless they stick closely to a Vata-pacifying routine. Spring moisture produces an increase of secretions in everybody, but Kapha types will have problems with congestion, water-retention, and a resurgence of acne following the drier days of winter.

In a study of the effects of seasonal weather changes on health, Western scientists recently "discovered" a related phenomenon. By analyzing local hospital records, geographers at the University of Delaware found that hospital admissions of asthma sufferers doubled, from about four hundred in early September to more than eight hundred in early October, peaked again in the spring, and declined in the summer. This finding is totally consistent with Ayurvedic predictions. September marks the shift from summer to fall, when the increase in Vata aggravates asthma conditions due to wind-borne allergens. The natural increase of congestion in the spring, when Kapha is highest, tends to precipitate the more severe variety of bronchial asthma.

Obviously, unless we counter this external influence of the doshas through changes in our lifestyle and routine, our own doshas inevitably become imbalanced. With so many different types of therapies to choose from in Ayurveda, this is easy to do. Finding the right combination of therapies to balance your constitution and fit your circumstances is necessarily an idiosyncratic process, however, and we will describe the way to approach it below.

Nevertheless, climatic changes have a universal influence: Each season, one of the doshas naturally accumulates to some degree in each individual. As we said, accumulation is the first step in the disease process, when symptoms of imbalance are still localized in specific organs and easy to treat. Consequently, Ayurveda recommends a professional internal cleansing therapy for everyone during each seasonal change, or *at least once a year*, to clear out the accumulated doshas and prevent progression of disease.

This purification and rejuvenation treatment is known as pancha karma, which means "five actions," as we said earlier. The five actions refer to the five cleansing methods: emesis (therapeutic vomiting), laxatives, enemas, nasal cleansing, and blood purification. Not all five treatments are prescribed for everyone. Generally speaking, emesis reduces congestion due to Kapha imbalance; purgatives clean out the bowels to balance Pitta; and enemas clean out the colon to balance Vata. These treatments are actually the cornerstone of a three-part therapy that starts with a preparatory step of massage and sweating to liquefy the toxins and move them into the gastrointestinal tract, where they can be eliminated by the five actions from the body, and ends with post-treatment care, including dietary modifications and rejuvenation therapies. As you will see, pancha karma is also used as a detoxification program to treat existing skin disease and other disorders.

In fact, the full pancha karma treatment is a powerful medical therapy that is used to cure imbalance and disease, as well as to prevent it. Therefore, only trained practitioners and Ayurvedic physicians can prescribe and administer it properly. Since it is not intended as a home remedy, we are not going to include instructions here. Nevertheless, we recommend professional pancha karma treatment to prevent disease and slow down the aging process to everyone. However, since most of you will be unable to find a local Ayurvedic practitioner, we do include instructions for a modified cleansing and balancing therapy that you can do at home as part of your seasonal beauty program. This Tej Home Spa Treatment, as we call it, includes some aspects of pancha karma therapy, such as self-massage, oil treatments, sweating, and light fasting. You will find this purification and beauty treatment in the chapter on massage therapies.

THE LIFE CYCLES

According to Ayurveda, our development from birth to death also follows a cycle of the doshas. Infancy through adolescence is high Kapha period; early adulthood through middle age, Pitta; the senior years, Vata. The influence of the dominant dosha is evident in our experience at each stage.

Kapha governs growth in general, and the growth of bodily tissue and bones in particular. Of course, this is the type of development that characterizes childhood. The soft, smooth skin and "baby" fat, the need for lots of sleep, the capacity for learning, and the common childhood ailments such as coughs, colds, asthma, congestion, sore throats, earaches, and even adolescent acne are all typical Kapha manifestations. After the growth spurt, we enter the active career- and family-building years. This period is characterized by the development of Pitta qualities such as independence, ambition, confidence, sociability, and intelligence, as well as Pitta activities such as the pursuit of intellectual, creative, and professional goals, and the development of social networks and mature relationships. Pitta imbalances like heartburn, acid stomach, ulcers, hemorrhoids, and hypertension also typically manifest during this period. In the later years, we experience the effects of increased Vata, which brings dehydration and impedes movement and the flow of life in general. Accordingly, the skin dries out and becomes wrinkled, taste sensation diminishes, the bones get brittle, the teeth loosen, we get varicose veins, we eat and sleep less, become stiff and arthritic, forget more easily, and develop all sorts of aches and pains.

Chopra has likened the relationship of these three master cycles to "wheels within wheels"—and no doubt, the idea of trying to balance the overlapping daily, seasonal, and developmental influences of the doshas can be dizzying. For example, what do you do to maintain equilibrium during a daily Vata period in the middle of

Kapha season during the Pitta cycle of your life? The answer is, there is no answer—at least no answer that anyone else can give you—because there is no logical way to keep all the wheels spinning in perfect harmony all the time. (There *is* a way to accomplish this feat, as you will read later, but it involves direct experience of consciousness, rather than mental juggling.) The purpose of knowing the cycles is to understand the full range of environmental influences on your doshas, so you can make the best choices for your lifestyle from moment to moment, based upon your unique experience and constitution. The more you know about the causes and effects of imbalance, the more easily you can avoid it, the more quickly you can recognize when you are out of "sync," and the more ways you have to get back into the swing. So, if you are a middle-aged person with a hectic schedule of high-pressure meetings on a hot summer day, you can recognize the potential for Pitta problems and take steps beforehand to avoid them—such as eating certain foods, wearing certain colors and scents, listening to soothing music, meditating, going for a massage, or doing breathing exercises, all aimed at pacifying Pitta.

Except for the formula to be established in the Self—that is, to be attuned to your own nature—Ayurveda has no formula to determine the precise cause of every imbalance or to calculate exactly what you should be doing at any given time to prevent it. Once again, knowledge of the cycles, like all Ayurvedic knowledge, is a tool to help you live in harmony with your environment and your inner nature. It is a guideline, not a rule. The "routines" of Ayurveda follow nature's timetable—and in following them, we follow our own nature. This is the secret of how to enjoy a life of inner and outer harmony, a life of absolute beauty.

THE IMPACT OF MODERN CULTURE

We cannot properly address the idea of living in tune with nature without also addressing the ways in which our culture creates disharmony in the first place. Today, many of us spend more time in "civilized" environments than we do out "in nature." These man-made worlds have their own energetic influence, which we cannot ignore if we want to achieve inner balance. Disconnected from the cosmic rhythms, yet not exempted from cosmic laws, modern culture (like every culture in history) has its own constitution—its "collective consciousness"—

which reflects the energies and activities of the individuals and groups who compose it. Because the totality is always greater than its parts, however, the makeup of the culture also influences each member in turn.

In Ayurvedic terms, many aspects of modern life are vatogenic in nature. Contemporary society is characterized by constant movement, speed, change, and mobility—the attributes of Vata. The age of air and space travel (Vata, of course, *is* air and space), computers, computer games, electronics, mass media, communications, and information, is a Vata age. "The acceleration of change in our time is, itself, an elemental force," Alvin Toffler wrote in *Future Shock* almost thirty years ago—and the pace is not slowing down as we go into the next century. We produce more goods, consume more resources, create more garbage, cover more miles, meet more people, explore more places, and change jobs, homes, and partners more frequently than in any other time in history. In addition, we are exposed to a greater variety and volume of stimulation than ever before: we have many new ways to reproduce not only visual images, but also sounds, smells, and tastes. This constant barrage of sights and scents, the incessant, often jarring noise of the electronic age, all aggravate Vata. On the other hand, we rarely make time anymore for nurturing touch, which is the sense that most helps to balance the elements of space and air. Instead, we annually consume billions of dollars worth of antidepressants, tranquilizers, and other mood-altering drugs—purchased over the counter, by prescription, or on the street—most of which are vatogenic as well. Meanwhile, the West's aggressive, competitive work ethic aggravates Pitta. Its materialism and acquisitiveness, along with its diet rich in fats, carbohydrates, and sugar, and poor in nutritive value, all aggravate Kapha—a fact that is evident in the alarming rate of obesity among Americans of all ages.

These imbalances in our collective lifestyle contribute to the high levels of stress and disease in the industrialized nations of the world. You do not have to escape to the mountains and go "off grid" in order to escape their effects, however. By following the principles of Ayurveda and using some of the stress-reduction techniques described in the coming chapters, you can enjoy all the benefits of modern life and still have a natural, balanced lifestyle no matter where you live. An Ayurvedic lifestyle does not mean renouncing material comfort, achievement, or wealth. To the contrary, it means having the clarity and energy to attain all your goals and still have the health and longevity to enjoy your success.

STARTING ON THE JOURNEY
TO ABSOLUTE BEAUTY

The journey to absolute beauty is a "pathless path" because perfection always lies in the present. It is only the mistake of the intellect, as we have said, that we do not know our own perfection at all times. All the techniques of Ayurveda are simply a means to remember the truth of who we are. They offer something for everyone, so that no matter where in the journey this knowledge finds us, we can benefit from it.

As you read about the many balancing therapies, however, you are apt to say to yourself, "Since I cannot possibly do everything Ayurveda teaches, then it is useless to try anything." This would be an unfortunate decision indeed. Clearly, no one could incorporate all these activities into a day and still have time left over for life itself—and it is not the intent of Ayurveda that we should.

The basic approach to absolute beauty is very simple—seven simple steps appear on the opposite page. When you are ready to try some additional balancing techniques, pick one and stick with it for a week or so to assess how it influences your body and mind. Aroma therapy, music, and color therapy are easy techniques to do anywhere, and can be enjoyed while you are doing other things. Dietary changes, massage, and meditation are more demanding in terms of effort and time—and indeed, I find that many people are resistant to such lifestyle changes, either due to the inertia of old habits, philosophical differences, skepticism, or laziness. Nevertheless, my clients who do make the attempt find that the effects of these activities on their health and appearance are the most profound and enduring. So I encourage you try these therapies, especially if you have any skin problems, or want to avoid developing them.

Whatever therapies you choose, keep in mind that balance is the key in all aspects of Ayurveda—too much, or too little, of anything is never good in the long term. The Ayurvedic injunction in all treatment is to avoid or reduce those things

THE SEVEN STEPS TO BEAUTIFUL SKIN

- Know your basic skin type and choose every treatment accordingly.
- Cleanse, nourish, and moisturize daily.
- Balance yourself internally as well as externally.
- Introduce additional therapies gradually.
- Avoid the elements and attributes you have in abundance; favor the ones you lack.
- Keep a regular routine suited to your makeup and the natural cycles.
- Enjoy the present moment.

that imbalance your constitution, and favor or increase those things that balance it. The circumstances are rare when a person must eliminate something from her life completely in order to have beautiful skin. A moderate, comfortable, royal pace is enough to get you to the goal.

The principles of Ayurveda are merely signposts on the journey to radiance—they are there to give you a sense of direction, not to dictate how you must proceed. In the same way that the science of physics, for example, does not tell us how to live, but only describes the natural laws of matter we must live by, this ancient science of life describes the natural laws of the bodymind and offers suggestions, given these immutable laws, of what we could do to make life's journey more enjoyable. The goal is inner and outer balance, and such harmony can never be gained by imposing a fixed set of rules upon ourselves. The ever-changing nature of life itself demands a certain amount of flexibility. Indeed, fanaticism has no place in Ayurveda, which is essentially the knowledge of how to "go with the flow." An Ayurvedic lifestyle is not the goal, but it is the process itself—a lifelong experience of self-discovery through attention to our body, behavior, mind, and soul.

WHAT YOU NEED TO REMEMBER ABOUT YOUR SKIN TYPE AND BALANCE

	DRY SKIN		SENSITIVE SKIN		OILY SKIN	
	Is	*Needs*	*Is*	*Needs*	*Is*	*Needs*
Balancing doshas	Vata	Pitta+Kapha	Pitta	Vata+Kapha	Kapha	Vata+Pitta
Balancing elements	Space+Air	Fire+Water+Earth	Fire+Water	Space+Air+Earth	Water+Earth	Space+Air+Fire
Key attributes	Dry	Oily	Hot	Cold	Heavy	Light
	Cold	Hot	Light	Heavy	Cold	Hot
	Light	Heavy	Sharp	Slow	Oily	Dry
	Rough	Lubricating	Acidic	Bitter	Slow	Sharp
	Dispersing	Dense	Fluid	Static	Dense	Dispersing
	Quick	Stable	Slightly oily	Astringent	Static	Mobile
	Astringent	Salty	Pungent	Sweet	Sweet	Pungent

CHAPTER 7

NUTRITIONAL THERAPY

What is food to one, is to others bitter poison.

LUCRETIUS

One of the most striking contrasts between modern medical science and Ayurveda has to do with the different value they place on food and diet in the health and life of an individual. Anyone who has ever seen the meals served in American hospitals—cheese ravioli, overcooked frozen vegetables, white bread, butter, and jello or ice cream is a typical dinner—has to wonder how such lifeless food could possibly restore the health of an ailing patient, and how little allopathic medicine is concerned with nourishing the body and mind.

Until very recently, doctors in the West were not required to take more than an introductory course on nutrition to complete their medical training. Even then, the approach they learn is fundamentally mechanistic: The body is a car, and food is the fuel that runs the engine. For a smooth-running machine, so the wisdom goes, use the premium gasolines—selections from the four basic food groups. If the engine fails, you cut certain fuels from the "good" category and add them to the "bad," depending upon the part that has broken down. If the pancreas goes, sugars are out; if the gall bladder goes, fats are out; if the heart goes; cholesterol is out; and so on. According to Western medicine, these rules apply to everyone equally. That is, we can all fill up on the same gas because fuel is fuel, and in any case, no matter what kind you use, you can only get so many miles out of the human engine before it fails.

FOOD IS THE
PERFECT MEDICINE

In Ayurvedic medicine, diet is not an ancillary aspect of training. It is a central concern of the vaidyas, the Ayurvedic physicians, whose term for the physical body—annamaya kosha—literally means the *food sheath*. In their viewpoint, food *is* medicine. An inappropriate diet is the number-one physical factor in the cause of imbalance and disease, and a proper diet is essential in preventing and treating it. This is equally true for the body and the mind. Food not only builds, fuels, and repairs every cell in the body, but by means of the subtle energy of the doshas it also fuels and heals our emotions. As you will see, food makes or breaks our mood. Consequently, what you eat is as important to your mental health as it is to your physical condition.

In contrast to conventional Western wisdom, Ayurveda does not believe there is one menu perfect for everyone. Orange juice, cold cereal, and skim milk, the staples of an all-American breakfast, may give energy and a "Special K" figure to some types of people, but Ayurveda predicts that the same meal will leave others with an upset stomach or late-morning fatigue. *No food is intrinsically good or bad according to Ayurveda, but each person depending upon his or her constitution reacts differently to it.*

To treat a person's health without careful regard to diet is, in Ayurvedic terms, tantamount to tending a garden without checking soil quality or the amount of moisture and sunshine it receives. Even in the West, we would instantly fire a gardener who failed to attend to such a basic principle of life; yet for many years we have continued to support a medical system that regularly ignores it. Thanks to the work of people like Dr. Dean Ornish, the cardiologist who reversed the symptoms of advanced heart disease in forty patients through nonpharmaceutical treatment including meditation and diet, the broader medical community is beginning to recognize the therapeutic potential contained in ordinary foods. Nevertheless, modern science still seems to be wiser about how to grow perfect flowers than how to grow healthy, happy, beautiful people.

WHEN EVEN CHOCOLATE IS OKAY

How does Ayurveda know which foods are okay for an individual? And if no food is intrinsically bad, why do some people claim to break out just looking at french fries or chocolate?

As you read in Chapter 5, Ayurveda classifies all foods and herbs according to six tastes, or rasas. These are: sweet, sour, salty, pungent, bitter, and astringent. The taste of each food derives from its constitution—that is, its own balance of elements. Like all of nature, the animals, vegetables, and minerals we eat each contain their own unique proportion of space, air, fire, water, and earth. Different combinations of elements create different tastes. Foods with more earth and water, like grains, naturally taste sweet, for example, and those with more air and space, like green leafy vegetables, naturally taste bitter.

Of course, each taste affects the mind and body according to the attributes of the elements in it. Thus, sweet tastes soothe, lubricate, ground, and nourish the psychophysiology, just as you would expect the dense, viscous earth to do; bitter tastes lighten, stimulate, and dehydrate, just as you would expect a burst of cold air to do; and so forth. The table on page 188 lists each taste and the common foods associated with it.

The best diet for you is one that is heavy on the tastes whose attributes balance your constitution and light on the ones whose attributes imbalance it, according to the law of "like increases like." Pitta types, for example, are better off to reduce consumption of sour, salty, and pungent foods, all of which contain *fire* and aggravate sensitive skin. For these hot-blooded types, a bowl of salsa with salty chips will leave its mark in red on the complexion; however, they usually can enjoy a bowl of ice cold, sweet chocolate sherbet without regret. The Kapha person with a sweet, heavy "earth" constitution and oily skin is the type that tends to get pimples from anything chocolate. There is no reason to feel sorry for her and her kind, though. *They* can indulge in hot salsa and pretzels (but not greasy chips, Kapha!) with relative abandon once in awhile.

So, who gets to eat the greasy stuff? Dry-skinned Vata types can indulge in oily foods without breaking out, although too much spicy sauce or too many dry,

THE TASTES OF COMMON FOODS

TASTE (RASA)	ATTRIBUTES	FOODS
Sweet (Earth + Water)	Cold, oily, heavy.	Sugar, all fruit juices, honey, rice, wheat bread and other complex heavy carbohydrates, milk, cream, butter, beef, fish, lamb, pork, fats, oils, beets, cucumber, potato, apple, fresh figs, grapes, melons, peaches, pears, plums, licorice, red clove, saffron, cardamom, cinnamon, most nuts, ghee.
Sour (Earth + Fire)	Hot, heavy, oily.	Yogurt, cheese, green grapes, lemons, orange, spinach, tamarind, banana, tomato, vinegar, all fermented foods, pickles.
Salty (Water + Fire)	Hot, oily, heavy.	Sea salt, rock salt, salted nuts, potato chips, fast food, canned food, seaweed, kelp, seafood.
Pungent (Fire + Air)	Hot, light, dry.	Onion, garlic, radish, ginger, chili, asafoetida, clove, cayenne pepper, hot mustard sauce, chicken, eggs, safflower oil, black pepper, pumpkin seeds.
Bitter (Air + Space)	Cold, light, dry.	Dark leafy greens, bitter greens, turmeric, goldenseal, dandelion root, fenugreek, gentian root, tonic water, alkaloids (caffeine, nicotine).
Astringent (Air + Earth)	Cold, medium.	Unripe banana, pomegranate, cranberries, beans, lentils, broccoli, cabbage, cauliflower, celery, potato, spinach, cinnamon, alum.

cold dishes can send their rarefied constitutions into a state of frenzy.

Keep in mind, however, that even someone with oily skin, for instance, can have an occasional taste of chocolate as long as the complexion is normal. There is usually no need to eliminate any food from your diet completely unless a skin condition already exists, and even then, you only have to stay away from those foods until your doshas are balanced again. When you are healthy, you can enjoy just about anything *in moderation* without dire consequences for your skin.

In fact, although your daily diet should always be richest in those tastes that balance your skin type, Ayurveda recommends that *every meal should contain all six tastes in some proportion no matter what skin type you are*. This is vital not only to nourish all five elements in your makeup (*lack* of a taste imbalances the doshas just as surely as too much). It is also an effortless way to control your appetite. What is the secret? Having the full range of tastes in every meal completely satisfies the senses, so even a moderate portion of food relieves not just your stomach hunger but also the psychological hunger that leads to cravings and overeating. The one stipulation is that the meal *includes each taste in its proper proportion for your makeup*. (When family members have different constitutions, the simplest way to insure that everyone eats a balanced diet is to cook with a large variety of foods. In my own family, we eat different vegetables, grains, and spices every day. By rotating foods in this way, no one eats an overload of one taste. If a family member already has an imbalance, then he or she takes herbs to help restore equilibrium [see Appendix D] and does additional Ayurvedic therapies.)

On the next page, we have indicated the best ratios for each skin type by showing the three tastes to favor and the three to "avoid" *in order*, from strongest to weakest effect on the leading dosha. For example, look at the list for dry skin. Salty tastes—first on the favor list—have the strongest *balancing* action on Vata, while sour has less, and sweet, the least. Similarly, bitter tastes—first on the avoid list—have the strongest *imbalancing* effect on Vata; astringent, less; and pungent, the least. In other words, the best diet for Vata or dry skin types includes salty tastes in the highest proportion and bitter tastes in the lowest proportion; the best diet for Pitta or sensitive skin types includes bitter tastes in the highest proportion and sour in the lowest; and for Kapha or oily skin types, pungent in the highest proportion and sweet in the lowest.

	TASTES TO FAVOR	TASTES TO AVOID
Dry skin (Vata)	*Salty,* sour, sweet	*Bitter,* astringent, pungent
Sensitive skin (Pitta)	*Bitter,* sweet, astringent	*Sour,* pungent, salty
Oily skin (Kapha)	*Pungent,* bitter, astringent	*Sweet,* salty, sour

Of course, this does not mean that dry skin types should be sprinkling a cup of salt on their plates or that Pitta types have to live on escarole. On the contrary, Ayurveda recommends that we avoid the pure forms of a taste—or what we might call the active ingredient, such as pure sugar, table salt, hot peppers, pure bitters, and so forth—because of their potency. Rather, we can and should get all six tastes that we need from the delicious variety of complex foods, which nature has provided conveniently for our enjoyment—that is, sweet from fruits and grains, for example, salty from vegetables and seaweeds, pungent from spices, bitter from greens, and so forth. In Appendix H, you can see the benefits of each taste for the skin and body when we eat them in the right proportions, as well as the problems that develop when we don't.

DO YOU WANT IT HOT OR COLD?

The theory of tastes reflects the Ayurvedic understanding of the metabolic processes, which start with the act of chewing and end weeks later when all the nutrients have been absorbed and assimilated by the body tissue and cells through the process of dhatu transformation. With each stage of metabolism, as the nutrients are chemically altered, they produce new "tastes," so to speak, or new influences on the physiology. (Rasa, the "first taste," is the only one that actually produces a sensation on the tongue.) Nutrients produce a second taste, or biochemical effect, called *vipak,* as the food is further broken down in the stomach. Before the nutrients enter the stomach, however, they also create an immediate *heating* or *cooling* and *drying* or *lubricating* effect on the body known as *virya.* Every food has a hot or cold potential energy that is only partly offset by how the food has been stored, cooked, or prepared, as you will see. Together with rasa, virya is the most important intrinsic factor in how a particular food will influence your doshas.

	VIRYAS TO FAVOR	VIRYAS TO AVOID
Dry skin (Vata)	Hot, moist	Cold, dry
Sensitive skin (Pitta)	Cold, moist	Hot, dry
Oily skin (Kapha)	Hot, dry	Cold, moist

Again, Ayurveda determines which taste is right for each type according to its action on the doshas. The dietary rule of thumb is: Know the predominant attributes of your constitution, then eat mainly foods with attributes that are *not yours*. (Later we will tell you how to adjust your diet to the seasons.) When the doshas are in balance, you will notice that you are naturally satisfied with the tastes and qualities that balance you. Conversely, when you are *out of balance*, you will notice yourself craving *more* foods that imbalance you. When that happens, you can start the process of regaining your constitutional equilibrium by reincluding foods into your diet whose rasa and virya *pacify* your doshas.

Once you are eating the six tastes in right proportion again, you even may notice that you are satisfied completely with a smaller plate of food. So listen to what your body is telling you. This is the essence of living in tune with nature.

FOOD FOR THE SOUL

You can understand Ayurvedic nutrition more easily if you expand your view about the nature of food beyond the chemistry of proteins, amino acids, minerals, vitamins, and the like to include the concept of subtle energies. In fact, the Ayurvedic theory of tastes is a theory of energetics. In the broadest sense, Ayurveda perceives all human experience as *food*: The world around us is digested by the five senses; ideas are digested by the mind and intellect; feelings and foodstuffs are digested by the body. Energetically, a pungent taste, a steaming bath, the spicy scent of clove, the friction of a vigorous massage, the vibrancy of the color red, and even a critical word or an angry mood, for example, all exert the exact same kind of fiery influence on the psychophysiology. All of this energy—all of this food—transforms our biochemistry, each in its characteristic way, and ultimately becomes the stuff of our tissues and cells.

Therefore, from the Ayurvedic standpoint, our personal lifestyle is a *mind-body diet*—we nourish ourselves, or imbalance ourselves, with every thought and deed. *We eat life*, and the practical question posed by Ayurveda is: What quality of life do we want to feed to ourselves?

Western physics has its own version of this energetic principle. It states: Every action has an equal and opposite reaction. The difference between the two ideas is only in the extent of their application. Modern science is still unconvinced that the laws of matter have anything to do with the laws of mind, whereas Ayurveda is premised on the indivisibility of the two. In the Ayurvedic view, the patterns of intelligence inherent in all things resonate with each other at all times on the level of unified pure consciousness; and in this way every interaction, depending upon the nature of the energies that come together, produces either harmony or discord—balance or imbalance—on the subtle intelligence we call the doshas. This is equally true of food's effect on the mind and the mind's effect on food.

THE EFFECT OF FOOD
ON MOOD

According to Frawley in *Ayurvedic Healing*, the ancient sages describe the effects of diet in this way:

> *The food that is eaten is divided threefold. The gross part becomes excrement. The middle part becomes flesh. The subtle part becomes the mind. . . . The water that is drunk is divided threefold. The gross part becomes urine. The middle part becomes blood. The subtle part becomes the life force.*

Indeed, this is what happens to food as it is transformed by the seven dhatus. The "gross part" is the malas, the roughage and other waste that is the natural by-product of digestion. The "middle part" is the extracted nutrients that build and repair tissue. The "subtle part" is the *taste* of our food and drink—the hidden intelligence—that literally feeds the qualities of our mind and emotions. In general, sweet foods produce sweet, satisfying moods, sour foods produce sharp or sour moods, and so forth. As you will see, we are what we eat.

TASTE'S EFFECTS ON MIND AND EMOTIONS

	IN PROPORTION	IN EXCESS
Sweet	Love, satisfaction.	Desire, attachment, need, passivity.
Salty	Mental ease.	Mental rigidity, greed, addiction.
Sour	Mental acuity.	Envy, regret, resentment.
Pungent	Ambition, motivation.	Hate, anger, jealousy, aggression.
Bitter	Mental clarity, insight.	Grief, disillusionment.
Astringent	Optimism, well-being.	Fear.

This understanding of tastes rings true in terms of ordinary experience. Every day, we speak of such things as sweet love, bitter tears, the sour grapes of losing out on what we want, and the taste of fear, which is actually the dry sensation on the tongue that comes from the fight-or-flight response. On the deepest level of awareness, the bodymind already knows that tastes and emotions are one and the same in terms of their effects. Indeed, Ayurveda speaks of both by a single name: *rasa*, or essence. The six rasas are the flavors of our emotions as well as our food. Likewise, our emotions are the flavors of our experience.

When we eat a balance of tastes, the mind is balanced. When we have peace of mind, life is balanced. When we have a disproportion of any taste, the mind gets swept up in emotion, and harmony is lost. Thus, eating a bowl of chili when Pitta is dominant or when you feel hot and irritable, is like pouring gasoline on a fire; but when the natural fire is low, a pinch of pungent ginger gives a needed spark. Similarly, loading up on carbohydrates, sugar, and cream-rich dairy—the classic "comfort" foods—literally adds weight to your troubles when Kapha is dominant or when you feel depressed; but a breakfast of hot cereal is stabilizing and strengthening when Vata is dominant or when you feel ungrounded or afraid.

All these flavorful effects, unlike the nutritive effects of food, are relatively immediate and short-lived. As soon as we have a meal, we experience the impact of rasa, vipak, and virya on our awareness. This is true whether or not we are awake to notice it, since tastes affect us consciously and subconsciously. In fact, long before Freud, vaidyas analyzed dream images to assess the balance of elemental energies—the balance of tastes—in the body. According to Ayurvedic theory, violent, hostile dreams or images of burning objects indicate excess fire and are a common experience after eating a large spicy dinner. Dreams of being bogged down in the mud or physically stuck or burdened in some way indicate too much earth and water and typically follow a night of bingeing on comfort foods. Images of running, fleeing, and fear signal excess space and air and often result from late-night partying with popcorn, chips, and ice cold beer. So if you wake up in the morning with similar dreams, think about what you ate the day before.

At the same time, remember that a rose is a rose is a rose—and a rasa, a rasa. At the subtle level of awareness where all experience is ultimately digested, the essential intelligence of the object itself—its taste—is always the same whether you eat it, think it, see it, smell it, touch it, or say its name. Therefore, whether a taste originates in your thoughts or your food, it always has the same effect on *consciousness*. Of course, that does not mean we can get pimples just thinking about chocolate. A sweet memory, or even a thought of sweet food, for example, does *not* add pounds to the hips. However, it will produce a subtle influence on the mood.

THE EFFECT OF MOOD
ON FOOD

The "flavors" of our consciousness also transform food. You already have some understanding of how this occurs from the discussion of stress in Chapter 3. Emotions stimulate the production of hormones, which regulate most of our physiological processes, including digestion, absorption, and assimilation. Fear, for example, stimulates the release of adrenaline, which affects the kidneys and in turn causes dehydration and a lack of digestive fluids. Anger—what Daniel Goleman in *Emotional Intelligence* calls the "the fight wing of the fight-or-flight response"—tenses the muscles and gets the blood pumping to the extremities in preparation for a fisticuffs, consequently diverting blood from the abdominal area. Grief, a low-arousal state, slows the body metabolism. In every case, we disrupt digestion, and the undigested food becomes toxins, or poisonous waste that blocks the colon and prevents further absorption of nutrients. As we discuss below, improper digestion and a "dirty" colon are two of the leading diet-related factors in aging and disease. Indeed, if we are not metabolizing nutrients properly, we can eat all the right foods and still be undernourished, or even eat a low-calorie, low-fat diet and retain weight.

This is only part of the picture, however. According to energetic principles, our mood transforms food even as we prepare, cook, serve, and sit down to eat it. Thought itself is a subtle energy whose influence extends beyond the physical limits of the nervous system and its messenger molecules by means of our touch and attention. Changes in our mental state alter the biochemistry of the skin, including such properties as its galvanic resistance and pH balance. These physiological changes, along with other shifts in body energies, alter the foods we handle: a hot, sweaty palm and a cool, dry one will have different energetic effects on the vegetables you cut and cook.

At the same time, thoughts themselves have a subtle vibratory effect on the physical world. As we said, each one is an impulse of energy—of consciousness—that sends a ripple of influence through the universe like a grain dropped in a cosmic pond. Indeed, we pour attention onto the world like spice wherever we direct our awareness. The taste of that spice—sweet, salty, sour, pungent, bitter, or astringent—depends upon the flavor of our emotions. If we prepare or eat a meal when we are angry, for example, the pungent energy of our thought is added to the food like a dash of cayenne. We not only feed our anger to the ones we serve, but in effect, feed it back to ourselves as we eat the meal. Anger literally turns sweet food sour in the stomach, transforming nourishment into poison—and so we say, anger eats us up inside and creates heartburn.

Now, imagine what we add to the taste of our food when we cook a meal with joy and care and sit down to eat with a peaceful mind and grateful heart. Right attitude transforms mere nourishment into nectar. A perfectly balanced diet of pure-grown foods eaten with mindfulness and an uplifted mood not only supplies all necessary nutrients to the seven dhatus, it also stimulates the production of ojas, which strengthens immunity and gives brilliance to the skin.

Later in this chapter, you will read the story of mealtime in my parents' home in India and the tips for right cooking and eating, which are the essence of Ayurvedic etiquette. For the moment, however, simply spend some time with the idea that we nourish ourselves and others through our own thoughts. It is one of the central principles of Vedic philosophy and the quintessential principle of absolute beauty. Your state of mind as you eat is the most crucial ingredient of your diet, besides the balance of tastes for your skin type, because of its effect on digestion and on the subtle energy of foods. In fact, Ayurveda teaches that right attitude ultimately outweighs right tastes as the primary factor in good nutrition, since even a perfect meal becomes toxic in the body when our emotions are discordant. In Dr. Frawley's words: "Anger can damage the liver as much as alcoholism. So herbs and diet are not enough if the taste of the mind has not changed."

THE SECRETS OF GOOD DIGESTION

The first body tissue is known as *rasa*dhatu. A "taste" of it exists in every body tissue since it is the one from which the other six sequentially arise. Plasma, blood, fat, muscle, bone, bone marrow and nerve, and reproductive tissue are all essentially rasadhatu that has been transformed to perform these specialized functions. In this transformation process, each dhatu gets its raw material from the one that precedes it, except for the first—rasadhatu—which gets *its* raw material only from our food. If that raw material is inadequate, the first dhatu to the last will not function properly.

Clearly, right diet is necessary for healthy tissue and beautiful skin. However, right diet alone does not guarantee it. As we said, the purest food in the world is worth little to our health if the body cannot digest and assimilate the nutrients in it. Consequently, Ayurveda has much to say on the subject of digestion. We have already considered some of the psychological factors in digestion above, and we will have more to say about this toward the end of this chapter. Now, however, we will look at the key physical and dietary factors in healthy tissue transformation and metabolism: the strength of the digestive fire, known as agni; the accumulation of toxins in the gastrointestinal tract; and toxic foods.

All the functions of life require energy—agni—to keep going. The physical body produces this required energy through the process of metabolism, or the bio-chemical transformation of substances, among which Ayurveda counts both thoughts and food. Metabolism includes not only the digestive function but also the absorption and assimilation of the energy it makes available, and the elimination of bodily wastes. When the metabolism is impaired as the result of wrong diet or stress, the body cannot properly digest or absorb the nutrients from food. Whatever is not digested or excreted through normal channels accumulates as ama, or toxins in the weak areas of the body. All disease is due to the accumulation of ama, according to Ayurveda. However, ama builds up in the body only when agni is disturbed as a result of imbalance. Consequently, we need to have good digestion, absorption, and elimination—that is, strong agni—if we want to derive full nutritive value from food and keep the body toxin-free.

REKINDLING
THE DIGESTIVE FIRE

Agni is the biological fire (also called tejas) that regulates metabolism and fuels digestion. Agni's transformative principle, along with the moving force of prana and the stabilizing force of ojas, makes up the triumvirate of life forces. When these forces are imbalanced, so are the biological functions—the doshas—which they govern. Thus, a healthy metabolism depends first on the balance of doshas.

A sluggish metabolism (improper digestion indicated by symptoms such as weight gain, indigestion, constipation, gas, distention, stomachaches, diarrhea, acidity, and heartburn) is a sign of low fire and air, or low Pitta and Vata—which is the equivalent of *high* earth and water, or *high* Kapha. In either case, you can restore balance to your constitution by going on a Kapha-pacifying diet, which includes predominantly pungent, bitter, and astringent tastes. The light, dispersing and hot, stimulating qualities of these tastes increase the forces of prana (movement and elimination) and agni (metabolism), which are needed to clear out ama from the body and get the digestive fires burning again. Consequently, a Kapha-pacifying diet also enhances the body's natural process of detoxification.

A DIET TO REDUCE TOXINS AND REKINDLE AGNI

This purification routine for body and mind includes a few days of light eating on a Kapha-pacifying diet, plus mental rest. The main staple of the diet is *khichadi*, a nourishing mixture of rice and lentils (see recipe below). The duration of the routine depends upon your skin type:

Dry (Vata)	1–3 days per month
Sensitive (Pitta)	1–4 days per month
Oily (Kapha)	1–5 days per month

No Sweet fruits, juices, or vegetables; breads, cookies, pastries, sweets, dairy, cheese, fried food, canned food, nuts, alcohol, oils, salt, sugar, legumes, or grains (except khichadi).

Yes Herbal teas: Ginger, cardamom, fennel, cinnamon, aloe.

Water: At least 6–8 glasses per day.

Fruits or fruit juices: Lemon, lime, grapefruit, pomegranate.

Steamed vegetables (optional): Beets, carrots, fennel, kale, spinach, broccoli, cauliflower. *Limit*: ½ cup per day total.

Vegetable juices: Celery, parsley.

Herbs: Triphala (½ teaspoon with warm water at night).

Khichadi: Have up to one bowlful whenever you feel hungry throughout the day. To make, wash 1 cup *each* yellow mung dal beans and rice and put aside. Put 3 Tbsp ghee in a steel saucepan. Heat on low flame. Add 1 inch fresh peeled ginger + ¼ inch cinnamon bark + 5 whole cardamom seeds + 5 whole cloves + 10 black pepper or coriander or cumin seeds + ¼ tsp turmeric powder, and sauté lightly. Add rice and dal to pan, and sauté lightly with the spices. Add 6 cups water and cook until the rice and dal are soft.

MENTAL FAST:

- Rest as much as possible. Avoid work, TV, arguments, discussions, meetings.
- Do light reading, simple breathing exercises, and take leisurely walks.
- Take a warm bath daily with ginger or eucalyptus oil; scrub the body thoroughly.
- Spend time in meditation or prayer.

OUT WITH THE OLD, IN WITH THE NEW—PROBLEM SKIN AND THE PRINCIPLE OF DETOXIFICATION

All skin disease, including acne, eczema, and psoriasis, is due in part to accumulated ama. Once toxins have collected in the intestinal tract, dietary changes alone cannot restore balance to the doshas because the toxins themselves inhibit the absorption of nutrients. If you try to "clean up" your diet without first cleaning out the body of toxins, the "good" food you consume literally goes to waste in the digestive system. As we have said, it is like pouring clean water into dirt—all you end up with is mud. Therefore, *if you have an existing skin problem, it is essential to detoxify and cleanse the body internally before you try to balance your diet.*

The traditional Ayurvedic method for eliminating bodily toxins and balancing the doshas is the medical treatment known as pancha karma, which we mentioned in Chapter 6. If you cannot do a professional treatment, however, you can do a modified purification program at home. This three-stage process of preliminary oil and sweat treatments, an internal cleansing and detoxification regimen, and post-treatment nutritive therapy is described fully in the chapter on massage therapies. If you have an imbalance at this time, you should go through this treatment before you begin the Kapha balancing diet described above. The entire treatment, including the detoxification diet, takes about two weeks to clear out ama and reset agni. Once the process is completed, you can begin your normal Ayurvedic diet to maintain balance.

Even if you do not have a particular problem or imbalance, Ayurveda recommends pancha karma treatment for everyone three times of year at the season's change when the doshas naturally tend to be aggravated.

AVOIDING TOXIC FOOD

The dietary principles for staying young and beautiful are universal: Keep the colon clean and the digestion strong. Western nutritionist Paavo Airola writes that many people around the world have discovered these rejuvenating secrets and claim to have found the perfect diet to do the job: Scandinavians favor rye and whey; Germans advocate mineral water, fermented foods, and the use of juice fasts; Mexicans cook with lime, papaya, and hot pepper; Asians eat a diet low in fat and high in fiber-rich grains, for example. In theory, most of these regimens do help either to avoid constipation or promote digestion, but in practice they do not work for everyone. This is Ayurveda's unique insight into diet and nutrition. To keep *your* colon clean and digestion strong you must eat a diet compatible with your own constitution, because wrong foods imbalance the doshas, and imbalance itself leads to poor metabolism and toxic buildup.

Nevertheless, certain foods are best excluded from all diets, or at least drastically reduced, because of their intrinsic toxic effects. These include:

- Chemically fed, chemically treated, genetically altered or irradiated foods.
- Chemically preserved, processed, canned, and frozen foods.
- Artificial sweeteners, artificial coloring, and "no fat" or "low fat" foods made with fat substitutes.
- Deep-fried fast foods and other foods cooked in reused oils.
- Hydrogenated oil, animal fat, shortening, and margarine.

These toxic foods are low in pranic energy—that is, they are lifeless—according to Ayurveda. In fact, all stale foods and old leftovers (more than a day old) lack prana and create toxins in the body.

Other foods that should be avoided or eaten only occasionally because of their imbalancing effect on all doshas include: coffee, black tea, caffeinated drinks, hot peppers and pungent spices, iodized salt, white sugar, and alcohol.

In addition, certain foods that may be otherwise perfectly good for you will create toxins in the body when eaten in the wrong combinations or at the wrong temperatures, because they will not be properly digested. Therefore, avoid eating: dairy products with animal foods, including fish; dairy with salty, sour, or pungent foods; hot (in temperature) with cold foods; sweet with salty foods; raw fruit with any other food; ghee and honey in equal amounts; cooked honey; and ice cold milk.

FOODS FOR YOUR SKIN TYPE

Now that you have the basic principles of Ayurvedic nutrition, it is time to figure out exactly what to eat. In Appendix I, you will find a list of foods for each skin type. (As many of my clients do, you can make a copy of the list and carry it in your wallet for reference until you become familiar with its contents.) Each chart includes a variety of vegetables, fruits, grains, legumes, dairy products, animal proteins, condiments, nuts, seeds, oils, sweeteners, beverages, spices, and supplements that balance the dominant dosha. In general, you will take most of your foods from this list.

Keep in mind, however, that each food chart reflects the ideal for a *single dosha*, while each of us is a combination of all three doshas, with a strong tendency toward one or two. *No single list represents the perfect diet for anyone for all time.* Rather, each one is there to draw your attention to the kind of influence each food has on the mind and body—either vatogenic, pittogenic, or kaphogenic. Keeping the doshas balanced is a constant give-and-take process with ourselves and the environment, not a once-and-for-all, I-am-doing-it-this-way diet resolution. The trick is to add and subtract items to and from the food list for your skin type until you arrive at a dietary program that is tailored to the condition of your doshas *at the present time*. (For example, if you are Vata but have a Pitta imbalance, add some Pitta-balancing foods to your regular Vata diet and cut down on foods from the Vata list that happen to be aggravating to Pitta.) Since nature does not stand still for the convenience of any one dosha, everyone's condition fluctuates with circumstances. To adjust your diet to a change in climate, season, routine, or life stage, for example, or to adjust to a particular physical or emotional imbalance, you will have to consider doshas other than your dominant one in order to come up with the perfect mix of foods for you.

Consequently, memorizing a food chart or pulling it out every time you eat is not ultimately the easiest or even the best way to make your dietary choices. In truth, because of differences in cuisines, the year-round availability in recent times of produce from other climates, and certain variable qualities of the foods themselves, Ayurvedic practitioners do not always agree as to the effects of a particular food on the doshas—and if you compare lists from book to book, you are sure to find occasional discrepancies. Food charts are a useful and necessary place to start your Ayurvedic diet, but they are not the final authority. In the final analysis, no one can know better than you what foods leave you feeling balanced, healthy, alert, and glowing, and what foods leave you feeling out of sorts.

This self-awareness is the essence of Ayurvedic self-care. To live in harmony with nature, you only have to know the attributes of space, air, fire, water, and earth; and since their essential qualities are already very familiar from ordinary life, you do not need to learn anything new. You simply need to look at the old, familiar things in a new way. If you start to open your senses and awareness to the elemental qualities of foods—how their individual colors, shapes, textures, smells, tastes, and energies reflect their own constitution—and pay attention to their subtle effects as you eat them, you will start to recognize *by direct experience* which foods are best for your constitution.

Indeed, just as humans have distinctive physical features by which we can guess their constitution, so too do the foods in nature's garden. The chart opposite shows various physical characteristics of the elements as they appear in foods. By observing these qualities in various foods, you can begin to detect which ones are best for your skin type. To give you just a handful of examples, a vegetable such as lettuce that has a *translucent* quality is high in space, which is intrinsically clear. If you see this quality in a vegetable and you know that your constitution is high in Vata (which is space and air), then the translucent food is probably not the best for you. Similarly, Pitta types should avoid most red and orange foods since they are naturally high in fire. Kapha types should avoid sticky, heavy foods, such as cooked rice and pasta, which are naturally high in water and earth. A prepared food that is predominantly sweet and solid, such as an oatmeal raisin cookie, for example, is high in earth. Ice cream, which is cold, sweet, and heavy (as in heavy cream), is also high in earth. Crackers and cold cereals, however, are dry and rough in texture and are high in air. The description of the major food categories following the chart will also help you to recognize the elemental content of foods.

THE ELEMENTAL ATTRIBUTES OF FOODS

SPACE	AIR	FIRE	WATER	EARTH
Hollow	Quick	Intense	Cool	Solid
Resonant	Rough	Hot	Dense	Dense
Translucent	Hard	Medium-size	Heavy	Heavy
Blue	Dry	Sharp	Large	Large
Cold	Variable	Light	Moist	Oily
Astringent	Fresh	Fluid	Smooth	Sour
	Wiry	Oily	Cloudy	Sweet
	Light	Fetid	Sticky	
	Compact	Red	White	
	Dark(gray/green)	Orange	Clear	
	Bitter	Pungent	Sweet	
	Astringent	Sour	Salty	

A NEW LOOK AT THE BASIC FOOD GROUPS

Every food group contains good choices and bad choices for every dosha. The descriptions below highlight a few choices within each category in terms of their elemental makeup, but they are by no means complete.

Vegetables. If you know what to look for, you will find that the constitution of natural foods is often evident. Think of a garden. The roots of all vegetation grow downward into the soil, absorbing minerals and holding moisture to nourish the plant. Closest to the earth—indeed, buried in it—they are naturally rich in earth and water. Therefore, tubers and other subterranean vegetables, including carrots, potatoes, beets, and onions, and ones that grow near the ground, such as cucumbers and zucchini, generally balance Vata types, who need grounding, moist elements to counter their naturally light, dry, dispersing natures.

Now consider leaves. They grow upward and furthest from the ground. Structured with a web of hollow veins to take in oxygen, they are naturally astringent and light. Consequently, leafy greens, along with most small-seeded hollow vegetables (peppers excluded), make a poor food choice for airy Vatas, but a good one for Kaphas, who need to lighten up.

The stem or stalk, the middle portion of the plant, draws water up from the roots, and oxygen down from the leaves, to nourish the whole plant, making it both diuretic and cool in quality. Thus, celery, broccoli, and asparagus are good for the fire and water in Pitta, and acceptable for the water in Kapha, but not so good for cool, dry Vata.

In general, vegetables tend to be sattvic—purifying—and nutritive. Along with fruits, they are best eaten in their season and should account for 20 to 30 percent of the daily diet for Vata and Pitta types, and 40 to 50 percent for Kapha types. All vegetables when uncooked tend to be somewhat astringent and therefore cool and drying. Thus, everyone should favor *cooked* vegetables in the wintertime, and Vata types should eat them year round.

Fruits. In botanical terms, fruit is the developed ovary of a plant—it carries the seeds of life, protecting and nourishing them. Fruit pulp is rich in water and air, so it is cool, light, purifying, harmonizing, and sattvic in nature. Dense but juicy sweet or sour fruits like berries, cherries, plums, oranges, grapefruits, and bananas, when ripe, are better for Vata, while cooler, sweeter fruits like sweet grapes, melons, and mangos are good for Pitta. However, Kapha types should stick to dried fruits, apples, pears, and other compact, astringent fruits to avoid edema (water retention) due to the high water content of most fruit. Sour fruits are better in the morning to liquefy Kapha, and energy-giving sweet fruits are better in the afternoon. To ensure proper digestion, do not eat fruit in combination with other foods. If you drink fruit juice and have dry or oily skin, dilute it with water and "heat it up" with pungent spices such as cardamom, clove, cinnamon, or nutmeg.

Seed foods. Dense, mineral-rich grains, beans, and nut kernels are all types of seeds. Grains are the seeds of grass-like plants, which tend to grow in very wet soil, giving grain its sticky quality. Beans, or legumes, are pod seeds, which grow encased in a hollow shell. Nut kernels are actually the large seeds of one-seeded dry fruits. Naturally, they are all rich in earth, where they are planted and spring to life, so they are basically nutritive and sweet.

Grains tend to be sattvic and neutral in energy, so they are generally good for all types. In fact, the average daily diet should be 30 to 40 percent grains if you have oily skin, and 50 percent grains if you have dry or sensitive skin. Diuretic grains such as barley, corn, and rye are better for Kapha, nutritive grains such as wheat and Basmati rice are better for Vata and Pitta.

Beans are more astringent than grain, containing high amounts of air. They are a good protein source, but they are rajasic in nature and therefore cause emotional irritability. Because of their lighter, cooler, drying effect, they are a good choice for Kapha types and a poor one for gassy Vata types. You can mitigate the astringent effect of beans through proper cooking. First soak them overnight in water, rinse, and add fresh water. Bring to a boil and rinse again. Cook them in a fresh pot of water and serve them with a little oil and spice, such as cumin, cayenne, asafoetida, and onion, to balance their cooling action.

Nuts are generally warm, heavy, and oily, as well as sattvic in nature. They help to build muscle, enrich the blood, and strengthen memory. Their tonic, nutritive properties are excellent for Vata types, especially in the morning, but only in small amounts since they are hard to digest. Kapha types should avoid all nuts except pumpkin seeds, which are lighter. The best way to eat nuts is first to soak them in water, remove the skins, and lightly roast them with pungent spices so they will be lighter and easier to digest.

Oils. Cooking oils are extracts made from the fatty acids of fruits, nuts, and seeds, such as avocado, olive, coconut, corn, safflower, soya, sesame, and sunflower oil. They are generally sweet, heavy, and warm, which makes them an excellent food for Vata. Pittas can use the cooler, fruity oils such as coconut and olive oil; Kaphas should avoid all but almond and corn oils. Hydrogenated oils, margarine, shortening, and other processed fats are inherently toxic because they are produced with industrial solvents. *All* oils, even pure organic oils, become toxic when they are overheated or reused. Toxic fats, which are common in packaged, processed, fried, and fast foods, are a major factor in obesity.

Dairy. As you might expect, milk is sweet, cooling, and sattvic. It builds all seven tissues, increases plasma, fat, and reproductive tissue, and calms nerves. However, it is also damp, sticky and heavy, and in excess, increases ama. Whole warm milk and milk products balance Vata, and after the body-building years of childhood, aggravate Kapha. Milk, butter, and ghee are good for Pitta, but the sour taste of buttermilk, yogurt, and cheese is not.

The best way to drink milk is heated on the stove (but not boiled) and flavored with saffron or cardamom to help digestion and increase ojas. If you prefer, Kapha can substitute a pinch of cinnamon, Vata and Pitta can substitute vanilla, and Vata can substitute honey as well. This drink makes an excellent sedative before bedtime. Do not have milk in combination with meat, fish, yeast bread, sour fruit, nuts, or pickled vegetables, all of which cause milk to become toxic in the body.

Animal protein. Generally speaking, foods from animal sources are nutritive and strengthening, but they are also tamasic in nature and therefore tend to breed toxins, feed infections, and dull the mind as well as our capacity for compassion and love. Still, there are degrees of difference within this broad range: warm-blooded animals native to cold climates have the heaviest, most dulling effect, while poultry and cold-blooded fish are lighter. Therefore, red meat is a poor choice for all skin types, and because of its high oil content, seafood is a poor choice for Pitta and Kapha types. Poultry, particularly turkey and egg whites, are good all-around choices.

If you do eat animal protein, try to use products from organically fed sources to reduce your consumption of toxic chemicals and chemical residues found in high concentrations in the meat, milk, and eggs of most commercially bred animals. However, the tamasic property of all animal foods inhibits the production of ojas, so eating a vegetarian diet, or at least reducing your intake of animal foods, is the best choice for youthful, glowing skin.

Beverages. Ayurveda teaches us to eat when we are hungry and drink when we are thirsty, so we do not dampen the digestive fires. All beverages, including water, are best when taken hot or at room temperature, although Pitta types can tolerate cooler drinks. Stimulants, including alchohol and caffeine in any form, naturally aggravate Vata and Pitta. However, Kaphas, who sometimes need an extra kick, can tolerate them in small amounts—but no cream and sugar in the coffee, to avoid weight gain. Carbonated drinks—*avec gas* as the French aptly call them—will wreak havoc on naturally bloated Vata types, but all types are advised to avoid them. A chilled glass of milk or coconut milk is a cooling drink for Pitta, and aloe vera juice is a perfect blood purifier. Warm milk at night is a perfect drink for hyperactive

Vatas, as we said; they can also tolerate an occasional milk shake or sweet lassi. Kaphas should take all fruit juices diluted with water.

Herbs and herbal teas. As we discussed in Chapter 4, Ayurveda values herbs both as foods and as medicines. Each herb has a distinctive flavor as well as specific short-term therapeutic effects. They include digestives, laxatives, diuretics, blood purifiers, stimulants, sedatives, expectorants, astringents, emmenagogues (to regulate menstruation), diaphoretics (to induce sweating), antipyretics (to expel heat), antiseptics, tonics, and rejuvenators. In Appendix D, you will find a list of herbs and their uses for each dosha. Herbal teas are fine for all types, although many, because of their taste and potency, are not suitable except as medicinals.

Sweeteners and spices. Vata types can tolerate all spices and sweeteners except white sugar. Kapha types can tolerate all spices except salt, but no sweeteners except honey. Pitta types can have all sweeteners except honey and molasses, and some spices including cinnamon, cardamom, and turmeric. Again, this does not mean, for example, that Vata types can never eat white sugar. It simply means that white sugar imbalances Vata and therefore should be eaten only in small amounts under normal conditions and avoided when Vata imbalance exists.

FOODS HIGH IN VATA	FOODS HIGH IN PITTA	FOODS HIGH IN KAPHA
Aggravate dry skin	*Aggravate sensitive skin*	*Aggravate oily skin*
Balance sensitive & oily skin	*Balance dry & oily skin*	*Balance dry & sensitive skin*
Most leafy greens and lettuces	Pungent spices	Oily and fried foods
All types of cabbage	Sour or pungent fruits and vegetables	Sweet, juicy fruits
Bitter vegetables	Pickles, vinegars, salts	Cool, creamy foods
Hollow vegetables with tiny seeds (peppers are also high in fire)	Acidic foods/medicines/stimulants	Sticky, cold foods
Eggplant, peppers, potatoes, and other "nightshades"	Animal foods	Sweets
Most dry, compact legumes	Nuts	
	Red-colored foods	

FINE-TUNING YOUR DIET

In this high-tech culture, we tend to go through the day at breakneck speed. We live in the fast lane, work in the fast track, talk fast, walk fast, drive fast cars, crave instant gratification, and of course, eat fast food. Perhaps that is why so many people balk at the suggestion that they take time to think about their diet or prepare healthier meals. In my years of practice, I have found that the average American is very resistant to changing her or his eating habits. Everyone wants a perfect complexion, but they want it in a quick fix.

It is not out of the question to achieve ageless beauty. It does take some time, however—and some willingness to try something new. After all, if our old habits had worked, we would not still be looking for miracle creams or the fountain of youth. While Ayurveda offers "instant" treatments like color and aroma therapy, in many cases these methods can only provide temporary relief of symptoms. The causes of every skin problem from acne to age lines lie deep within consciousness and physiology, and until a person is willing to go beyond the external remedies—to the thoughts and foods that feed the body cells—there can be no permanent change.

Many people mistakenly believe that they see the effects of their diet on the skin within hours of a meal. They point to an emerging blemish on their face and announce, "This is the pizza I ate for lunch today," or "This is that chocolate bar I had on my break." In actuality, the body takes days to break food down, absorb its nutrients into the bloodstream, and completely assimilate them into the seven dhatus. Other than allergic reactions, which can show up almost immediately, diet-related problems take at least three to five days or more to appear on the surface of the skin, because that is how many days it takes for nutrients to pass from one dhatu to the next. Cystic acne, for example, a Kapha problem that affects the fourth dhatu, or adipose tissue, takes about twenty days to emerge after eating Kaphogenic foods. An imbalance in the seventh dhatu takes thirty-five to forty days to appear. By the same token, the effects of a lifetime of poor eating habits cannot be undone by cutting out a portion of french fries. You must consistently cleanse, nourish, and lubricate the body *internally* as well as externally if you want to have truly flawless and radiant skin.

I know that some people with chronic skin problems do not like to hear that news, since they regard any sort of diet program as one more punishment on top of their affliction. However, these are the people who often have the most to gain from Ayurvedic nutritional therapies, which are frequently effective in cases that have resisted years of conventional treatment. Indeed, I have had the pleasure of seeing many clients over the years say good-bye to lifelong skin problems and say hello to a vibrant new look after a short course of detoxification and some simple dietary changes.

Whether you have a problem condition or just want to counter the effects of time, eating the right diet for your skin type can have a great payoff with just a minimal effort. You do not have to make a lifetime commitment to Ayurvedic nutrition right now—or ever. Just try the diet prescribed for your skin type for a few weeks—start with one or two adjustments in your regimen every few days if that is all you can handle—and see if you do not notice a change—not just in the quality of your complexion, but in your sense of well-being, too. Below you will find a number of practical ideas for fine-tuning your diet to your doshas. You will see that there are many ways to achieve balance without giving up the foods you love.

If you do *not* want to make a change in your diet right now, you can still enjoy many increased benefits from your food simply by eating more mindfully. In the section on Ayurvedic etiquette, you will learn some simple techniques to enhance the nourishing properties of your food and to promote good digestion, which require nothing more than a shift in attention. Ayurveda teaches the value of bringing our awareness to all things, including the preparation and enjoyment of our meals. As you become more conscious of how you eat, you naturally become more conscious of what you eat and what influence it produces in your body and mind. Then you will find the transition to an Ayurvedically balanced diet an effortless and even enjoyable next step towards absolute beauty.

ACHIEVING BALANCE MEAL BY MEAL

Eating right does not necessarily mean depriving yourself of the things you like. More often than not, it simply means adding more tastes to your food palate, moderating your consumption of others, and taking the time to savor your food.

Here are a few ideas to help you achieve balance and still enjoy an occasional helping of the foods on the "avoid" list for your leading dosha.

Combine foods. One way to "have it all" is to combine foods, or tastes, that counteract each other's effects. Just think of the opposite pairs of attributes and tastes: heavy and light; moist and dry; heating and cooling; sweet and bitter; pungent and astringent; and salty and sour.

For example, dressing a salad with sweet, heating oil mitigates the cooling, drying property of green leafy vegetables. Adding the pungent taste of ginger or cinnamon to a dish of ice cream balances its heavy, cooling effect and makes it easier to digest. By the same token, the sweet taste in carbohydrates is a good antidote for pungent tastes. Contrary to common belief, the best remedy for an overdose of hot red pepper is not a gulp of water but a bite of bread or grain. Similarly, the hops in beer, not the chill, is probably what makes it such a satisfying beverage with spicy cuisines. But use this method judiciously: Salad for dinner every day will imbalance Vata, no matter how much oil you add to it.

Cook your food. In some cases, you can mitigate the action of a food by cooking it. For example, you can counter the cooling action of many fruits and vegetables by steaming or baking them, and the drying action of dried fruits by stewing them. You will find many such suggestions in Ayurvedic cookbooks.

Eat smaller servings. Ayurveda teaches that every meal should include *all six* tastes in proper proportion to your doshas—notice the emphasis on "all six." One of the most common mistakes in diet is that we eat certain tastes in excess and never eat other tastes at all. Creating balance means altering the proportion of tastes to complement your doshas. It does not necessarily mean eliminating foods from your diet completely. If you habitually eat too much of one food that is not recommended for your skin type, you do not have to sacrifice it completely. Simply reduce your consumption until you achieve a more balanced proportion. If you find it hard to do this in one swift blow, try reducing your consumption gradually over time.

Our tastebuds naturally become inured to certain flavors from constant overstimulation. We have all had the experience of suddenly wanting a second dash of salt or a third spoonful of sugar where one or two used to do. In the same way that we become habituated to craving more, we can habituate the senses to being satisfied with less—the only difference between the two processes is that the second requires conscious participation.

For example, try reducing the amount of cream cheese on your morning bagel by a half teaspoon a day—which is not much more than the size of small gumball. At the end of one week, your serving will be a tablespoon smaller and you will probably not even notice the change. If you continue cutting back a little each day, you eventually end up with just a teaspoon or less of the cheese spread thinly over an entire bagel, and yet you feel just as satisfied with the taste. Try this with any food: put one less slice of cheese or meat on a sandwich; one less shake of salt on your potato; one less teaspoon of sauce on your pasta; one less pat of butter in the frying pan. You can also try this with entire meals. For example, reduce an eight-ounce serving of protein to seven ounces and make up the difference with a bigger serving of a side dish that balances you.

If you cut back and add in tastes in this manner a little every day, you will end up with a perfect balance of tastes on your plate with little feeling of deprivation. Used in conjunction with other mindful eating methods we describe below, this technique is also great for dieters looking to lower their calorie or fat intake, since a few bites of your favorite food is eventually all that it takes to feel satisfied.

EATING IN TIME WITH NATURE

Eating in time with the daily, seasonal, and lifetime Vata, Pitta, and Kapha cycles also helps to keep your doshas in balance. As we described in Chapter 6, there are six daily cycles (that is, two Kapha-Pitta-Vata cycles) of four hours each, starting with the morning Kapha cycle at sunrise. The three seasonal cycles will vary depending upon your geographical location, but generally speaking, Kapha season spans the cold, damp months of late winter and early spring; Pitta season spans the humid, hot months of late spring and summer; Vata season spans the cool, dry windy months of fall and early winter. You will see how to adjust your diet during these cycles below.

The life cycles include the Kapha years of childhood and adolescence; the Pitta years of early and middle adulthood, and the Vata years of maturity and old age. Naturally, during the growth years of Kapha, the appetite and digestion are strongest; and during the mature Vata years, they are much less. For children, reduce Kapha-aggravating foods such as sweets and cheese to prevent colds and congestion, which are common childhood problems. In the active middle years, watch your intake of Pitta foods to avoid ulcers, high blood pressure, acid stomach, and other digestive problems typical of this active Pitta period. Eat more Vata-pacifying foods during the mature years when the body typically becomes dehydrated.

Seasonal changes. If you have dry skin, stick closely to your Vata-pacifying diet during the dry, windy days of autumn and the early winter chill, when the complexion tends to get dehydrated and the feet and hands develop cracks and fissures. If you have sensitive skin, stick closely to your Pitta-pacifying diet during the hot and arid days of late spring and summer to avoid irritating your complexion and your moods. If you have oily skin, stick closely to your Kapha-pacifying diet in the late winter and spring when the cold, damp air tends to weigh down your spirits and stimulate the production of mucus in the body. During the alternate periods, add to your normal diet some foods that pacify the seasonal dosha and cut down on ones that aggravate it.

In general, eat more lightly in hot weather when the digestive fire is naturally weakest and more heartily in the cold when agni is strongest. At all times, try to eat the foods from your list that are locally grown and in season. During the twenty-day period at the change of seasons, do a pancha karma treatment plus a short detoxification and cleansing diet to clean out the morbid doshas and reset your metabolism.

Daily changes. On a daily basis, select foods according to the time of day. In the morning, eat a small breakfast of sattvic foods such as cereals, seeds, nuts, bananas, pears, raisins, and coconut milk. Have your largest meal at midday, when agni peaks. During this rajasic period, favor protein and other high-energy foods as well as foods with high water content (especially in the summer). In the mellow evening hours, eat a light meal such as soup, steamed light vegetables, breads, beans, and greens. Citrus fruit, root vegetables, yogurt, high-protein foods, and fried foods are too hard on the digestion, which slows down considerably during this Kapha time. Eat dinner at least two to three hours prior to sleep, because if you go to bed on a full stomach, the undigested food produces ama.

Every constitution tends to experience an energy slump during the Vata period, from 2 to 6 P.M. For a quick pick-me-up, have hot herbal tea with a few roasted nuts or a warm bagel with "lite" cream cheese for dry skin; juicy cold watermelon in summer or a pear or apple in winter for sensitive skin (no hot beverages at this time for these fiery types); and tea with some raisins and pumpkin seeds or sliced apple dipped in honey and cinnamon for oily skin. For a more substantial treat, try one of the special snack recipes created for us by Ayurvedic chef Richard LaMarita of the Natural Gourmet Cooking School in New York City (see pages 214–215).

Changes in routine. Whether nature changes the calendar or you do, any alteration in routine affects the balance of doshas. For example, putting in a lot of long days and late nights at work aggravates Vata, while sleeping late every morning aggravates Kapha. Traveling, particularly air travel, also aggravates Vata. Visiting a hot climate aggravates Pitta; visiting a cold one aggravates Kapha; a few autumn days in the Windy City will surely aggravate Vata.

No matter what type of skin you have, if you find yourself in any of these situations, increase your consumption of foods that pacify the aggravated dosha for the duration of the change. If you can plan ahead, start to modify your diet a day or two in advance of the change. Eating a Vata-pacifying diet the day before and during a plane flight, for example, helps to alleviate the symptoms of jet lag.

ADJUSTING FOR IMBALANCES

The diet prescribed for your dominant dosha is always good to follow whether or not you have a skin condition. However, if you do have a problem, we recommend that you first go on a detoxification regimen (see page 251) and then modify your diet according to the nature of the imbalance. If you have not done the Skin Problem Quiz in Chapter 3 to determine what your imbalance is, please do so now.

When the leading dosha is imbalanced. If your leading dosha is aggravated, stay on the normal pacifying diet for that dosha and veer from it as little as possible for two to four weeks, or until the symptoms of imbalance are gone. You can enhance the recovery process by adding one or more additional balancing therapies into your daily routine, such as meditation, massage, aroma therapy, color therapy, sound therapy, or breathing exercises. These techniques help to heal the emotions while your diet repairs the bodily imbalance.

When other doshas are imbalanced. When a dosha other than your leading dosha is out of balance, stay on your normal diet, but also avoid foods that aggravate the imbalanced dosha (see the "no" lists under the dietary suggestions for each dosha in Appendix I). In addition, treat the psychological and emotional factors related to the imbalanced dosha (as shown below) with the appropriate herbal remedies, meditation, and other balancing therapies.

For Vata: Anxiety, worry, fear, nervousness.

For Pitta: Frustration, anger, hostility, jealousy.

For Kapha: Depression, grief, attachment, possessiveness.

If your leading dosha *and* another dosha are imbalanced, balance the leading dosha primarily through diet, and the second dosha primarily through calming the emotions. In my experience, whenever a dosha other than the leading dosha is out of balance, the cause tends to be emotional rather than physical. It is also true that these imbalances are usually easier to correct.

MIDAFTERNOON TREATS

FOR VATA:

SESAME-COCONUT CANDY

Yields 20–24 pieces

½ cup almond butter	¼–½ cup maple syrup
¼ cup sunflower seeds, roasted	2 Tbsp cocoa powder
½ cup dates, chopped	coconut, lightly toasted
1 Tbsp apple juice	sesame seeds, toasted

Mix together the coconut and sesame seeds and set aside. Mix all other ingredients together, roll mixture into small balls and coat with sesame-coconut mixture. Refrigerate

FOR PITTA:
CILANTRO-WHITE BEAN DIP
WITH RAW VEGETABLES

Yields 1½ cups

1 cucumber, peeled and cut into
    lengths ¼" by 2"

¼ bulb fennel, cut into strips

2–3 celery stalks, cut into
    lengths ¼" by 2"

1 cup cannellini (white) beans,
    soaked overnight

¼ cup fresh cilantro, chopped

1 tsp olive oil

salt and pepper

Prepare raw vegetables as instructed, and set aside. Rinse soaked beans thoroughly and cook in water for about 1 hour or until tender. Drain the beans, *saving* the water. Place cooked beans, fresh cilantro, and olive oil in a food processor and puree until creamy. Add the bean water for desired consistency. Salt and pepper to taste. Use as a dip with raw vegetables.

FOR KAPHA:
SPICY DRIED FRUIT COMPOTE
Yields 6–8 servings

½ lb dried figs, chopped

1 cup fresh apple, peeled, cored,
    and roughly chopped
    (about 2 apples)

½ lb prunes, chopped

½ lb raisins

½–1 cup water

¼ tsp dry ginger powder

3–5 cloves

½ tsp cardamom powder

½ tsp anise or almond extract

juice from one orange

Put all ingredients in medium saucepan and cook uncovered over medium flame for 20 minutes or until soft. Stir regularly and add more water if the fruit is absorbing water quickly. Add orange juice a few minutes before it is done.

MINDFUL EATING

America is known as a melting pot of cultures, and it is true that cuisines from nearly every nation are served on our tables. Yet, in the past century, no one country has done more to change the world's eating habits than the United States, which has exported fast food to Paris and Beijing, baby formula to tribal societies in Africa, and soft drinks to desert nomads in the Mideast.

Ironically, most Americans have plenty to eat, but few eat well. We not only eat the wrong kinds of foods, but we often eat for the wrong reasons. Influenced by advertisers who promise us sexier, happier, more sociable, and exciting lives if we buy their brands or serve their products, we eat to look good, feel good, have a party, soothe our hurts and loneliness, or distract ourselves from the problems of life. Many times we eat just to get it over with and get on to the next appointment on our busy schedules. Rarely do we eat *consciously* for the one reason we *must* eat: to supply nutrients to our cells. It seems that the more privileged we are, the less mindful we become of the gift of life contained in every morsel of food. In so doing, we starve the body *and* the spirit, no matter how many calories we actually consume.

In the beginning of this chapter, you read about the influence of our thoughts on the quality of our diet and digestion. Here you will find some additional food for thought about the value of awareness in the act of eating, including practical suggestions on how to cook, serve, and eat your meals to bring out the healing, nourishing qualities of your food as well as your own pleasure in its tastes. These ideas originate in Ayurvedic customs that have been practiced in India for thousands of years, but they are offered to you with the recognition that we live in a culture that seems light years away from the ancient Vedic civilization. I was fortunate to grow up in India at a time when cultural values and the pace of life were not yet as Westernized as they are today, and it was easier to raise a family according to these customs. The brief story of my childhood may help bridge the centuries and continents that separate these worlds and give you a sense of the graciousness and beauty we gain by our mindfulness.

As a young girl, I lived in the countryside surrounded by my extended family. My aunt had a farm a few miles from our home, and every morning my sister and I had the responsibility of awakening before sunrise and walking the distance to fetch the day's supply of fresh milk and vegetables. We often stopped to pray at a temple along the way, and we always arrived at the farm past daybreak, just in time to milk the cows, help with chores, share a breakfast with my cousins, and return to my mother's kitchen by ten o'clock. Then we did homework until noon, when it was time to go to school.

Dinner was the meal that everyone ate together—grandparents, aunts, uncles, cousins, and my own parents and siblings. The routine was the same every day of my childhood. Although it was a light meal, its importance to the family was great. Everyone had to be in the house before sundown. In accord with Ayurvedic tradition, before the meal we always washed our feet and hands and changed into clean clothes in order to remove the subtle influences we had accumulated and carried through the day. Refreshed in body and mind, we were then prepared to say our evening prayers and pay respect to our elders. This was a wonderful custom in which we children touched the feet of our grandparents and asked for their blessings. At the age of five or six, my youngest brother began a habit that endured through childhood of putting his head on our grandfather's feet and refusing to get up until grandfather blessed him with the words, "You will be a very famous doctor." Of course, grandfather always obliged so we could get on with our meal, and today my brother is a renowned ophthalmologist in India. For the rest of us, this was also the time to tell our grandfather if anything from the day was troubling us. If no one had a problem to be solved, my grandparents might sit with us for a while to go over our lessons or recite our multiplication tables. The benefit of this daily ritual was that no one ever sat down to dinner burdened with unspoken frustrations, worries, or hurts that could turn even the pure-grown foods from our farm into poisons in the body.

Our family ate in Indian fashion, sitting in a large circle on the floor. No serving platters were set out; and except for the elder women who dished out each course to us individually, no one else handled the food until it was time to eat. First, each element of the meal was arranged ceremoniously on a large steel dish called a *thali*. The order was always the same: a little bit of salt and a piece of lemon, then salad, then vegetables, then lentils and curry in small cups, then rice, then ghee—the full complement of salty, sour, bitter, astringent, pungent, and sweet tastes on every plate. Once the ghee was served, we prayed aloud to offer this nourishment to the Divine within. In fact, the act of eating itself was considered a *yagyna*, a Vedic ritual, in which we were feeding food to agni, the transformative life force inside us. When this was done, each of us took a pinch of rice and water and sprinkled it on the ground around our plate, returning a portion of food in gratitude back to the earth. About an hour and a half after we had begun to prepare for dinner, it was at last time to eat. Then, of course, we scooped the food up using our fingers and bread.

This we did with great delight and very often in silence. We were free to speak at the meal, and we did when the whim struck, but all conversation centered on the satisfying tastes of the food itself or on the happy events in our day. Otherwise, we ate in a comfortable, comforting quiet, mindful only of the nourishment we were enjoying and the honor we were paying to our bodies, which provide a dwelling for the soul. We ended with another prayer, this one for the givers of the food, asking that God grant them happiness. I am sure that no one ever got up from a meal at home with a case of indigestion.

This is the wisdom of Ayurveda—to eat without distractions or in the company of those we love, attentive to the food, and in an uplifted state of mind. Mindful eating—indeed, mindfulness in all activity—is a necessary condition for balance. After all, balance is not a fixed state; it is not something we can set like an automatic pilot. It is a continuous process of conscious give-and-take; it is dynamic. To be balanced is to *be* natural, and nature exists in a constant state of flux. Thus, we cannot achieve balance by resisting. The tree that is still standing after a storm is the tree that bends with the wind; it is also the tree with deep roots.

My own family living in New York City does not eat farm-fresh produce anymore, my husband and I each have demanding careers which make it difficult to coordinate a leisurely dinner by sundown every evening, and our children and their cousins are scattered around the globe. We have not abandoned the traditions of my parents—we pray and eat together as often as we can. But we also adjust our activities to be harmonious with life here. Like a typical American family, we sometimes

go out to eat or even take dinner in front of the television if there is a program we want to watch. It is only natural that we make some accommodations to the culture and time in which we live; the principles of Ayurveda are there to keep us balanced and rooted, not rigid.

AYURVEDIC ETIQUETTE:
TIPS FOR HEALTHY DIGESTION

The goal of Ayurvedic etiquette is not good table manners (although grace in action is a hallmark of beauty). Rather, it is the nourishment of body and soul.

Some of the suggestions you will find here, like the idea of silent eating, may be completely new to you; some of them, like the idea of blessing your food, may be familiar from your own religious tradition. Ayurveda is a *science*, however, not a religion, and its principles for right eating are not commandments to be followed out of duty or fear or guilt. To the contrary, pressure, fear, guilt, depression, envy, anger, or any stressful feeling we carry to the dining table only impedes proper digestion and turns food rancid in the stomach. Ayurveda offers these guidelines to reawaken our appreciation of food on a very fundamental level, teaching that the act of eating is an act of attunement with our own nature and an act of respect for life. Perhaps some of these ideas will give added meaning to religious or spiritual rituals you may already practice at home.

Cleanse the body and mind before you begin to cook or eat. Besides the choice of food on your plate, nothing is more important to proper digestion than your state of mind when you prepare, serve, and eat a meal. Energetically, the emotions we carry into the kitchen to cook affect the food we eat as much as those we carry to the dining table. Thoughts get added to the mix like extra herbs and spices in a recipe: the cook's care and attention enhance the life force in food; anger, worry, or thoughtlessness destroy it. For this reason, Ayurveda recommends that we cleanse the body by washing the hands and feet, or bathing, and cleanse the mind by meditating before preparing (or eating) a meal, and that we nourish the spirit while cooking by offering blessings for the food and love and benefit to those who will eat it.

Choose a manner of doing this that suits you. You can find some prayers or make them up and say them as you start to cook or serve. You can chant or sing as you work. You can simply choose to occupy your mind with uplifting thoughts. Whatever way you decide, easily give yourself over to your task. Open your awareness to the bounty and beauty of nature as you handle and prepare her foods for your nourishment. This does not require a lot of effort or concentration. You do not have to force all other thoughts from your mind or berate yourself for having them; simply let unpleasant thoughts go in favor of the pleasant ones. The purpose is to increase the joy of activity, not the burden of your labor. More than anything, it is a matter of your intention and the quality of your attention as you work.

Do not eat when you are upset. Take a few minutes before you sit down to eat to address any worries or upsets. You can do this together as a family, as we did in India. Or, if you are alone, ask yourself if anything is bothering you and write down whatever comes to mind. You do not necessarily have to resolve the problems on the spot, but at least acknowledge them, decide on when or how you will address them, and then set them aside with the understanding that you will return to them at the appropriate time. When you bring your concerns into your conscious awareness in this way, you diminish their power to control you. You may also discover that merely *committing* yourself to a time when you will resolve a problem relieves your mind of its weight. Then you are free to eat your meal without angers or fears eating at you.

Begin your meal with a blessing. Beyond any religious significance, saying grace at meals is a natural aid to digestion: it shifts the attention away from negative thoughts, settles the mind and body, and uplifts the emotions—all conditions that promote healthy metabolism of nutrients and prevent the production of ama.

If you are not comfortable with prayer, try just taking a quiet moment—the time it takes for one or two full breaths—to appreciate the beauty of the meal on your table and your own capacity to savor it. Actually *look* at the different foods on your full plate before you pick up your fork. Use all your senses: smell the aromas, notice the array of colors and what is hot or cold, feel the steam, be still and listen to your breath. Bringing your conscious awareness in this way to the simple joy of eating also helps to relieve compulsive eating habits, which are by definition unconscious patterns of behavior.

Chew your food until it becomes liquid, and savor the flavor. The digestive juices in the mouth and the gastrointestinal tract each have a different job to do in breaking down nutrients for the body—the former acts on carbohydrates and the latter on proteins. If the job is not done adequately at the front end, it cannot be completed at the back end, and the undigested food becomes ama. So, chew your food until it is liquid.

As you do this, let all your senses savor the meal. Eat at a modest pace and eat with the whole mouth. Taste the food; notice its different flavors on the tongue (sweet, salty, sour, and bitter in order from tip to back); feel its changing textures; pay attention to changes in your mood and physiology; listen to your body. Eating in this way is actually a remarkable experience. You will discover sensations you probably never noticed before, and a level of pleasure and satisfaction from your food that you probably never imagined. One of the great insights of the Veda is that we can perceive the wholeness of life through any one of the senses when we direct our full awareness through it to the object of our perception during meditation.

In fact, there is a Vedic practice to develop sensory control that includes a technique to master taste. It entails sitting with the eyes closed and chewing mindfully on a *single* raisin for *hours* at a time with absolutely no other intention but to savor it. For fun, I urge you to try the technique now, but only for one minute. Don't worry if you find it challenging to keep the raisin in your mouth more than twenty or thirty seconds. No one has become a yogi on the first try.

However, you may get a glimpse of infinity through a raisin as you experience it dissolving layer after subtle layer upon the tongue. Indeed, it is almost possible to imagine how one could achieve utter fulfillment and bliss from a single taste.

Eat silently or have good conversation. Ayurveda recommends that we take meals in silence simply because mindful eating requires our thoughts, mouth, and tongue to be busy with things other than speaking. In all activity, Ayurveda teaches that we should not divide the mind, but give our attention fully to what we are doing. This is called living in the moment, or being present—or simply, wholeness. When we split our attention between our food and our company, or even between our food and a newspaper or TV, one or the other experience does not get the attention it needs to be fully digested and enjoyed. This is a life half lived.

Does this mean you must always eat alone, or in stoic silence, and concentrate on every bite as if life depended on it? Not at all. In every circumstance, *be natural.*

If you are alone, open your mind and senses to your experience; what happens, happens. If you are in company, put down your fork, speak, and enjoy the conversation. Whenever you have a choice, eat in the company of those you love, and speak with them about what you love and about the pleasure of sharing your delicious meal. In this way, you make every meal a feast for body and soul.

Eat a modest portion: Cup your hands side by side, and imagine them filled with your favorite food. According to Ayurveda, this is approximately the size of the total serving one should eat at each meal. As you can see, it is not much more than what fills an average bowl. In fact, it is only half a handful *less* than what fills the average stomach—and this is as full as Ayurveda suggests the stomach should get. The rule of thumb is: Fill up one-third with food and one-third with water (not juice), and leave the remaining space for air. *Sip* the water along with your meal and drink it hot, warm, or at room temperature to avoid dousing or cooling the digestive fire. Never gorge yourself because it is bad for digestion and only expands the stomach and increases the appetite with every meal.

In theory, this may seem like a spartan regimen, especially by American standards, where bigger means better. However, if your meal includes the right balance of tastes for your constitution, and if you eat it mindfully, you will find in practice that you are satisfied completely well before you are sated.

TIPS FOR WEIGHT CONTROL

Oral gratification is a meal balanced with all six tastes eaten mindfully. In a nutshell, that is Ayurveda's secret to weight control, and it has been used successfully for thousands of years. Any method short of that only *creates* imbalance and ultimately leads to further weight gain once the problem exists.

According to Ayurveda, there are three reasons why people gain weight: They are born with a Kapha body type and have a natural tendency to put on pounds; they develop a Kapha imbalance as a result of stress or wrong diet; or they have a sluggish metabolism due to low agni. Consequently, there are only three healthy ways to lose weight: Pacify Kapha; rekindle agni; or both. In a word: balance.

As we said, once you have an imbalance, you tend to desire foods that cause the imbalance. That is the reason dieters frequently find themselves in a self-perpetuating cycle of craving the very foods they are trying to give up. Even when they find the willpower to abstain, the sense of deprivation itself creates new stress and further imbalance, and so the cravings continue. This cycle of imbalance-desire-imbalance occurs whether or not the precipitating factor is diet-related. Any stress, including the stress of kicking a nicotine habit, can set off this chain reaction. Consequently, weight problems are not diet problems alone; they are stress-related problems, and if you do not address the stress issues that underlie the weight gain, eventually the weight will creep back on. Therefore, in addition to the specific diet recommendations given here (including the detoxification regimen, the balancing diet for Kapha and agni, and mindful eating), we encourage anyone who is concerned with weight control or dieting to incorporate into her or his daily routine the practice of meditation, massage, breathing exercise, physical exercise, or any of the stress management techniques we describe in this book.

In effect, all the knowledge in this book is Ayurveda's answer to dieting. When we are totally balanced in body, breath, mind, and spirit, psychological hunger, which is at the root of all eating problems, is not an issue. Balance is a state of total satisfaction—it is the ultimate gratification. We *think* we crave chocolate or chips or ice cream or donuts or pizza. In truth, it is not a taste we crave at all. It is wholeness itself.

Guilt-free satisfaction for cravings. To satisfy cravings on the spot, we also have a secret weapon for you that is easy to use and completely guilt-free. In the chapter on aroma therapy, you will find instructions for making mood oils—essential oils that are scented with specific herbs or spices to balance the doshas. Mood oils work on the principle that the scent of a food or herb arises from its taste: a pungent taste has a pungent aroma, a sweet taste, a sweet aroma, and so forth. When we have a craving, we do not really want the taste so much as we want the sense of balance it gives. Since the scent of a food has the same balancing effect on the mind as its taste, you can get the satisfaction you are seeking by smelling it rather than eating it. In fact, according to energetic principles, you can "eat" a taste with any of the senses to satisfy psychological hunger. Here is an idea of how this principle works:

- *For "comfort" cravings.* The desire for comfort foods—sweets, creamy foods, and carbohydrates—is really a desire for something warm, nurturing, calming, and grounding, and usually accompanies feelings (conscious or subconscious) of anxiety, insecurity, restlessness, fear, worry, or upset. In other words, it is an expression of Vata imbalance and calls for Vata-pacifying techniques. To satisfy the craving, you can dab a sweet oil such as orange blossom or neroli on your pulse points, or use the oil to give yourself a massage; recall a "sweet" memory; have a loving conversation with a friend, or listen to sweet music, or drink fennel tea.

- *For "energy" cravings.* The desire for quick pick-me-ups—a caffeine rush from a coffee, a cola, or a chocolate bar, for example—is really a desire for something warming and stimulating and usually accompanies feelings of dullness, lethargy, sadness, and depression. These are all symptoms of a Kapha imbalance and call for Kapha-pacifying techniques. To appease the desire, use pungent oils such as bergamot, clove, or rosemary, listen to some loud, lively music; get a vigorous massage, or drink ginger or spice tea with honey.

- *For "I-have-to-settle-down" cravings.* Feelings of extreme excitation, frustration, anger, and annoyance—the craving for fast finger foods like salty chips and nuts—indicates a Pitta imbalance. To satisfy yourself, use sweet scents such as jasmine or sandalwood; take a moonlight walk; get a calming massage; look at vibrant scenery or artwork; drink licorice or cardamom tea.

Help for compulsive eating. We have mentioned briefly the fact that mindful eating is an effective way to counter compulsive eating. If you have a history of eating problems, however, you may find that you feel resistant to the idea of mindfulness altogether, or that you feel agitated or uncomfortable when you first try the technique. One client with a lifelong weight problem noticed when she began to practice mindfulness at meals that she had a strong impulse to steal food from the plates of strangers at a nearby table in a restaurant. Although it upset her at the time, she later realized that her mother, who had been mentally ill, had abused her as a child because of her weight, and that she had grown up with a constant dread that she would not have enough to eat.

Although the specifics of the story are unique to this woman, this kind of emotional response is not uncommon in individuals who have spent a lifetime being "unconscious" around food. Indeed, when you are "present" at the table with your food instead of using food to stuff down feelings or cope with stress, forgotten emotions naturally start to come up. If this happens to you, just recognize the feelings for what they are. Put down your fork and breathe easily. Be with the feelings instead of the food. Take as long as you need. If you prefer, get up from the table altogether and sit down in another room until the emotions start to subside. Avoid resuming your meal until you feel calm enough to stay present with some of your emotions even as you eat. Like undigested food, these undigested emotions need to be cleansed from the body to restore balance to the doshas.

Changing lifelong eating habits is a challenge for most people, even those without eating problems. Nevertheless, we cannot hope to change how we look if we do not change how we eat because, on the cellular level, our bodies are actually composed of food. As we said in the beginning of the chapter, the Ayurvedic term for the physical body is annamaya kosha—literally, the food sheath. All body tissue is formed from the nutrients we consume. Is it any surprise, then, that stale, processed, lifeless foods create aged-looking, lifeless skin? What we feed it is what we get. Therefore, eat a diet rich in fresh foods, as well as enlivening thoughts and loving feelings, if you want to create the radiant look of absolute beauty.

CHAPTER 8

MASSAGE THERAPY

Where touching begins, there love and humanity also begin.
ASHLEY MONTAGU

Touch is so fundamental to life that newborn infants deprived of any direct contact with others will die even if all their other basic needs are met. Conversely, infants who are regularly caressed develop stronger immunity to infection and disease. Despite this evidence, once we outgrow our parents' laps and mature into adults, we in Western society have only limited opportunities for nurturing touch. While we would consider it cruel and unusal punishment to deprive a seeing person of visual stimulation, or a hearing person of sound and music, or a hungry person of the taste and smell of his food, many of us will let ourselves go for days and weeks without another person's reassuring hug or gentle stroke. Except for the few minutes daily we spend in the bath or shower, most of us do not even take time on a regular basis to touch or pamper our own skin—even though it is the only body organ that is totally exposed to view and always at our fingertips.

Touch nourishes life just as surely as food does. Yet, rich as we are materially in the Western world, we seem to be a society starved for tactile experience. Montagu in his book on *Touching* suggests that the West's "frenetic preoccupation with sex . . . is, in many cases, not the expression of a sexual interest at all, but rather a search for the satisfaction of the need for contact."

This is not generally the case in India where, in accord with the teachings of Ayurveda, many traditional families practice massage at home regularly. Infants, for example, receive a massage at birth and every day thereafter for the first three years of life, and new mothers get a special massage every day for forty days following childbirth. Children continue to get frequent massages until adolescence.

The lack of nurturing touch received by older children and adults in fast-paced Western societies truly is unfortunate, given how much of life depends upon the functions of the skin. As the organ of touch and sensation, the skin is the point of contact and defense between us and the rest of humanity, as well as between us and the environment. Its nerves, blood vessels, and glands carry vital information and nutrients to every other organ, and help to regulate critical bodily functions, including water and temperature control, absorption, secretion, and excretion. New research shows that it also plays a key role in the body's immune response. While we can survive to a ripe old age without sight, sound, taste, or smell, we cannot last more than five hours once the skin fails.

A section of skin the size of a quarter contains three million cells, a hundred or more sweat glands, and a yard of blood vessels. With more than six hundred forty thousand sensory receptors overall, the skin is in constant communication with the brain, even when we sleep. It is the source of our experience of heat, cold, pressure, pain, tenderness, and the range of human emotions. Skin not only feels sensations, it knows the difference between a caress and a slap, a prick and a push, a scratch and a tickle, a flutter and a goose bump. Of course, skin is also a source of sensuality and erotic pleasure, particularly the specialized zones of the lips, nipples, and genitals, where the nerve endings are most abundant. "[I]n no other relationship is the skin so totally involved as in sexual intercourse," notes Montagu.

As we have said, skin is our second brain: It discriminates, it thinks, it knows, it communicates, it feels, it makes and receives love. Is it any wonder, then, that the stimulation of this largest of the sensory organs has far-reaching effects on our health and well-being?

PURIFICATION, OLEATION, AND ELIMINATION

As you recall from Chapter 3, the seven layers of skin are each connected functionally to one of seven dhatus, or bodily tissues; so any time we stimulate or nourish the skin tissue we simultaneously balance and nourish the blood, muscle, fat, bone, nerve, and reproductive tissue as well. Ayurveda refers to this phenomenon of nutrient supply between skin and dhatus as the *rasasara*, the essence of the skin that rises to the top, like cream in milk.

In view of this understanding, Ayurveda naturally stresses the value of regular oil massage, or *snehana*. In fact, together with an herbal cleansing process known as *lepas*, and sweat treatments known as *swedana*, massage therapies using herbal oils to help detoxify the body constitute the first vital stage of pancha karma, the curative treatment described in Chapter 6 as a basic tool of Ayurvedic medicine. As we said then, only trained clinicians can administer the full pancha karma treatment properly. However, facets of this balancing therapy can be done effectively at home, including a head-to-toe self-massage known as *abhyanga*; a mini self-massage; specialized massage for the stomach, eyes, and nose; another for a natural face-lift; and even a "heavenly" oil treatment for the "third eye," the door to consciousness.

In this chapter you will learn how to do these oil massages as well as the before-and-after treatments of herbal cleansing and sweating. All together, these three external processes of purification, oleation, and sweating, along with your daily skin care regimen of cleansing, nourishing, and moisturizing, constitute a complete home spa treatment. For those who cannot find or afford professional pancha karma therapy, you will also learn an easy *internal* cleansing treatment, including purgatives and light fasting, that may be used in conjunction with the spa treatment as a do-it-yourself seasonal regimen to balance the doshas. Finally, you will learn two simple sets of yogic exercises—or what I call "inner massage"—one to maintain flexibility and a toned, youthful appearance; and the other, called "the fountain of youth," to rejuvenate the main energy centers of the body.

As you are about to see, Ayurvedic massage therapies are much more than just a means to relax tense muscles and calm the mind. They are, first and foremost, powerful balancing treatments that stimulate the body's energy channels and open the mind to the experience of inner wakefulness. In effect, Ayurvedic massage is a technique for cultivating blissful awareness, which is the ultimate basis of radiant beauty.

A TOUCH AWAY FROM IRRESISTIBLE SKIN: THE SECRET OF AYURVEDIC MASSAGE

If you want skin that is irresistible to the touch, the secret is to touch yourself. Ayurvedic body massage is one of the most effective means of slowing the skin's aging process and achieving overall softness and luster, because it works to purify, nourish, and tone the body on a deep cellular level. Your regular skin care routine is sufficient—and necessary—to clear away the daily accumulation of dirt and dead cells. But only the deep detoxifying and rejuvenating action of a massage can eliminate ama, the stored-up cellular waste beneath the skin that Ayurveda believes precipitates genetic breakdown and disease.

Ayurvedic massage is also a powerful stress management tool that reopens blocked energy channels and balances the psychophysiology. One client came to me for a single massage, and reported back the next day that within hours of the treatment she had gotten her period for the first time since starting a high-pressure job three months before.

Actually, Ayurvedic massage works on several levels to reduce stress, improve immune functioning, revitalize the body, and increase your Tej factor. First of all, massage relaxes and tones the muscles, and stimulates glands under the skin to produce hormones, including seratonin, that calm the mind and emotions. It also promotes drainage of the lymph system, which functions as a filtering mechanism for the blood plasma and plays an essential role in immunity. Lymphatic fluid delivers nutrients to cells and then carries away cellular debris and foreign particles. As the fluid passes through the lymph nodes, the waste products are flushed out, and the purified plasma is recirculated through the heart and back into the bloodstream. Unlike the blood system, however, which is kept circulating by the pumping action of the heart, the lymph system is not self-propelled. Instead, lymphatic flow depends upon muscular contractions to pump the body's network of lymph vessels. When the muscles are inactive, the unfiltered fluid stagnates in the vessels and nodes, weakening the body's line of defense against disease. Massage is vital to health because it works the muscles and drains the nodes. It is the only way most of us have in today's sedentary work world, other than regular physical exercise, to keep the lymph system flowing properly.

Several benefits of Ayurvedic massage derive directly from its use of herbal oils. These enhance the purification process at the same time they nourish the skin and relax the mind. As you read in Chapter 4, we make Ayurvedic body and face oils with pure herbal essences diluted in organic oil or ghee, both of which are selected according to skin type. These essential oils, as we call them, have natural antioxidant, antibacterial, and anti-inflammatory properties due to the medicinal herbs, and a molecular density similar to skin, so they are able to penetrate easily. You can see proof of this penetrating action for yourself in the visible traces of oil found in the urine following massage. In fact, the oils used in Ayurvedic massage completely permeate the dermis within five minutes, and all seven skin layers within eight to ten minutes more. Consequently, they do not leave the skin feeling greasy; and although bathing after massage is recommended to help relax the body and promote penetration of the oil, it is not for the purpose of washing off excess oil.

Once absorbed into the body, the massage oils have numerous effects besides softening and soothing the skin. Their nutrients go directly into the bloodstream through the many capillaries in the skin. When we massage oil into the skin, it lubricates the internal environment, loosens stored-up toxins, and carries them back to the digestive tract for elimination. It also stimulates *dhatu-agni*, the metabolic fire of each tissue, and thus enhances ojas. To put it in Western terms, the herbal oil improves blood circulation, heats up the body, and gives the skin a natural softness and glow. Of course, its aroma also works to balance the emotions, relax the mind, and otherwise add to the blissful effects of healing touch.

Contemporary scientists have found that Ayurvedic massage has the added benefit of reducing free-radical activity, a key factor in cellular aging, cancer, and heart disease. Free radicals are unstable oxygen atoms that randomly combine with other molecules—consequently "oxidizing" them—usually to the detriment of the host cell. Research indicates that the proliferation of these oxidant molecules leads to increased stress and poor immunity. In studies by Dr. Hari Sharma of the Ohio State University College of Medicine, subjects showed a significant decrease in lipid peroxide, a free radical found in the blood, after only three days of Ayurvedic detoxification treatment and oil massage, indicating improved immune response.

MARMAS AND CHAKRAS: THE VITAL ENERGY CENTERS OF THE BODY

Perhaps the most profound effect of Ayurvedic massage comes from its unique balancing procedure, called *marma therapy*, which focuses on the body's vital energy points. Ayurveda describes these marma points as junctions of matter and consciousness—that is, the points where the subtle life forces of Vata, Pitta, and Kapha converge and therefore where the body's organizing intelligence is most concentrated. Ayurveda identifies 107 major marma points throughout the body, including seven *mahamarmas*, or great marmas, which are also known as the seven *chakras*. These centers are especially vital, and because they are close to the surface of the skin, they are extremely sensitive to touch. All the marmas are connected by invisible energy currents—or pulses—called *nadis* that run throughout the body. The navel, which is the seat of the second chakra, is particularly rich in nadis—all seventy-two thousand of the body's subtle currents converge at this point, because it is the life-giving channel from mother to unborn child.

The marmas are located physically over the lymph nodes and at the joints where five anatomical structures intersect: blood vessels, ligaments, muscles, nerves, and bone. Consequently they play a key role in balancing the doshas and in stimulating the three circulatory systems of the body—lymph, blood, and nerves.

The mahamarmas, or chakras, center on the seven ductless glands of the endocrine system, which is the main producer of hormones, the chemical messengers that regulate physiological functioning and emotions. These are, starting with the first chakra, the reproductive glands; the pancreas and abdominal region; the adrenals; the thymus; the thyroid; the pituitary, also called the "third eye"; and the pineal gland, also called the "crown chakra." The chakras are linked by three intertwining channels that undulate like a snake from the base of the spine up to the crown and carry the *kundalini* energy, the latent spiritual force residing in the first chakra. As you will see, by opening up these energy centers through massage, yoga exercise, and meditation, we naturally become more focused and vital.

The energy channels known as meridians, which are familiar to the West through Chinese acupuncture, are similar but not identical to the Ayurvedic system, which actually predates it. Indeed, the knowledge of the body's vital energy centers originates from *Dhanurveda*, the branch of Veda dealing with the science of warfare. Dhanurveda categorizes the marma points according to which ones, if injured or pierced by a foreign body (an arrow, for instance), would result in instant death, in death over time, in painful death, or disability. Later in history, one of the great vaidyas, who also invented surgical medicine, borrowed this knowledge of vital energy points from the ancient warriors and began to apply it to the art of balance and healing.

Of course, many Westerners are surprised to learn that the culture that brings them meditation and swamis is also expert in warfare. However, that response is emblematic of the Western misunderstanding of Vedic science, which is nothing if not practical. After all, knowledge of how to achieve purity and bliss would not be whole without knowledge of how to vanquish evil. Indeed, the *Bhagavad Gita*, one of the epic works of Vedic literature, is essentially the tale of Lord Krishna preparing the great archer, Arjuna, to go into the battlefield.

Together, knowledge of the marma points and the use of medicinal herbs and oils make Ayurvedic massage a profound technique for balancing subtle energy of the body and mind. The specific techniques of massage appear below.

THE TEJ HOME SPA TREATMENT: EXTERNAL CLEANSING AND OLEATION

Snehana, the Sanskrit term for oleation, comes from the root word *sneha*, which means "love"; the literal meaning of *snehana* is "loving your own body." This, then, is the essence of Ayurvedic self-massage. It not only helps the skin to look younger, smoother, clearer, and more radiant, but also helps the body to overcome tension and fatigue, and the mind to feel peaceful and refreshed.

The Tej Home Spa Treatment is a three-stage therapy including herbal cleansing, oleation—in this case, self-massage—and sweating. Although we recommend that you give yourself this loving treatment at least once a week—it takes about sixty to ninety minutes to do the full regimen—you can do it more or less frequently, or you can do just one aspect of the treatment alone. In fact, we recommend that you do the full self-massage as often as possible within the parameters set for your skin type.

CLEANSING

Herbal cleansing, phase one of the Tej Home Spa treatment, prepares the skin for oleation by cleaning and opening the pores using herbal pastes called lepas (see below). It should be done before a massage, or at least once a week. However, you can also use the herbal paste every day in place of soap for your regular cleansing routine. It not only kills bacteria and removes foul smells from the body caused by perspiration, it also improves circulation, provides nourishment, and stimulates the metabolic processes of the skin tissue for a healthier, more glowing complexion.

LEPAS (HERBAL CLEANSERS)

To make:
- *For daily use:* Mix ¼ tsp turmeric powder + 1 Tbsp chickpea flour. To make a paste, add:

  For dry skin: 1 tsp almond oil + *water.*
  For sensitive skin: *Milk* only.
  For oily skin: *Water* only.

- *For monthly use (deeper exfoliation):* Mix 2 Tbsp red lentil powder into ½ cup whole milk. Soak overnight. Add to mixture 1 tsp almond paste + 1 tsp cashew butter + 1 tsp dried milk + ½ tsp wheat germ oil. To make paste, add some rosewater.

To use: Apply paste all over the body. Lie down and relax for 15–20 minutes while paste dries. Using your hands or a dry towel, gently rub off dry paste in a circular motion. Finish with oil massage.

STEP-BY-STEP SELF-MASSAGE

Self-massage, or abhyanga, includes oleation of all the joints of the body and more than fifteen main marma points. It is a totally balancing treatment because each energy point you massage has a mirror image: Points on the hands and legs correspond to the seven chakras; points on the head correspond to those on the feet and fingers; points on the neck, to ankles and wrists; on the chest, to the calves; on the stomach, to knees and elbows; on the lower back, to the thighs; and points on the pelvic girdle, to shoulders and hips. Whenever you massage one point, you stimulate the corresponding point, too.

The procedure for self-massage is broken down into nine steps, literally from head to toe. In order, they are: the scalp, the feet, the lower limbs, the upper limbs, the back, the stomach, the upper torso, the neck and face, and the face-lift. For special purposes, or if you are short on time, you also can do a variety of mini-massages based on individual aspects of the full version. These include the mini-ahbyanga, or the three-minute scalp and foot massage (Steps 1 and 2); the stomach massage (Step 6) and the natural face-lift (Step 9). As you will see, there is a different value to be gained from each one. Before you begin, read the general directions below.

Getting ready. Regular massage is beneficial for all constitutional types and vital to healthy, young-looking skin. However, not all types of massage are suitable for everyone because each of us has a different proportion of elements in our makeup, and each element requires different herbs, oils, fragrances, and touch to balance. A light, allover massage with warming, soothing sesame oil may benefit a person with Vata skin, for example. But that same warming oil when rubbed into sensitive skin will aggravate Pitta and probably cause a rash. Therefore, each component of the massage procedure—including the frequency and time, the type of herbal oil or powder used, the method of application, and the amount of pressure and stimulation—must complement your skin type. Otherwise, you may find that the treatment does more harm than good.

When and where to massage: The best time to do this massage is before bathing in the morning, or at bedtime. Always empty your bladder before massage, and never massage directly after a meal.

Dry skin types, who tend to get ungrounded, benefit most from frequent touch (remember, skin is the seat of Vata). They can do self-massage comfortably up to two times a day, although evening massage is most useful for Vata disorders such as insomnia and anxiety. Sensitive skin types generally require less frequent massage, and may enjoy its soothing effects especially in the afternoons during high Pitta time. Oily skin types can do with the least frequent massage—no more than one or two times a week—and tend to benefit more from the treatment during the morning Kapha period, when they need some stimulation to get going.

The best way to do a massage is seated on a big towel or mat (one that can be easily laundered), preferably in a warm, quiet, comfortable place. If you feel like it, play soft music and burn incense nearby.

How to use the oil. Always use warm, medicated body and face oil made with ingredients suited to your skin type (see Chapter 5). You can warm the oil by running hot water over the jar—just be sure the lid is closed tightly. Use face oil, *not* body oil, on your face. To avoid the need for shampooing after a massage, also use face oil on the scalp points; it is less greasy.

Use at least a few drops of oil to massage each marma point and each joint. Add some to your hand or directly onto the skin at each new massage point. Apply *more* oil for dry skin, *less* for sensitive skin, and the *least* for oily skin. For *very oily* skin, *do not apply oil.* Instead, use dry chickpea flour and a loofa or body brush.

How to massage yourself. To do self-massage correctly, you need to know the type of stroke to use and the type of pressure to apply. There are three basic strokes in this massage, depending upon which part of the body you are working on:

- *On marma points and joints*, use the fingers or palms in a clockwise circular motion, except where otherwise indicated. For example, on some delicate areas of the skin, we only press the point; in some cases, we tap the marma points to awaken the subtle energy.
- *Between joints*, stroke in an up-and-down motion on the limb to create friction.
- *On the hands and feet*, stroke in an upward direction only.
- *On the neck and face*, stroke gently upward or sideward only.

Generally speaking, Ayurvedic massage does not involve deep muscle massage, because it aims to stimulate the subtler energies of the marma points and nadis. For this purpose, we use a light but firm touch. How much pressure to use depends upon your own skin type:

- *Sattvic touch* is light, settled, and slow. It generally is used on *dry* skin types, or in situations when a person needs calming. Too much pressure aggravates air, and a massage that is too deep and vigorous can do a Vata constitution more harm than good, causing feelings of ungroundedness and agitation.

- *Rajasic touch* uses moderate pressure and speed, and is good for *sensitive* skin types. Too much stimulation also aggravates Pitta.

- *Tamasic touch* is the deepest and most vigorous. It is good for *oily* skin types whose thick, slow Kapha natures need more stimulation.

There is no precise way to measure these differences in pressure—it is literally a matter of feel.

How long to massage. Traditionally, Ayurveda advises to do a massage stroke seven to eleven times on each point. The complete massage at that rate takes about fifteen to twenty minutes. If you do not have time, do fewer strokes. If you are very short on time, do a mini-massage of just the scalp and feet. This is especially good to do at night if you have insomnia, and in the morning if you are anxious. Its calming effect is always beneficial if Vata or Pitta is high. Kapha types, of course, rarely need such regular soothing.

For good results, however, what is more important than the quantity of strokes is the *quality*. By this, we mean not just the type of pressure applied, but also the *attention* paid. Your awareness of the bodily sensation of each stroke facilitates the processes of stress release and purification. For the greatest effect, therefore, *always keep your eyes closed and your awareness gently focused on each point as you massage it.*

In the case of disorder or disease, Ayurvedic doctors often administer massage with medicated oils every day for months. As a health maintenance measure, however, frequent massage is usually enough. To stay fit and balanced, do the full self-massage a minimum of three times the first week and once a week thereafter. Do more frequently if there is an existing imbalance, *except* in those circumstances listed below when some or all massage therapies are contra-indicated.

Caution!

- To avoid pushing toxins deeper into tissues, *never* massage when you are menstruating or have severe weakness, emaciation, constipation, or fever.

- *Never* massage the *stomach* if you are pregnant or have intestinal ulcers or heart problems.

Now, here are the step-by-step instructions for doing a full-body self-massage, along with illustrations showing the numbered marma points and the direction of the massage strokes:

STEP 1: THE SCALP

Massage each point below in three steps: First, put oil on your palm and *pat* the point, applying light pressure. Then, massage the point clockwise. (Use your middle finger instead of your palm on the medulla oblongata.) Finally, grab the hair over the point, curl it around your finger, and pull once firmly to stimulate the nerves. Release.

- *Brahma Randhra (1)*. Located in middle of the scalp about eight finger-widths above eyebrows. It is the soft spot remaining on the scalp from infancy, when the skull has not yet fully formed around the brain. According to Ayurveda, it is the seat of bliss, or samadhi, and also the place where prana, the life force, leaves the body. This is where we begin every Ayurvedic massage. (Relieves pressure headaches.)
- *Adhipati (2)*. The "crown" chakra, located on top of head midway between the ears. (Relieves hypertension.)
- *Manya Mula (3)*. The medulla oblongata is the deep indentation at the base of the skull bone just above the hairline on the back of the neck. It is also the spot where the prana enters the body before birth. (Good for pancreatic dysfunction and tension headaches.)

STEP 2: THE FEET

Massage the right foot and leg completely, and then repeat on left. Massage the foot as follows:

- *Talahridaya (4)*. With your thumb, massage the point located in the middle of the arch of the foot. (Good for the heart.)
- *Big toe*. With your thumb, massage the point located in the middle of the underside of the big toe. (Regulates hormonal activity.)
- *Toes*. Starting with the big toe, massage the underside of the base joint of each toe. Then go back and massage each toe all around by gently pulling upward

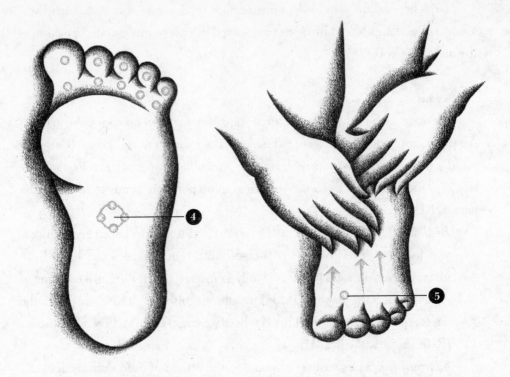

from base to tip. (Each toe, from biggest to smallest, corresponds to a major organ: brain, lungs, intestines, kidney, heart.)

- *Kshipra (5)*. This marma point is located on the topside of the foot, in the groove just between the base of the big toe and the second toe. Place the tips of both thumbs between the big toe and the first toe, and massage about one inch towards the ankle until you feel the bone.
- *Foot*. With both hands, massage foot in upward stroke from toes to ankles. Do top and bottom.

STEP 3: THE LOWER LIMBS

- *Ankle*. Wrap your hands around the joint and massage clockwise.
- *Lower leg*. Massage up and down from ankle to knee on all sides.
- *Knee*. With palms, massage clockwise around the whole kneecap and massage in circular motion gently behind the knee joint.

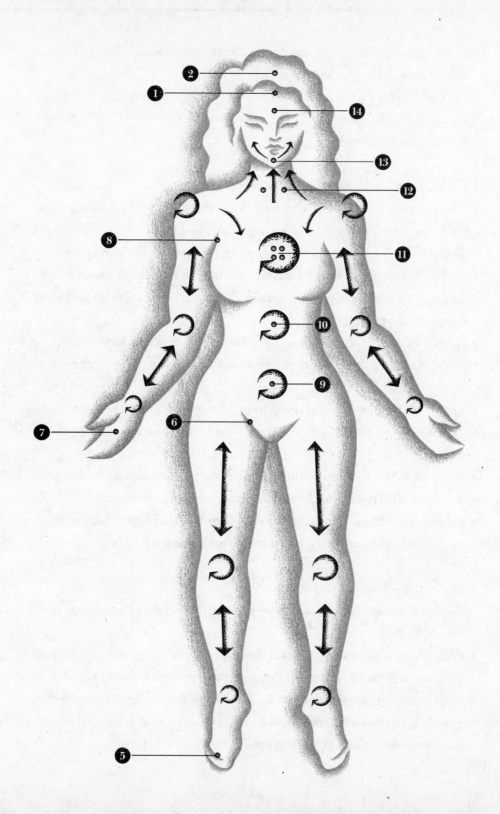

- *Upper leg.* Massage up and down from knee to groin on all sides. If you have cellulite, also knead the thigh muscles.
- *Lohitaksha (6).* This is the marma point at the top of the leg, located in the crease midway between the hip and groin above the lymph node. With middle finger, massage clockwise.

STEP 4: THE UPPER LIMBS

The procedure is similar to that for the feet and lower limbs. Massage right hand and arm completely, and then repeat on left, as follows:

- *Talahridaya (7).* Massage this point, which is in the middle of the palm.
- *Fingers.* Starting with thumb, massage the base of each finger one at a time. Then go back and massage each finger all around by gently pulling upward from base to tip.
- *Hand.* On topside of hand, massage upward from fingers to wrist.
- *Wrist.* Hold the joint lightly between your thumb and fingers; massage clockwise.
- *Lower arm.* Massage up and down from wrist to elbow on all sides.
- *The elbow.* Wrap your thumb and fingers around the joint and massage clockwise.
- *Upper arm.* Massage up and down from elbow to shoulder on all sides.
- *Shoulder.* With palms, massage clockwise around the whole joint.
- *Kashadra (8).* Raise your arm and with your middle finger, gently massage the point in center of underarm where a lymph gland is located.

STEP 5: THE BACK

The back is difficult to massage by yourself. Do as much as you can without straining, as follows:

- With your palms, massage up and down the whole lower back area and up the spine as far as you can reach. Then massage shoulders and upper back as far down as you can reach. Or twist a towel into a thick rope and apply the oil directly to it. Hold it behind you like a short jump rope, with one end in each hand, and rub the towel back and forth across your whole back.

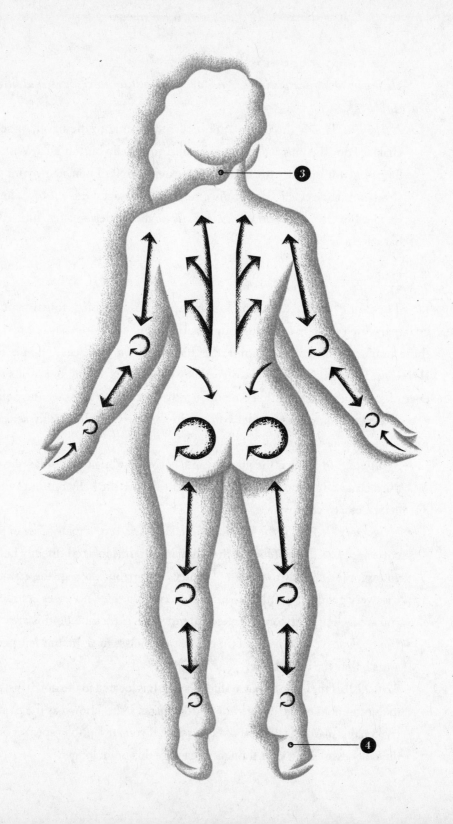

STEP 6: THE ABDOMEN

Skip the abdominal massage if you are pregnant, or have intestinal ulcers or heart problems.

- *Nabhi (9).* This is the marma point corresponding to the navel, or second chakra. Pour the massage oil directly into your belly button area. With your fingers, gently massage the navel in a clockwise circle. Then, using your palm, slowly make the circles bigger until you have massaged the entire abdominal region. Reverse direction, and slowly make the circles smaller until you are back at the navel.

STEP 7: THE UPPER TORSO

The points on the upper torso correspond to the third, fourth, and fifth chakras, which are connected to our emotions and self-expression. Massaging these points (or any marma point, for that matter) may release the energy of stored-up feelings. Consequently, you may experience strong emotions as the energy goes out. If feelings do come up, gently continue to massage the point for a minute or two longer as you let your awareness be with the bodily sensation. Keep breathing easily.

- *Agrapata (10).* This is the marma point corresponding to the solar plexus, or third chakra. It is located four to five inches above the belly button. With fingers, massage clockwise.
- *Hridaya (11).* This is the heart, or fourth chakra. It is actually a set of points covering an area about four inches in diameter. It is located directly between your nipples, slightly to the left of center. With palm, massage the area clockwise very gently. (Stimulates love and passion for life and eases emotions. If you are feeling angry or frustrated, spend some extra time on this area.)
- *Upper chest.* With one hand on either side, massage right and left pectoral areas below the collarbone.
- *Neela (12).* This is the throat, or fifth chakra. It is located in the notch just above the sternum. Massage using your middle finger. (Also known as the Saraswati chakra, this marma is the seat of expression. If you are a public speaker or have difficulty expressing your thoughts, massage this point longer.)

STEP 8: THE NECK AND FACE

When massaging these delicate areas, use face oil for best results.

- *Neck.* With both palms, gently massage neck in upward direction from base to chin.

- *Hanu (13).* This marma point is located in the crease between the chin and lower lip. With right palm facing you, place your index finger on this point on top of the chin, and your middle finger under the chin so the two fingers are cutting your chin like scissors. Using these fingers, massage chin and jawline in an upward direction all the way to the right ear. Do the same with the left hand on left side of face. Alternately massage right and left side a few times in this way.

- *Laugh lines and cheeks.* Using your index fingers, massage the laugh lines in an upward direction, starting at the chin and going up to the base of the nose. Then, with a flat palm, continue to massage upward from the base of the nose across the cheeks all the way out to the temples.

- *Eyes.* Using the ring finger, place your fingertips on the underside of your eyebrows where they meet at the nose. Then massage outward along the eyebrow area, down and inward along the lower lid area, and back up towards the nose, as if you were finger-painting circles around the eyes.

- *Sthapani (14).* This is the sixth chakra, also known as the "third eye." It is the point at the center of forehead above the eyebrows where Indian women wear the red dot known as a *bindi.* With your middle finger, massage each side of nose upward to this point.

- *Forehead.* Place your middle and ring fingers together across bridge of nose and massage upward to hairline. Repeat six to ten times, alternating right and left hands. Then place right fingertips on left temple and stroke across forehead with palm to right temple. Place left fingertips on right temple, and with palm, stroke across forehead to left temple. Repeat six to ten times, alternating. (Reduces lines on forehead.)

STEP 9: THE NATURAL FACE-LIFT

Do this to complete the full body massage, or any time you want a face-lift. Massage each point below in clockwise direction for twenty to thirty seconds. Use

middle finger(s) unless directed otherwise:

- *Center of chin.* (Relief for head colds.)
- *Both corners of mouth.*
- *Center of bone between the nose and upper lip.*
- *Outside corners of nose* where nostrils flare. (This point corresponds to the small intestines. It relieves the sinuses.)
- *Center of cheekbones.* Push up on underside of bone and massage.
- *Lower lids* just above cheekbone. *Press gently* with ring finger. *Do not massage.*

- *Brows*. Use bottom of thumbs to press upward on the inside corner of eyebrows at bridge of nose. Then, with thumb and forefinger, pinch each brow across the whole arch from inside to outside corner. Repeat. (When done correctly, this may cause mild soreness due to stored tension. Good for headaches and bladder problems.)
- *Temples*. Use flat fingers to massage gently.
- *Third eye*.
- *Crown*. Place both hands on top of head and move them rapidly back and forth to vibrate the scalp.

THE MINI-MASSAGES

SCALP AND FOOT MASSAGE

If you do not have time for a full self-massage every day, the best alternative is to anoint the scalp and feet (Steps 1 and 2 above). These two steps alone are good for balancing the subtle energies of the body, because the marma points on the head mirror those on the feet, and vice versa. Massaging the Brahma Randhra and other marmas on the scalp also vitalizes the mind and body, increases the flow of cerebro-spinal fluid, helps to improve eyesight and memory, and balances the pituitary and pineal glands. Herbal oils rubbed on the feet pacify Vata and also penetrate to the brain and nervous system through nerve endings in the skin to provide a soothing, grounding effect. This quick and easy self-massage is especially useful for dry skin types or anyone with vitiated Vata.

Anointing the feet is an ancient practice in many cultures. As you massage your feet in the morning, consider the fact that you depend upon them to carry you through the day, and that by anointing them you consecrate your day. It is an expression of kindness to yourself as well as gratitude to life.

ABDOMINAL MASSAGE FOR STRONGER ABS

A stomach massage twice a day, first thing in the morning and again at bedtime, is the easiest and healthiest method for flattening the tummy besides dieting and regular exercise. It not only reduces fat deposits and cellulite around the abdomen (for even greater results, use Tej or Bindi Medodhara Oil), but also strengthens and tones the muscles, especially following childbirth. (Avoid stomach massage during pregnancy, however, or if you have ulcers or heart disease.)

Stomach massage is also an excellent method for reducing the effects of stress. The stomach and solar plexus (the second and third chakras) are major depositories for toxins, since most of us hold unexpressed emotions in our guts. In addition, seventy-two thousand nadis converge at the navel, so abdominal massage stimulates energy flow throughout the body. It also balances the elements of water and fire, which are seated in the second and third chakras. Other benefits include relief from chronic constipation, improved digestion, and improved kidney function.

Follow Step 6 above. Take one minute to massage clockwise, and one minute, counterclockwise.

THE NATURAL FACE-LIFT

This mini-massage (Step 9 above) tones the cheeks and smooths away laugh lines, crow's-feet, and wrinkles on the forehead. It is very gentle on these delicate areas of the skin, so there is no reason you cannot do it as often as you like, no matter what your skin type. Remember to use your face oil for this massage.

THE BLISS COMPRESS

One of the most enjoyable aspects of pancha karma is an oil treatment called *shirodhara*. This therapy requires special equipment, including a vessel designed to pour a continuous and controlled stream of medicated liquid directly onto the third eye, while the patient lies face up on a table that has a drainage system at the head. The purpose of the treatment is to relieve mental tension and induce a blissful state of mind. Unfortunately, this is not a therapy that can be done easily at home. However, you can enjoy some of the same blissful effects of shirodhara by using an oil compress, or *pichu*.

For the best results, do the full self-massage and steam bath to prepare the body for this treatment. Or, if time is short, just do a neck, face, and scalp massage with oil, followed by a warm towel compress. To make the pichu, soak the middle part of a folded cloth in the appropriate liquid. Lie down and apply the oil compress to the marma point in the center of the forehead, just above the brow line. Relax with eyes closed for ten minutes. If you like, listen to soft music and burn incense.

For Vata: Warm sesame oil.

For Pitta: Equal parts cool milk + coconut oil + ghee + 2–3 drops rose water. Also good for cooling off in the summer.

For Kapha: 3 parts warm sesame oil + 1 part ginger tea.
Also good for relief of winter head colds or sinus conditions.

In general these treatments are good for all dry conditions, constipation, migraines, nosebleeds, lower back pain, kidney problems, and nervous disorders. Of course, it is also blissful.

EYE AND NOSE WASH

Pancha karma also includes an oil treatment for the eyes called *netra basti,* which helps to improve eyesight and reduce wrinkles, and another for the nose called *nasya,* which opens the breathing channels, enhances the sense of smell, and purifies the pranic energy. Like shirodhara, these therapies customarily are administered by professional practitioners; however, you can do a modified version at home. Again, prepare for these oil treatments either using the Tej Home Spa Treatment or more simply, by massaging the neck, face, and scalp followed by a warm compress.

For the eye wash, apply one drop of melted ghee to each eye using an eye-dropper. Lie down and relax with eyes closed for five to ten minutes.

For the nose wash, clean the inside of nostrils with a wet finger or cotton swab, and then massage the outside in an upward direction. With a warm compress over the nose, lie down with the head tilted back. Add seven to eight drops of melted ghee in each nostril, and take a deep breath. Let the drops drip down the nasal passageway, and swallow. Do not spit out. Relax for few minutes.

I encourage you to use all these Ayurvedic massage therapies regularly, not only on yourself but also on those dear to you. Administering snehana is a truly loving, healing act, and a wonderful way to care for yourself and others. Moreover, the long-term health benefits of regular massage often are profound. Susan's experience is a case in point—and although it is a dramatic example, it is indicative of the kind of healing that Ayurveda can bring about through its knowledge of the doshas and its understanding of how the body communicates through its illnesses.

A woman in her fifties who suffered from terrible arthritis and had long been trapped in an unhappy marriage, Susan came to me the first time, like many women, just to relax and pamper her skin. Since she had a Vata constitution that thrives on the touch and nourishment of oil massage, she naturally enjoyed the first session, and for the next two years she returned every week, driving four hours round-trip just for the pleasure of the one-hour treatment. To her surprise, her arthritis pain vanished completely within a year.

Of course, as a student of Ayurveda, I was not surprised at all. Arthritis is a Vata imbalance that has entered the bone and connective tissue, which together hold up the body. Susan's arthritic body, unable to carry its own weight without terrible pain, was simply expressing what she had been unable to say to her husband for years: that she felt unloved and unsupported. Along with dietary changes and herbs to improve her digestion, the nurturing, strengthening touch that is part and parcel of snehana—the experience of loving the body—finally gave her what she needed and the arthritis disappeared, even though the status of her marriage had not changed.

SWEATING

Oleation, or snehana, is one half of the two-step preparation process of pancha karma; the second half is swedana, or sweating. Ayurvedic sweat baths can be done using wet heat, such as steam baths; dry heat, such as saunas or hot stones; or self-generated heat, such as exercise or body wraps. Sweating not only enhances detoxification by liquefying ama, it also promotes the elimination of wastes by activating the sweat glands. Sweating also stimulates agni and relieves coldness, stiffness, and heaviness due to bloating. Caution: *If you have a Pitta imbalance, do not do sweating*.

Do this treatment following the self-massage if you have time, or at least once a week as desired. Spend five to ten minutes in the bath, or until you perspire.

- *Steam heat*. Fill the bathtub with warm water and add a few drops of herbal decoction: Use dashamala or bala for dry skin; comfrey, chamomile, or nettle for sensitive skin; and rosemary or sage for oily skin. Sit in the tub and close the shower curtains to collect the steam.

- *Dry heat.* Wrap yourself in hot towels and lie down under a blanket to induce sweating.

Always finish the Tej Home Spa treatment with your regular cleansing, nourishing, and moisturizing routine. When you have completed these external purification techniques, you are ready for the internal cleansing and oleation treatments of pancha karma.

PANCHA KARMA AT HOME: INTERNAL CLEANSING AND OLEATION

Ayurveda prescribes professional pancha karma therapy as a preventive measure for everyone at least three times a year at the change of seasons. Even if you cannot find or afford professional treatment, however, you can still do a cleansing and oleation therapy at home to help reestablish proper balance and functioning of the doshas, dhatus, and malas.

Like all pancha karma therapies, this home version has three phases:

- the preparatory phase of internal and external oleation, plus sweating to bring the toxins into the gastrointestinal tract and ready them for elimination;
- the eliminative procedure itself;
- a post-treatment management phase that focuses on diet to rekindle the digestive fire and rejuvenate the body.

The time to do pancha karma is always during the last ten days of one season through the first ten days of the next. Within that twenty-day span, you will need about nine consecutive days to complete the therapy. During those nine days, you will need to eat lightly and take as much rest as possible. On the seventh day, when you do the eliminative treatment, you will need to do a partial fast and take a *full day of rest.* Consequently, we suggest that you plan the therapy when you can take time off from your usual routine.

The same cautions that apply to regular massage and sweat treatments apply here. Do not do the therapy if any part of it is contraindicated by existing problems or symptoms. If you have an imbalance or illness, consult your physician first. This home therapy is a palliative. It is not a substitute for either the curative action of professional pancha karma therapy or any other medical treatment.

THE NINE-DAY ROUTINE

THE NIGHT BEFORE YOU BEGIN

The night before Day 1, begin the preparatory phase of the treatment by taking 1 tsp of melted ghee in ½ glass of warm milk before bedtime.

DAYS 1–3: INTERNAL OLEATION

Daily diet. The traditional diet during pancha karma is khichadi, a dish made from lentils and rice (see recipe on page 198). Eat about two cupfuls at each meal. Throughout the day, drink at least six to eight glasses of water (no colder than room temperature) and as much ginger or lemon herbal tea as you desire.

Daily activity. As much as you can, relax, take short, quiet walks, read, listen to music, meditate, and sleep. Try to keep work and vigorous activity to a minimum. Lights out no later than 10:30 P.M.

Oleation treatment. Before bedtime, take:

- Day 1:   2 tsp melted ghee in ½ glass of warm milk.
- Day 2 & Day 3:   3 tsp melted ghee in ½ glass warm milk.

DAYS 4–6: EXTERNAL OLEATION AND SWEATING

Daily diet. Same as above.

Daily activity. Same as above.

Oleation treatment. Stop taking ghee. To prepare the body for eliminative treatment, do the full self-massage (snehana) followed by a sweat treatment (swedana) each day in the morning or evening. *If you have sensitive skin or high Pitta, do not do the sweat treatment for more than two to four minutes a day.* As an alternative for Pitta types, you can wrap yourself in a warm towel and relax.

After the full self-massage is an excellent time to do pichu, the bliss massage, or any other mini-massage described above.

DAY 7: PURGATION AND LIGHT FASTING

This is a day of cleansing, light fasting, and *complete rest.*

In the morning when you wake up, take a purgative of pure castor oil (the odorless kind is okay) with ginger or lemon tea: Use 2 Tbsp castor oil for dry or oily skin (Vata or Kapha) and 1 Tbsp for sensitive skin (Pitta).

Diet. Follow the castor oil with a cup of herbal tea every half hour. *Take no solid food until you have moved your bowels about four to six times.* In the evening, eat a light dinner of rice and dal.

Activity. No work or vigorous activity today. Relax, meditate, read, listen to soft music, and sleep as often as you like. Again, it is a good day to give yourself the bliss massage.

DAYS 8–9: TONIFICATION AND REJUVENATION

Adjust your diet for the new season. Eat lightly and rest as often as possible for at least one to two days more.

INNER MASSAGE:
YOGIC POSTURES AND EXERCISE

Physical exercise is an inner massage. Like cutaneous massage, it works the muscles to relieve tension, tone the skin, improve circulation, and stimulate lymphatic flow. As such, it is a vital part of any health and beauty routine.

Long before the advent of step aerobics and weight training, the rishis taught a system of body/mind exercise known as *Hatha Yoga,* which is one of the eightfold paths of yoga, or *union.* This ancient spiritual discipline involves a series of body postures, or *asanas,* that not only develop physical flexibility, strength, and grace, but also enliven the subtle energies, balance prana, steady the emotions, and increase the vigor of the mind. Unlike most of its modern counterparts, yogic exercise integrates all aspects of existence: body, breath, mind, and spirit.

The three sets of yogic exercises we have included here are suitable for all skin types, and are particularly beneficial for maintaining a youthful appearance, vibrant sexuality, and strength and flexibility of the mind as well as the body. The first set balances and enlivens the seven chakras, or vital energy points; the second set develops vigor of the mind and body; the third set maintains flexibility of the spine. You can do these yogic exercises at any time, but only on an empty stomach. An ideal time, however, is in conjunction with the breathing exercises and meditation described in Chapters 12 and 13. Wear loose-fitting cotton clothes and use a carpet or mat on the floor under you.

REJUVENATING THE CHAKRAS

The chakras are the seven vital energy centers of the body, where the subtle life forces are most concentrated. The flow of energy means life; stagnation means aging and disease.

This set of yogic exercises is called the "fountain of youth" because it opens the energy channels and helps to keep them continually vibrating. Since the seven chakras also correspond respectively to the elements of earth, water, fire, air, and space, plus the aspects of mind and soul, this inner massage balances every level of life.

Do these exercises together as a group in the order given. The complete set takes about twenty minutes. However, if you have an imbalance of an element, for two to three days do the exercise for the corresponding chakra(s) only. Then resume the full routine.

Caution: Do not do any of these chakra exercises if you are ill.

First chakra. This exercise has two parts:
- Sitting erect, inhale as you close the vaginal and anal openings by tightening your groin muscles and squeezing the buttocks. Hold a few seconds; exhale and release. This exercise balances male and female energy in the body. It also stimulates the sexual glands and pushes sexual energy up the spinal column so it circulates through the vital energy channels and back to the genital region. By keeping the energy moving without expending it, this exercise helps to keep you younger. It is especially revitalizing during periods of sexual *in*activity.

• Stand erect. Reach both hands out to either side, keeping your right palm up and left palm down. Twirl around in a full circle in a clockwise direction. Begin with three to four repetitions a day. Gradually increase to ten to twelve a day. This exercise rotates the energy of all seven chakras and helps to balance the earth element. It may cause temporary dizziness.

Second chakra. This yogic posture is called the Shoulder Stand. Lie face up on the floor with your hands under your hips. Inhale deeply as you tighten the sphincter muscle and lift your legs straight up overhead with feet pointed to ceiling. Bend your elbows and use your hands and arms to support the hips and back; tuck your chin into your chest. Hold the posture a few seconds with your back as straight as possible. Exhale as you slowly lower your body and head back to the floor. Repeat ten times. *Do not do when constipated*.

This exercise is said to give eternal youth. It stimulates the thyroid gland, which strengthens tejas, or the metabolic fire, and the immune system. It helps to balance the water element.

Third chakra. Kneel on the floor with your head, back, and buttocks erect. Grab the back of your thighs for support, and gently drop the head forward until your chin is resting on your chest. Slowly inhale as you lift the head, arch the spine backward as far as possible and let your head drop gently behind you. Slowly exhale and return to the starting position. Repeat three to four times.

This exercise helps to balance the fire element.

Fourth chakra. Sit on the floor with your legs straight out in front of you. Place your hands next to your hips, palms down, with fingers pointing behind you; then bend your knees so your feet are flat on the ground and twelve inches apart. Supporting your weight on your hands and feet, carefully lift your buttocks off the floor as you push your hips and stomach up towards the ceiling, and let your head drop back, making a sort of table out of your torso and limbs. Inhale and tighten the muscles as you lift up; exhale and relax as you return to the seated position. Repeat two to three times.

This exercise may be a little difficult at first. Do not strain to get it perfect immediately. It helps to balance the air element.

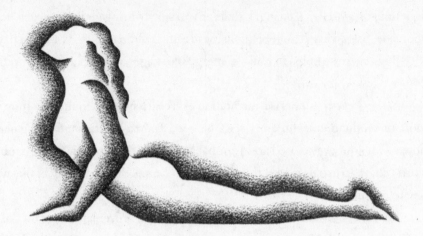

Fifth chakra. This posture is called the Cobra. Lie face down with hands on the floor at shoulder level and feet spread two feet apart. Supporting your torso with your arms, slowly lift the head and chin off the floor as you arch your back. Inhale as you lift your chin up to the ceiling; then lower it and press it into your chest. Exhale and slowly return to start position.

This exercise stimulates the brain centers and helps to balance the space element.

Sixth chakra. Lie face up on the floor with your hands at your side. Inhale as you raise your arms straight overhead and point your toes away from you, stretching the spine. Hold this position for a few seconds as you visualize energy spinning like a horizontal wheel through each chakra and then radiating out from the body like

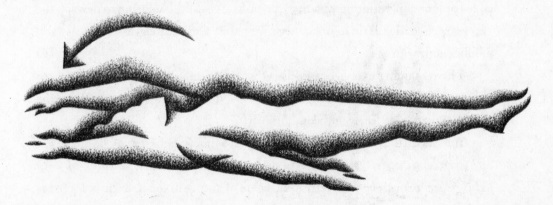

light until every cell is glowing. Exhale as you slowly bring your arms back down to your side. Start with three repetitions, gradually building up to six times a day.

This exercise helps to remove energy blockages and balance the mind.

Seventh chakra. This is a meditation exercise using the sound "so-hum." (For more on meditation techniques, see Chapter 13). In a quiet, comfortable place, sit down with your eyes closed and your back straight. Slowly inhale as you think the word "so," and exhale as you think the word "hum." Breathe easily in this way for one to two minutes.

So-hum is the universal breath: We take in the universal energy on the word "so" and let go of the ego on the word "hum." In this way we remove the barrier between individual and universal Self, and help to establish the experience of unbounded awareness, which is perfect balance.

STRENGTHENING MIND
AND BODY

Yogic exercise flexes mental "muscle" as well as physical muscle. The set below begins with four exercises to strengthen the powers of the intellect: concentration, analytical thinking, perseverance, remembrance, and retention. These qualities are vital to mindfulness, or what we call applied consciousness. In the final chapter, we will speak at length about how you can use volition, intention, and attention to transform thinking and behavior patterns that create stress and unhappiness, and lead to aging and disease. As you begin to practice these principles of action in a mindful way, these yogic exercises will help you to develop the necessary mental muscle to break negative habits and create greater success and fulfillment in your life.

The remaining exercises in this set work directly on the face and body to reduce wrinkles, tone and strengthen muscles, maintain flexibility, and keep the vital energy moving freely and continuously.

You can do these exercises as a set or one at a time, whenever you desire. *Begin each exercise sitting either cross-legged or in the lotus posture, or standing up with feet together, whichever you prefer.* The value of sitting in lotus or cross-legged is that it takes almost no effort to keep the back straight and the body steady when you are in these positions, so you do not become distracted or fatigued.

Yogic prayer. Press palms together, the four fingers pointing upward, and the tips of the thumbs resting on the sternal notch, which is the indentation at the base of the throat at the top of the sternum. Close your eyes, and breathe gently and fully as you focus your thoughts for one minute on the divine within or on any image of light, love, or tranquility. This develops concentration.

Thinking and perseverance. Slowly lift your chin and bend your head backward as far as possible without straining. Close your eyes, then breathe in and out through the nose quickly and forcefully about twelve times. Let your torso move in and out freely like bellows as you breathe. This increases the power of thinking and perseverance.

Remembrance. Place an object on the floor about five feet away from you. Look directly at the object, and then breathe in and out twelve times quickly as you focus your awareness on the crown of the head. This develops the power of remembrance.

Retention. Gently bend your head forward and tuck your chin into your chest. Close your eyes, then breathe in and out twelve times quickly as you focus your attention on the nape of neck. This develops power of retention.

Vigor of the cheeks and teeth. Touch the tip of your tongue to the roof of your mouth, pucker your lips as if whistling, take a deep breath, and close the mouth. With the inhaled air, blow up your cheeks, gently tuck your chin into your chest, and hold the breath as long as possible, until cheeks turn red. Gently lift your head and exhale through the nose. Repeat six times.

This exercise smooths out wrinkles, strengthens the cheeks and teeth, and helps to prevent and cure pyorrhea, a disease of the gums.

Vigor of the neck. Close your lips and pull the corners of the mouth up as hard as possible in a big grin. Breathe in through the nose as you tighten your cheek muscles, and at the same time stretch your throat and neck upward. Hold as long as possible, then slowly exhale as you release and relax your face and neck muscles. Repeat six times.

This exercise strengthens the thyroid gland, improves metabolism, and corrects a double chin, which results from poor metabolism.

Vigor of the jaw. Open your mouth wide and gently close your teeth with the bottom set thrust forward *in front* of the top set. Now put your right thumb and forefinger inside your mouth on either side of the teeth and use them to stretch the lips and cheeks apart from side to side as far as possible; then close the fingers to release. Quickly push and release the fingers this way like a spring thirty times. The mouth should be stretched into a wide-open grin with the neck and jaw stretched forward.

This exercise is good for strengthening and toning the cheek, jaw, and neck muscles. Together with the neck exercise above, it produces the same effect as a lymph drainage massage, which typically costs two hundred dollars or more. By contracting and releasing the series of muscles that goes from the chin, jaw, and neck to below the ears, these exercises help to drain the lymph fluids from these areas and to circulate them to the heart.

Vigor of the chest. Standing with arms down at sides, inhale slowly as you raise the arms out to the side and up overhead without bending the knees. Exhale slowly as you lower the arms and return to the start position. Repeat twelve times.

Vigor of the abdomen. Through the nose, exhale and draw in your abdomen; then inhale and release your stomach. Do *not* hold the breath. Repeat three times.

Vigor of the muladhara. This exercise is done in three sets: first with feet together, then four inches apart, then two feet apart. To do each set, stand up straight with legs stiff. Inhale as you tighten the rectal and vaginal muscles. Hold as long as possible, then exhale and release. Repeat three times.

Sexual energy flows from the first chakra, or the muladhara. It can flow down and out, as it does when we experience orgasm, or up through the chakras and out the crown, as it does when we are enlightened. By closing off the downward path in this exercise, the energy moves up the spine, balancing the mind and body and expanding consciousness. This exercise enhances sexual vigor and helps to relieve all sexual problems.

Salutation to the sun. The Sun Salute, or *surya manascar,* is a standard yogic exercise routine that maintains the flexibility of the spine and opens the three main energy channels that go up the spine, linking the seven chakras and all other energy channels of the body. When the spine becomes stiff, we become old. Thus, the Sun Salute keeps the entire body balanced and fit.

As the name suggests, this exercise is customarily practiced in the early morning as you stand facing east toward the sun. It consists of a sequence of six basic postures that are done in a double cycle, first in forward order and then in reverse. The short routine takes about a minute and flows rhythmically, almost like a dance, because you change postures with each breath.

You can find directions for the Sun Salute in almost any basic Hatha Yoga book. Rather than repeat them here, we refer you to your local bookstore.

WORKING OUT
TO BALANCE WITHIN

As we said, all the exercises above are suitable for every skin type. Of course, many of you may want to do other types of exercise as well, either for fitness or recreation, or both. Today, workout programs exist for every age and temperament, from baby exercise classes to "dancing with the oldies" for the senior crowd. We cannot comment here on every one, but we can offer some general advice on how to choose an exercise program that will help to keep your doshas in balance.

Oily skin types (Kaphas), who are prone to be inactive and overweight but who also have strong physical endurance, benefit most from vigorous, high-impact, aerobic workouts of up to an hour a day. Because of their innate tendency to retain water, they also benefit by working up a good sweat. Although their stamina is good once they are up and moving, Kaphas tend to have problems getting themselves started. For this reason, they are better off in a structured workout program where they have a personal trainer, an aerobics instructor, or a tennis partner, for example, who will motivate them to begin. Following a workout, oily skin types (as well as sensitive skin types) also need to be vigilant about cleansing and nourishing the face and skin in order to minimize acne problems that commonly result from exercise, especially in the warmer seasons.

High-energy sensitive skin types (Pittas) also benefit from fast-paced workouts where they can burn off extra energy and let off steam. However, they do not have Kapha's stamina. Although they like to push themselves hard, they should limit their workout periods to avoid getting burned out. Self-starters by nature, Pittas do not generally have problems getting motivated, so they can stick to an exercise program without coaching. On the other hand, they are naturally aggressive and enjoy competitive sports. Even if they choose a solitary form of exercise such as running or swimming, they are likely to turn their workout into a training program for a marathon or swim meet.

Dry skin types (Vatas) always benefit from a regular routine, including regular daily exercise. Their problem is the opposite of Kapha's—that is, they jump into activity enthusiastically but they have limited physical endurance and psychological tenacity. Therefore, Vatas are best off in a structured workout program of low-impact, moderately paced exercise such as ballet or modern dance classes, Hatha Yoga, hiking, swimming, or fitness walking. They should limit their workout to thirty to forty-five minutes a day.

For all skin types, the best time of day to exercise is during the morning Kapha period, when the metabolism is naturally slow and the mind and body can benefit most from the extra stimulation. Exercising during the afternoon Vata period tends to cause fatigue, and at night, tends to stimulate the mind and induce insomnia. Since vigorous exercise alters the metabolism, do not work out for a half hour before eating and for at least one to two hours after eating. On the other hand, a ten- to fifteen-minute stroll following meals is beneficial for healthy digestion.

"Touch," as one writer noted, "is the most personally experienced of all sensations." It is also the sense most closely linked to the skin—the organ that enables the most intimate human contact is the body's *largest* organ. Indeed, we seem to be creatures designed for tactile experience. Every inch of the skin is made for touching—and being touched. Therefore, touch would seem to play a large role in human experience. Yet, we have created a culture in which we live increasingly in physical isolation from each other, in which the only touch that many children know is in the form of violence and abuse, and in which many adults rarely experience loving physical contact outside of a sexual context. In addition, the speed and artificial stimulation of the "high tech" age has engendered a "low touch" society. We even use machines to exercise the body, and dream of having abs, pecs, and buns "of steel"—we want to *become* machines, as if we could outdistance or outsmart the aging process with technology.

The only way to conquer aging and disease is to reduce stress and expand awareness. The technologies of *consciousness* are the tools of the science of longevity. Machines wear out and run down; the intelligence of the bodymind is self-renewing. Exercise your biceps by picking up a baby or giving a massage to someone you love. If you want to press a couple of hundred pounds, hug some friends. A nurturing touch can undo a day's worth of tension and stress.

CHAPTER 9

AROMA THERAPY

How much better is thy love than wine!
And the smell of thine ointments than all manner of spices?
A garden shut up is my sister, my bride;
A spring shut up, a fountain sealed.
Thy shoots are a park of pomegranates,
With precious fruits;
Henna with spikenard plants,
Spikenard and saffron, calamus and cinnamon,
With all trees of frankincense;
Myrrh and aloes, with all the chief spices.
Thou art a fountain of gardens . . .

THE SONG OF SONGS

All of us practice aroma therapy twenty-four hours a day. The air we breathe is a potpourri of odors and fumes from soaps, lotions, sprays, polishes, paints, foods, flowers, garbage, gasoline, chemicals, and living bodies of every sort. Even while we sleep, the scent of night cream and laundered sheets is directly under our noses. Nevertheless, we tend to disregard all but the most obvious effects of this constant olfactory barrage.

This is an unfortunate oversight in the view of Ayurveda, which holds that all five senses are important to the balance of the doshas. Smells have a primal influence, as you will see. Anatomically, the nose is the gateway to the brain, and every scent we inhale exerts a direct influence on the mind and emotions. At the same time, all odors ride into the body on the breath, which also carries the life force, prana. In this way, the nose is also the gateway to consciousness. For these reasons, Ayurveda is concerned equally with *what* we breathe, which is the subject of this chapter, and *how* we breathe, which is the subject of Chapter 13.

SMELL: THE ANCIENT ROOT
OF EMOTIONS

The human olfactory organ has ten million scent-detecting cells, each with two hair-like nerve endings that pick up odor molecules dissolved in the mucous membrane. These twenty million *cilia* are sensitive to thousands of different odorants. According to Peter and Kate Damian in *Aromatherapy: Scent and Psyche*, a trained nose can identify hundreds of distinct odors, and even the untrained instrument can sense a single molecule of bell pepper in a trillion air molecules. While this gives humans an acute sense of smell compared to our other senses, our prowess in this area is nothing compared to the rest of the animal kingdom. Sheepdogs, for example, have twenty-two times more olfactory cells than humans, which gives them a sensitivity a *million* times more acute.

Indeed, for slithering reptiles and all four-footed creatures, who live with their noses close to the ground, a sharp sense of smell is still crucial for survival, and eons of evolution have not diminished their capacities to sniff out the most subtle scents. This is not the case for upright *Homo sapiens*. Along with most other primates and birds, we literally have broader horizons, and so rely more heavily on the distance senses of vision and hearing than on the so-called chemical senses of touch, taste, and smell. Furthermore, human reason, and the civilization it has wrought, afford our species other, more adaptive means of protection that have reduced even further the survival value of a good nose. As a result, by Darwin's rules of natural selection, the human nose, as sensitive as it is, just isn't what it used to be as a sensory organ. At this point in our evolution, once we mature past earliest infancy—when smell is for a brief time our preeminent sense—olfaction is less important to our brute survival than it is to our mental well-being and our pure pleasure.

In *Emotional Intelligence*, Daniel Goleman, writing on the development of the brain, calls the sense of smell "[t]he most ancient root of our emotional life." As he explains, the olfactory center began as two thin layers of neurons encircling the top of the reptilian brain stem, which was an outgrowth of the spinal cord and the earliest brain-like organ. These neuronal layers could react to surroundings and regulate functions—one determined what was "edible or toxic, sexually available, enemy or meal"; the other gave the signal to "bite, spit, flee, chase"—but they could not think or learn.

With the appearance of the first mammals, a primitive emotional center emerged as a new layer of neurons in a ring shape—or "limbus"—around the olfactory center and brain stem. In time, this *limbic system* developed capacities for memory and learning, and established a vital link between olfaction and emotion. "These revolutionary advances allowed an animal to be much smarter in its choices for survival . . . If a food led to sickness, it could be avoided the next time," Goleman states. Although the animal still decided "what to eat and what to spurn" solely on the basis of smell, the olfactory-limbic connection now enabled comparison of new and known smells, and discrimination between good ones and bad, so the animal could adapt its behavior to each new circumstance. This new apparatus was named aptly the "nose brain," or *rhinencephalon*. It gave us our primal emotional choice—the subjective like/dislike component of our bodily pleasure/pain response—and became the root of the thinking brain. Emotional nuances and "the ability to have feelings *about* our feelings" arrived later with the full power of human reason. Nevertheless, our wiring is such that the ancient brain always responds first in times of emotional crises or sexual desire and passion—the behaviors necessary for human survival. Says Goleman, "When we are in the grip of craving or fury, head-over-heels in love or recoiling in dread, it is the limbic system that has us in its grip."

Because of its ancient wiring, the human olfactory sense retains a privileged connection to the "emotional" brain as well. While other sense stimuli reach the limbic system by way of higher brain centers, olfactory stimuli take a direct route via the nasal opening and olfactory organ and nerves to the neural matrix of fear, passion, aggression, desire, pleasure, need, and instinct—the limbic structures that are literally beyond reason. As a result, scents and smell have a unique capacity to evoke vivid impressions and emotional "flavors," even ones that were formed at an age before we had words to name them. Ayurvedic aroma therapy takes advantage of this capacity by using scents selected for their known effects on different body types to balance our mind and emotions.

LOST AND FOUND:
THE REDISCOVERY OF AROMA
THERAPY BY WESTERN SCIENCE

Having unraveled these brain/mind mysteries and the neurochemical links between our senses and emotions thanks to advanced technologies, modern scientists have developed renewed interest in studying the psychophysiological and behavioral effects of aromas. In fact, the use of scents to stir the mind or relax the body dates back to ancient times, and spans continents and civilizations. Following the European Renaissance, however, Westerners generally dismissed the healing science of scent as a form of alchemy, and until the late nineteenth century left it largely to the province of "occultists" and perfumers. At that time, medical interest in aromatic oils was revived when scientists observed during an epidemic of tuberculosis that French flower growers and perfume workers showed an unusually high immunity to the disease as well as to other respiratory ailments.

In the 1920s, a French perfumer, Rene M. Gattefosse, brought new popularity in the West to the medicinal powers of essential oils after he inadvertently plunged his badly burned hand into pure lavender oil following a laboratory explosion. When the wound healed within hours with no infection or scarring, Gattefosse turned his professional attention to the therapeutic and cosmetic uses of pure essences. Westerners credit him with coining the term "aromatherapy," which was (in French) the title of his first book; but India, Egypt, Persia, Greece, Rome, China, and many other cultures, ancient and modern, have practiced aroma therapy in one form or another continuously in history.

Working independently but in the same period as Gattefosse, two Italian physicians experimentally demonstrated the sedative and stimulative effects of specific essences on physiological functions, and revived scientific interest in the emotional and behavioral effects of aromas. In the next decades, several scientists around the globe entered the field, and by 1990, a Yale psychologist, using the scent of chocolate, proved that odors can help evoke memory and promote learning. Today, a large research industry has grown up around the use of aromas to modify moods and behavior. Recent studies have found, for example, that foul odors foster antisocial behavior while flowery ones enhance learning ability and memory, leading to the conclusion that certain aromas could be useful in environments such as classrooms, offices, and even the subways.

Where researchers have used pure essential oils, studies generally have confirmed the therapeutic and medicinal value of essences, which Ayurvedic physicians discovered ages ago. However, most Western aromatherapists and "aroma-chologists"–as the industry scientists have named themselves—still have little or no understanding of the role of individual body type and complementary attributes in the effective use of scents. Without this knowledge, they inadvertently misuse their tools by prescribing the same fragrance for everyone. While research has proven that specific essences have specific effects, we know from Ayurveda that not every effect is desirable for all people. As you will see, a soothing, cooling scent that *reduces* stress in hot-tempered Pittas can *produce* stress in laid-back Kaphas because it aggravates their natural lassitude and sensitivity to cold. Likewise, a stimulating aroma that balances Kapha will aggravate Vata's already excitable nature. Effective aroma therapy requires knowledge of the essences as well as of the individual using them.

Dr. John Ryder is a New York psychologist who has begun to investigate the therapeutic uses of aromas according to Ayurvedic principles. Using my Tej oils, he has combined Ayurvedic aroma therapy with modern stress management techniques to help his clients relax. After he gives them a questionnaire to determine their individual body type, the clients select the most appealing scent from a set of three Tej aromas and then learn a relaxation technique with self-hypnosis. Once they are in a very deep state, Ryder guides them through a visualization process in which they inhale the selected aroma while they are given suggestions to promote healing and relaxation. Within three to five such sessions, this pairing of the fragrance—a pleasant external stimulant—with the experience of deep calm establishes a conditioned response so that his clients are able to relax at any time simply by inhaling the appropriate aroma from a vial.

"It is a powerful aid in treating emotional disturbances because it resets the level of arousal in the limbic system through direct neuropsychologic mechanisms," Ryder explains. "In one step, it alleviates the effects of stress and promotes self control."

Apart from such independent experiments, much of the current scientific research on odors and the related field of flavors is supported by corporations who see a big future in "ambient fragrancing" and "sensory engineering" to improve products and employee productivity and to affect consumer behavior, for example. The larger profit potential, by far, lies in the use of synthetic scents rather than costly pure essences; and for the most part, synthetics are what researchers and manufacturers are using. As mentioned, unless the ingredients specify "essential oils," the word "fragrance" on a label always refers to synthetic scents.

MAKING SCENTS: THE VALUE OF PURE ESSENTIAL OILS

In Chapter 4, we introduced the subject of herbal essences, their origin in the fragrance and flavor of plants, their role in the plant's immunity, and their many uses in Ayurveda, including aroma therapy. As we said, extracted essential oils carry the plant hormones, the biochemical messengers that, like our own hormones, regulate the plant's life processes. The essence itself—the fragrance—serves the plant in two ways: On the one hand, it attracts bees and other creatures necessary for pollination and propagation; on the other hand, it repels predators and kills invading bacteria. It is the life force of the plant, and adds its life and intelligence to our body when we inhale or ingest it. As the Damians state in *Aromatherapy: Scent and Psyche*, the essence creates balance biochemically, electromagnetically, and hormonally, working *with* the body's subtle intelligence and talking "chemical to chemical, charge to charge, phytohormone to human hormone." This shared language "should hardly be surprising," they add, "since our physical existence directly depends in various ways upon the plant kingdom."

Unlike scents that are synthesized or chemically isolated from their natural source, pure essential oils have up to hundreds, and sometimes thousands, of biochemical components. Many of these components have yet to be isolated or even identified in the laboratory, so the biochemical complexity of pure oils is impossible to duplicate. With necessarily simpler structures and no life force, synthetic essences can never replicate the holistic action and intelligence of essential oils. Moreover, when they are absorbed into the bloodstream through the lungs and skin, the lifeless chemicals become toxic waste in the body. When they are inhaled, artificial scents are more likely to create an imbalance in the central nervous system and precipitate allergic reactions. These are effects we know. We are only just discovering the long-term effects of continued and increased exposure to synthetic deodorants, sprays, perfumes, and other artificial odorants in our atmosphere. The newest research indicates that we are dulling the olfactory nerves by overstimulation, and actually distorting our ability to identify odors or flavors with their natural sources. The impact on our health is insidious.

The Damians advocate the use of pure plant extracts, as does Ayurveda. They describe studies in which baby boomers and younger subjects, when asked to identify aromas, associated lemon scent not with lemons but with household cleaning products, and confused pine scent with lemon for the same reason. In other research, individuals given the choice between artificial berry-flavored drinks and real berry juice preferred the drink with the man-made taste. "Actually," one flavor-lab scientist told the Damians, "if you squished fresh berries and blindfolded these tasters, they would say it wasn't fruit." The Damians note the dangers of living in this ersatz world: "Often, only because they smell and taste like those things that are good for us, we are duped into eating, drinking, and smelling things that are not good for us." Thus, we get the pleasure we want but none of the therapeutic or nutritional benefits we need.

In the final analysis, we cannot get real pleasure from artificial smells and tastes anyway. These fabricated molecules may satisfy the senses in the moment, but lacking the intelligence of natural foods, they eventually become toxins in the body and create imbalance in the doshas. Without balance, we cannot achieve life's ultimate and lasting pleasures—unbounded bliss and absolute beauty. Natural scents used according to Ayurvedic principles not only give instant sensual pleasure but also enliven the mind and balance the subtle energies of the body. In the following sections, you will see how to choose and use the scents that are right for you.

CHOOSING SCENTS:
A MATTER OF TASTE

Have you ever wondered why you are *not* attracted to everyone you meet? Or why you experience a special "chemistry" with one person and not another? Ayurveda answers that every person has a unique essence—our ojas—that is secreted by glands and is indeed a product of body chemistry. As Goleman writes, "Every living entity, be it nutritious, poisonous, sexual partner, predator or prey, has a distinctive molecular signature that can be carried in the wind." Like the essence of a flower, this molecular signature is detectable to our sense of smell, and like all smells, produces a specific psychophysiological response.

In this way, our "taste" in friends and mates is not unlike our taste in perfumes. Just as each of us prefers only a few fragrances out of hundreds on the market, we find ourselves drawn to the essence of certain people and repelled by others. Which ones attract, soothe, or stimulate us and which ones do not depends to a large degree upon our own essential nature—that is, the characteristic mixture of elements that forms us.

As mentioned in previous chapters, the smell of something originates from its taste: Whatever tastes sweet, smells sweet; whatever tastes sour, smells sour; and so on. There are six basic tastes, or *rasas*, and consequently six basic smells. Each one has the properties of the elements that make it up:

RASA	COMPOSITION	PROPERTIES
Sweet	Earth + Water	Cold, Oily, Heavy
Sour	Earth + Fire	Hot, Heavy, Oily
Salty	Water + Fire	Hot, Oily, Heavy
Pungent	Fire + Air	Hot, Light, Dry
Bitter	Air + Space	Cold, Light, Dry
Astringent	Air + Earth	Cold, Medium

A taste and its aroma, like water and steam, are just the same matter in different form, so they affect your constitution in similar ways. The fire in pungent food increases Pitta, for example, whether you *eat* it or *smell* it. However, a taste of food must be digested and absorbed through the intestinal tract before nutrients get into the bloodstream to effect pervasive change. On the other hand, odor molecules either pass via the air sacs and capillaries of the lungs straight into the bloodstream, or via the olfactory nerves to the limbic system and hypothalamus, where they stimulate the endocrine system and hormone production. Because these effects are immediate but short-lived, aroma therapy is ideal to ease feelings, enhance a mood, or reduce stress symptoms on the spot. Quick-acting and easy, it is a perfect adjunct to the daily skin care routine, especially if you have an imbalance rooted in emotional causes.

Essential oils are the basic tools of Ayurvedic aroma therapy, as we said. You can use them in any or all of the ways suggested here, but use only essences whose "tastes" and properties complement your skin type and constitution. These are:

For dry skin (*Vata*):	Sweet, warming, calming, hydrating
For sensitive skin (*Pitta*):	Sweet, cooling, soothing, hydrating
For oily skin (*Kapha*):	Pungent, warming, stimulating, drying

In total, there are only about 150 essential oils used in aroma therapy. You will find the most common ones for each skin type listed in Appendix E.

USING SCENTS

The "essence" in essential oils lends an extra aromatic value to preparations used in many Ayurvedic treatments, from facials to foot massage. You do not have to wait for these occasions, however, to enjoy an aroma's balancing effects. Here are several simple ways to get added beauty mileage around the clock from your sense of smell. You do not have to limit yourself to these forms. As long as you use the right scent, you cannot overdo aroma therapy, or do it wrong—so be as inventive as you like in utilizing your aromatic oils. For more information on how to prepare various formulations, refer to Appendix B.

PERFUMES

Any essential oil suited to your skin type makes an excellent perfume "as is." Just pick a favorite one—or mix a few—and dab on like cologne as often as desired.

These chemical-free, just-right-for-you fragrances are the perfect alternative for anyone with skin sensitivities or allergies to packaged perfumes. You can also make perfumes, which are blends of pure essences, usually in an alcohol base.

SAMPLE RECIPES

Sweet 8 drops rosewood + 4 drops *each* jasmine and ylang-ylang + 3 drops rose + 1 drop vanilla + alcohol base (80 drops).

Musky 10 drops patchouli + 8 drops sandalwood + 4 drops *each* ylang-ylang and jasmine + 2 drops clove + 1 drop cinnamon + alcohol base (80 drops).

MOOD OILS

No matter what our body or skin type, we all have times when we are overwhelmed by uncomfortable feelings. Mood oils are a blend of various essences to give you on-the-spot relief from bad moods or to spice up good ones. Cooling, sedating oils douse flaring tempers, for example. Warm, calming oils ease panic attacks. Stimulating oils lift the spirits. Massage mood oil on pulse or marma points, or add it to any Ayurvedic treatment that calls for essential oil.

SAMPLE RECIPES

Calming & Warming (Relieves anxiety; balances Vata)	3 drops *each* neroli & lemon + 2 drops *each* jasmine & sandalwood +1 drop vanilla + 1 oz pure jojoba oil base.
Calming & Cooling (Relieves anger; balances Pitta)	5 drops *each* sandalwood & vetiver + 1 drop jasmine + 1 oz pure jojoba oil base.
Stimulating (Relieves depression; balances Kapha)	4 drops bergamot + 3 drops *each* lavender & basil + 1 oz pure jojoba oil base.

SAMPLE RECIPES (CONTINUED)

Sedating (Relieves insomnia; balances Vata & Pitta)	6 drops rose + 2 drops *each* jasmine & chamomile + 1 oz pure jojoba oil base. (Or substitute Bindi Sleep Aroma.)
Grounding & Strengthening (Relieves fear; balances Vata)	4 drops patchouli + 2 drops *each* sandalwood & cardamom + 1 oz pure jojoba oil base.

APHRODISIAC OILS

Aphrodisiac oils stimulate the production of endorphins, the "feel good" hormones that are the body's natural painkillers. For romantic occasions, you can give yourself an added boost of pleasure by applying these oils to pulse points, such as the temple or wrist, where they are absorbed directly into the bloodstream through the skin.

TEJ LOVE OILS

For her: Mix 3 drops ylang-ylang + 2 drops *each* nutmeg & vanilla rose + 1 drop *each* jasmine & clove + 1 oz almond oil base.

For him: Mix 4 drops sandalwood + 2 drops *each* jasmine & ylang-ylang + 1 drop *each* cinnamon & clove + 1 oz almond oil base.

AROMATIC MISTS

Make an aromatic mist by putting four ounces of distilled water plus four to five drops of essential oil in a purse-size spray bottle. Carry it with you and spray on your face as desired. To freshen up on a hot day or balance Pitta, use a cooling, soothing scent like vetiver or sandalwood. To counter the drying effects of air travel, use a warming, soothing scent like rose geranium or orange. To lift your spirits and balance Kapha, use a warming, spicy scent like clove or juniper.

AROMATIC BATHS

An aromatic bath is a perfect way to cleanse body and mind simultaneously, and it is something you can enjoy alone or with your significant other. Take a soothing bath before bedtime or a stimulating bath to start your day. Or do it whenever the mood strikes. Pitta types will want to keep the water a little on the lukewarm side; Vata and Kapha types will want it hotter. Kaphas should not spend more than five to ten minutes once or twice a week in a hot bath, to avoid getting too lethargic.

ADD TO BATHWATER:

Soothing	8 drops neroli + 4 drops *each* sweet orange & geranium rose.
Detoxifying (Relieves aches & pains)	4 drops ginger + 3 drops *each* sage & rosemary.
Aphrodisiac	5 drops ylang-ylang + 3 drops lavender + 2 drops *each* geranium & cardamom.
Stimulating & Aphrodisiac (Relieves fatigue; promotes sexual energy)	3 drops *each* rosemary & bergamot & ylang-ylang + 1 cup white wine.
Rejuvenating	5 drops *each* rose & jasmine.
Tranquilizing	One handful of either dried valerian, lavender, linden, *or* chamomile wrapped in cheesecloth
Anti-inflammatory (Relieves itchy skin)	One cup vinegar + 2–3 drops sandalwood oil.
Warming & Soothing (Good for dry skin)	1 tsp honey + 10 drops rose oil.
Cooling & Soothing (Good for sensitive skin)	One handful of milk powder.
Warming & Stimulating (Good for oily skin)	5 drops lavender + 3 drops rosemary + 2 drops sweet orange *or* mint.

AMBIENT TECHNIQUES

There are many ways to use aromas in your environment so that everyone around you can enjoy their effects. Pick the scent according to the occasion. Use a stimulating fragrance for a party; an aphrodisiac scent for a romantic evening at home; a soothing scent for a relaxed intimate talk; and so forth.

VAPORIZERS

To lightly scent the air, add 5–10 drops of essential oil to a cup of hot water and leave uncovered in the room. You also can buy or make an aroma pot, which has a heat source (such as a votive candle, Sterno, or hot plate) to extend the vaporizing effect. Another easy way to vaporize an essence is to put a drop of aromatic oil on a lightbulb and *then* turn it on. Do not drop the oil directly on a hot bulb.

SCENTED CANDLES AND INCENSE

Scented candles do double duty as mood setters, because they are soft on the eyes as well as aromatic. Of course, they are also easy to buy and use.

Incense is an aromatic gum or other substance that is burned to produce a scented smoke. It is commonly sold in the form of a slender stick or small pellet. To use, place in an incense holder and ignite the incense with a match. After it is lit, blow out the flame and let the incense burn slowly like a hot ember.

CARE FOR THE NOSE

Humans have an estimated ten million scent-detecting cells. However, they cannot do their job adequately if the nasal passage is obstructed. The nose is lined with tiny hairs that act like brushes on a vacuum cleaner to pick the dust and dirt from the air before it enters the body. If the nose hairs are not kept clean, the air we breathe will not be cleansed. Moreover, perception will not be clear because the dirt dulls the sense and reduces our capacity to recognize strong smells, whether they are pleasant or offensive.

The technique to clean the nasal passages is an aspect of pancha karma known as *nasya* (for instructions, see page 250). It is beneficial particularly for people with oily skin, who are naturally more sensitive to smell because of their Kapha nature.

However, nasya is also a technique for "cleansing" the brain. The nose is the door to the brain, as we have said, as well as the entranceway for prana, so Ayurveda considers a clean nose to be essential to the clear experience of consciousness.

Even without using any special technique, you can find opportunities every day to enjoy the scents of life. For example, when was the last time you sat down to a meal with family or friends and invited everyone to *breathe* in, rather than dig in? When was the last time you savored the aromas yourself?

The sense of smell is a means we have to "digest" the world, and aroma therapy is a means to create balance and wholeness in our lives. Whenever you can, enjoy this gift. Sniff the garden flowers. Fill your lungs with ocean air. Take in the scents of a pine forest. Stop, and smell the coffee.

CHAPTER 10

COLOR THERAPY

To look at you gives joy; your eyes are like honey,
love flows over your gentle face . . .

SAPPHO

If you have ever taken a spring walk through botanical gardens, explored a coral reef in tropical seas, or watched a sunset over the Grand Canyon—or the Taj Mahal—then chances are you have experienced firsthand how a beautiful vista can make your spirits soar. The "high" you feel from such sights is not in your imagination. It is a symptom of actual hormonal changes in your body in response to stimuli received by the optic nerves and communicated electrochemically to the brain and the limbic system. If you were hooked up to the necessary instruments, scientists miles away could pinpoint your shift in mood just by measuring changes in your blood pressure, heart rate, brain wave activity, and the like as you take in the view.

Of course, this euphoric response, just like the stress response, depends to some degree upon our interpretation of an event. A person with a fear of heights will not feel relaxed and happy looking over the rim of the Grand Canyon, no matter how awe-inspiring a sight it may be to the rest of us. Nevertheless, Ayurveda teaches that everything the eyes behold affects our psychophysiology in specific ways because of the nature of light, color, and matter itself.

SEEING THE LIGHT:
THE SCIENCE OF COLOR

Light is radiant energy consisting of subtle fluctuations or vibrations in the electromagnetic field. All light waves travel through space at the same velocity of 186,000 miles per second; but the length and frequency of individual light waves varies astronomically. At one end of the spectrum, a single radio wave can be several miles long. X rays, gamma rays, ultraviolet rays, and infrared rays fill most of the middle range. At the other end of the spectrum, a cosmic ray can be a few trillionths of a centimeter short. Within this vast spectrum, only a narrow range of high frequency waves, between 3900 to 7700 angstrom units in length, produce the *subjective* sensation of light in the human eye.

The effect of the light depends upon its frequency. Just as different frequencies of audible sound waves give us different notes, different frequencies of visible light waves give us different colors. Split into its component frequencies, ordinary white light produces the visible spectrum we know as the rainbow: red, orange, yellow, green, blue, indigo, and violet. Each band of color—each frequency band—has specific vibratory effects.

Most color we see is light energy reflected by ordinary objects, then absorbed by the one hundred million light-sensitive cells of the retina, converted into a pattern of electrical impulses, and transmitted by optic nerves to the brain. There the mind assigns meaning to the pattern—for instance, big, red, round ball. We see white light when all colors, or frequencies, are present, and we see black when all are absent. (In fact, the colors we "see"—indeed, our entire "picture" of reality including shapes and forms—do not really exist in nature outside our sensory perception. There is no "red" or "white" in the phenomenal world, for example, but only a particular vibrational energy—a light wave—which the mind interprets as red or white when our awareness comes in contact with it through the machinery of the eye.)

Although ordinary light has its greatest impact through the visual sense, it also impacts the nervous system and the subtle energies of the body through the skin. We are all familiar with the physical effects of certain invisible radiation. X rays, for example, penetrate solid matter, transform gases into electrically charged particles, create a photographic image, and in large doses, produce radiation sickness in living things. UV rays damage the skin and take the fun out of too much sun. Along with X rays, gamma rays are the most penetrating form of light. Nevertheless, all light waves, including visible light, penetrate the entire body to some degree.

Of course, color also has a direct psychological impact. For example, Dutch scientists recently discovered that blue sleeping pills are more effective than red ones, even when they contain identical medicines, presumably because we associate blue with soothing, healing qualities.

The normal human eye can discriminate 150 hues formed by various combinations of the primary colors, red, yellow, and blue.

THE VISIBLE SPECTRUM

PRIMARY	SECONDARY	TERTIARY
Red	Orange (red + yellow)	
Yellow	Green (yellow + blue)	
Blue	Violet (red + blue)	Indigo (blue + violet)

Various colors have the capacity to vitalize, animate, heal, enlighten, energize, inspire, and fulfill the mind depending upon their characteristic vibration. Of the primary colors, red has the lowest frequency and is the *most* dense and powerful color; its hot, stimulating quality balances Kapha but aggravates Pitta and Vata. Blue, which has the highest frequency, is the *least* dense and intense; its soothing, cooling quality is too sedating for Kapha, but harmonizes Pitta. Yellow, which is warmer than blue and less intense than red, balances Vata. According to Ayurveda, we will and wish in blue, think in yellow, and act in red. When yellow (thinking) and red (action) are harmonized, we have balance in body and mind.

YOUR TRUE COLORS

Ayurvedic color therapies work to brighten your mood and bring out your inner glow. Like aroma therapies, they are "ready-to-wear" remedies that are convenient to use even on the most hectic days, since they require so little time and effort. They are especially advantageous for Pitta types whose light-sensitive constitutions respond most readily to visual stimulation. However, everyone can benefit from some added color in life—as long as the color is suited to our skin type.

DRESSING BY COLOR THE AYURVEDIC WAY

This is not a variation on the "color me beautiful" approach, which aims to match your clothes to your skin tone. Rather, dressing by color in Ayurvedic terms means livening up your wardrobe (or your home or office) with colors that will actually enliven your mood. When you feel happy, you look radiant—that is the essence of Ayurvedic beauty.

We are not suggesting here to match your wardrobe to your constitution. A monochromatic color scheme can be very tiresome. Rather, keep color in mind as an added means to create balance, and match colors to the mood you want to create for the occasion or the day.

COLOR BY MOOD

DOSHA/MOOD	DESCRIPTION	FAVOR	AVOID
Vata (Anxiety, fear, overstimulation)	Need warm, muted, calming colors.	Gold, orange, yellow, greenish or bluish white, deep purple, indigo; dark colors like brown are grounding in moderation.	Bright red.
Pitta (Anger, frustration, jealousy, hostility, "burnout")	Need cool, mild, soothing colors.	White, soft blues or greens, pastels.	Bright or dark colors, black.
Kapha (Depression, grief, lethargy, dullness)	Need warm, stimulating, bright, but light colors.	Bright red, orange, yellow.	Pink, white, blue, green, brown.

COLOR BY THE GLASSFUL

As we said, the body can absorb color in several ways, not only through the optic nerves. At the subtle vibratory level of existence, light waves affect everything they touch. Liquids exposed to light become infused with its faint vibrations. If the light is filtered through tinted glass, for example, the liquid absorbs the color in the form of its vibrations. In turn, our body tissue absorbs these vibrations when we drink the liquid.

Based upon this understanding, Ayurveda recommends "taking color," both externally and internally, by storing herbal preparations and drinking water in tinted glass containers and then keeping them in sunlight for a few hours before using. Green enhances prana, red stimulates tejas, yellow stimulates ojas.

PRECIOUS TEAS

Like colors and gems, metals have unique energetic effects on the mind and body. In addition to wearing jewelry forged from metals, you also can infuse liquids with their balancing properties, either by storing water or oils in gold, silver, or copper containers or by brewing precious metal "tea." To make gold or silver tea, take a piece of solid sterling or gold (22 karats or more) jewelry, boil it in water, and let simmer for an hour. Take two tablespoons of this tea once a day. Silver calms the intellect, gold stimulates the mind, copper reduces Kapha and fat.

You can also store your massage oils and other herbal preparations in metal containers to enhance their effects:

For dry skin:	Gold, brass
For sensitive skin:	Silver
For oily skin:	Gold, copper
For all types:	Stainless steel

JEWELS

Jewels are a symbol of power and royalty, but you do not need many to look and feel regal. Ayurvedic seers recognized that the structure and composition of gemstones and precious metals endows them with energetic properties that are helpful in strengthening the mind and balancing the emotions. Because of their ability to conduct and refract many forms of energy, gems and metals also protect against potentially harmful radiation in the atmosphere.

The therapeutic use of gems originates in Vedic astrology. Historically, astrology and Ayurveda were complementary sciences; astrology dealt primarily with healing mental disorders, and Ayurveda dealt with physical disorders. Vedic astrologers use specific gems to balance the subtle energetic influences of the sun, moon, and planets on the life force. To be effective, the gem must be worn directly against the skin, and be at least two karats in size if it is a precious stone, and four karats, if semiprecious. A tonic made by soaking a diamond in water overnight is good for the heart. Although Ayurveda also makes medicines with purified minerals, these remedies are not generally offered in the West.

The table on page 284 shows the balancing effects of various gemstones and precious metals. The first stone in each group (usually the precious stone) will always produce the strongest effects; the secondary stones produce similar effects, but to a lesser degree.

As Westerners, some of you may have difficulty reconciling your rational, scientific view of the world with the so-called occult science of astrology and the influence on human life of stars, planets, and rocks, no matter how precious they may be. From the Ayurvedic perspective, this influence is easy to explain in terms of the energetic nature of all matter and the effects of all types of vibratory action—electromagnetic waves, sound waves, the force of gravity, and even the imperceptible motion of molecules on the subtle energies and intelligence of the body.

In *The Re-Enchantment of Everyday Life*, author Thomas Moore addresses the West's skepticism of these ancient ideas, and argues eloquently for "the undeniable power the sky has over our moods and emotions" and the "benefits of an astrological worldview."

Moore writes: "Have you ever been stopped in your tracks at the appearance of a huge, yellow, egg-shaped moon rising on the horizon on a warm summer night? Have you ever commented on the thrilling sight of a purple-and-orange sunset casting its magic over thousands of people in a valley or flatlands? Have

THE EFFECTS OF GEMSTONES & PRECIOUS METALS

Ruby, red garnet, sunstone: Decreases Vata & Kapha; increases Pitta; improves circulation & digestion; strengthens heart.

Pearl, moonstone, milky quartz: Decreases Pitta & Vata; nourishes body tissue & nerves; improves fertility; reduces anxiety.

Red coral: Balances Pitta, decreases Vata; strengthens circulatory & reproductive systems; increases energy.

Emerald, peridot, jade, malachite: Balances Vata, decreases Pitta; promotes healing, strengthens lungs; regulates nervous system.

Yellow sapphire, yellow topaz, citrine, amber: Decreases Vata; increases energy & vitality; regulates hormones; increases ojas.

Diamond, cubic zirconia, clear quartz: Decreases Vata & Pitta, slightly increases Kapha; strengthens kidneys & reproductive system; enhances ojas.

Blue sapphire, amethyst, lapis lazuli, blue topaz, turquoise, aquamarine: Decreases Kapha & Vata; clears infections, heals wounds; protects against negative energy; counters tumor growth & weight gain.

Iridescent opal: Balances Vata & Kapha; increases creativity, promotes compassion & understanding in relationships.

Blackstone, black tourmaline, jet, smoky quartz: Increases Pitta & Vata; protects against negativity.

Gold: Balances Vata & Kapha; increases tejas & ojas; harmonizes & strengthens.

Silver: Balances Pitta; cools & calms.

you ever gone to the trouble of rising early to see the sunrise on the other side of a lake or over the peak of a mountain? If you have done these things, then in my definition you are an astrologer, or at least an aspiring one."

Of course, each of has known such moments, yet as individuals we may still have a problem making the conceptual leap from the breathtaking sight of a huge, yellow moon to the mind-soothing twinkle of a two-karat topaz.

The argument for the validity of both experiences is the same, however. The subtle structures of all things, large and small, reflect the same fundamental laws of nature. The same intelligence at the basis of our nature is at the basis of all nature. This is the all-encompassing wisdom and compassion of Mother Nature. She repeats herself at every level of existence, so that if we miss her perfection in one place we are sure to find it in another.

With similar insight, Moore defines the essence of astrology: "At base it is a form of relationship between human life and the world, a relationship in which we learn about ourselves by observing the sky. Inverting the idea that we have a sky within, we could see the heavens as our interiority turned inside out. In the mysterious dynamics of macrocosm/microcosm, the sky has a soul that to some measure overlaps with our own soul."

If we think of "it" as "Ayurveda" and substitute "the color blue" or "sapphire" for "the sky," Moore's statement is no less true. All things are a looking glass for consciousness if we know how to look. All things reflect back to us an aspect of our own nature if we open the doors of perception to nature's subtle intelligence and pay attention to its effects.

SIGHTS FOR SORE EYES

A major cause of crow's-feet and dark circles under the eyes is eyestrain and fatigue. As the organs of sight, the eyes work constantly for us during waking hours to sort through the infinite stream of visual stimuli in our environment and make some sense of the world. In fact, we absorb more energy through the eyes than through other senses because of the extremely high frequency of light waves. Despite its high saturation threshhold, the eye is sensitive to the most subtle changes in energy. Scientists working with subjects kept in a totally darkened room have found that the optic nerve is able to register even a single unit of light, an infinitesimal packet of waves known as a *photon*.

In terms of Ayurveda, the eyes are a seat of Pitta. They "digest" light energy, which is the fire element. When these windows to the soul are clouded by fatigue, our experience of life becomes clouded. In general, eye problems are Pitta problems. They are aggravated by anger, frustration, tension, and worry, too much hot, spicy, sour, and salty food, stimulants, exposure to heat, dust, and smoke, imperfect imagination and distorted thoughts, malnutrition, and an imbalanced routine. When the mind is tired or strained, nothing can rest the eyes; and when the mind is at rest, nothing can tire the eyes. Therefore the best remedy for eyestrain is meditation, stress reduction, and "cleansing" and relaxing the eyes by looking at pleasant sights. This last technique is particularly beneficial for visual Pitta types, but also helps every temperament. Vata relaxes by taking a walk in the morning sun, looking at the sea, and gazing at soft candlelight or logs burning in the fireplace. Pitta is soothed by moonlight walks, seeing the sky and trees, strolling through a flower garden, or looking at art. Kapha is enlivened by watching the sunrise and going on a nature walk.

A number of other techniques to avoid and relieve eyestrain appear in Chapter 5.

CHAPTER 11

SOUND THERAPY

There is geometry in the humming of the strings.
There is music in the spacings of the spheres.

PYTHAGORAS

Imagine the sound of fingernails moving across a blackboard, a jackhammer pounding concrete, a summer breeze rustling through a grove of trees, surf breaking gently on the seashore. The effects of certain sounds on the mind and body are so specific that even our thought of them causes a predictable response.

Ayurveda uses the known effects of particular sounds to bring balance to the psychophysiology. It does this in two ways: with music and with mantras, which are sounds used in meditation for their specific vibratory quality. As in all aspects of this ancient science, the balancing influence of these sounds, whether mantras or music, differs depending upon a person's temperament and body type. If you think of the five elements in terms of their relative densities—from the hollowness of space to the solidity of earth—you can begin to understand how different constitutions might respond to various types of sound.

Audible sound has an obvious palpable quality that we can experience directly, as when we hit a tuning fork on a hard surface and then hold it to the ear. We not only hear the sound waves, we can feel them reverberating. In fact, deaf people "listen" to music by feeling its vibrations.

My client Karen experienced a clear demonstration of sound's impact when she went with a friend to see an Off-Broadway show called *Stomp*. As the name suggests, this theatrical event features a group of performers who use ordinary objects such as brooms, garbage cans, stomping boots, and slapping hands to produce an inventive, often hilarious, and always very loud array of sounds. It has no words or music other than the ever-changing tones and rhythms created by these odd percussive instruments. Not long into the performance, Karen developed a headache. For relief, she put her hands over her ears. Although the sound was significantly muffled, Karen noticed that she could feel its powerful reverberations in her solar plexus. As the show continued her headache worsened, and afterwards she felt tense and agitated. Her friend, on the other hand, was totally exhilarated.

Ayurveda has a simple explanation for the extreme difference in their responses. Karen, whom I was treating for a dry complexion, is a typical Vata type with a Vata imbalance—that is, too much air and space. Repetitive loud, booming noises with changing rhythms are naturally aggravating to her slender, literally thin-skinned physique and jittery, erratic temperament. In contrast, her friend is a thick-set, earthy type with an easygoing disposition. His classic Kapha nature is not only unruffled by hard-driving sounds, such music actually gives him an invigorating get-up-and-go feeling.

What precisely is sound, that it has this ability to produce such noticeable and varying effects on the mind and body? In this chapter, we will answer this question from the objective view of modern physics, as well as from the subjective view of Ayurveda. Of course, we will also recommend the best kinds of music to create balance for each skin type.

According to Ayurveda, there are two kinds of sound: *struck* sound, which is produced by making something vibrate and is therefore open to investigation by the objective means of modern science; and *unstruck* sound, which is "soundless" and primordial, and therefore only experienced subjectively on the level of pure consciousness. The sound therapies using music obviously involve struck sound because they begin with audible vibrations. The "inner" experience of unstruck sound comes through the practice of meditation, which we will discuss in Chapter 13. While mantras have an audible aspect, they are the vehicles to experience the silence of pure consciousness in meditation, and in fact, have their greatest potency in their unstruck value as primordial sound, as you will see.

SOUND AND THE BODY

Sound, according to modern physics, is kinetic energy, or energy in motion. It is produced when energy moves through air or any elastic medium and sets the particles of the medium in motion themselves. This motion always occurs in the same direction as the sound energy. When sound travels through air, it sets the air molecules around it vibrating, causing some molecules to compress and others to spread apart or become rarefacted. This compression and rarefaction—differences in pressure—creates a sound wave. The frequency of a sound wave, which is the number of cycles or compressions per second, gives the sound its pitch or note. High frequency or fast vibrations—the compressions are more frequent—produce a high pitch, like the screeching of a subway car in the station. Low frequency or slow vibrations—the compressions are less frequent—produce a low pitch like the rumble of distant thunder.

In the human body we hear sound when the compressed air of sound waves sets the eardrums vibrating. However, the actual range of frequencies heard by the human ear—approximately twenty to twenty thousand cycles a second—is very small compared to the range of sounds that are present around us, just as the visible light is only a narrow band within the full spectrum of radiant energy. The eighty-eight notes on a piano span almost the entire range of frequencies we can hear. Many animals have a far smaller range of hearing than humans, and others, like dolphins and bats, can hear frequencies up to one hundred thousand cycles per second and higher. By definition, these ultrasound frequencies, as well as infrasound frequencies, do not register on our auditory sense; nevertheless, like all sound and all energy forms, they do have an impact on the body.

The auditory sense is similar to the visual sense in that both respond to certain vibratory frequencies. Whereas vision responds to a single octave of frequencies (the "notes" of the rainbow"), however, our hearing spans ten octaves and is better equipped for identifying patterns. On the other hand, sound itself has much less power than light. In sixty seconds, sound energy travels approximately twelve miles, while light energy travels nearly twelve million miles. To translate this comparison into ordinary experience, the sound of a full orchestra playing loudly produces about as much energy as the heat and light from a standard household lightbulb. All in all, the ears process far fewer stimuli than the eyes, and will overload at a much lower energy level—each second, the ear transmits about thirty thousand bits of data to the brain; the eye, about a hundred million bits.

To understand how sound energy affects the body as a whole, it is important to understand the property of *resonance*. Resonance is the state of amplified vibration that occurs when a medium is struck by an external sound stimulus with the same or similar frequency as the natural vibrational frequency of the medium. That is, the interaction of two identical frequencies intensifies the sound. Resonance demonstrates the fundamental Ayurvedic principle that like energies—or like frequencies—increase like.

Because it increases the intensity—the amplitude—of sound waves, resonance can be a powerful force. Radio receivers and many musical instruments, including acoustic guitars and violins, make constructive use of it, either to produce sound or increase sound volume. When the volume of an external stimulus increases too much, however, resonance can cause the vibrating medium to break up, as when a loud sound shatters glass.

Resonance also seems to be one mode by which sound affects the skin and body. In Russian experiments, the Pacinian corpuscles—the sensory receptors for deep touch, or pressure—have shown "very definite resonance properties" when subjected to acoustical stimulation, according to Ashley Montagu in *Touching*. Pacinian corpuscles are found around muscles, joints, ligaments, and tendons—the same locations where we find many marma points, the body's subtle energy centers described in the massage therapy chapter. Rich in "resonant" receptor sites, these energy points, which are linked together by a vast network of subtle channels in the body, would naturally be sensitive to sound stimulation, as Ayurveda predicts.

Other scientific studies of sound's effects on the body indicate that the skin itself is very sensitive to sound vibrations. In one study, subjects exposed to sound waves of varying intensities on the skin were able to pinpoint the location of the sensation with great accuracy. In a study of deaf subjects, researchers found that exposure to low frequency sounds decreases skin sensitivity to tactile experience, whereas exposure to high frequency sounds increases sensitivity.

Other studies show that very high levels of low frequency sounds and sounds below the human hearing range quickly result in physical effects such as vertigo and nausea. Apparently these discomforting effects are so powerful, the military has investigated using sound waves as a potential weapon.

Of course, sounds also affect our behavior and moods. The Muzak industry and "sensory engineers" have made sound an element of interior design along with space, light, and color. They have spent much money to determine which sounds make us more likely to buy and less likely to steal in shopping malls, and more productive and less tired at work.

Clearly, we "hear" and respond to sound through the body as well as through the eardrums. Moreover, we experience different effects depending on the type of sounds we hear. Ayurveda starts with the premise that there are particular sounds that resonate with particular body types to create balance and well-being. The question is, which sounds are right for which constitution? To answer, we will look at the science of sound from the Ayurvedic perspective.

THE ORIGINS OF SOUND IN VEDA

When the ancient rishis experienced the absolute silence of unbounded pure consciousness, they experienced a paradox: Their mind was at once completely still and empty of thought and at the same time contained a fullness and dynamism of an immense order. Although this flow within the unmanifest field of consciousness was beyond sensory perception, by an inner sense the rishis seemed to "feel" its force within the body; they seemed to "see" the shapes of its movement, and "hear" its soundless vibration. They experienced this vital energy as their very Being, and as it resonated infinitely within their consciousness, the voices of the sages suddenly burst forth into spontaneous chant. The sounds of their chant were the first human expression of the primordial sounds of nature.

All sound, according to Ayurveda, emerges out of silence, not a silence that is empty, but the silence that is full—the silence that underlies the diversity of creation.

This dynamic silence, although unmanifest by nature, creates a sort of vibration or hum within itself. The highly accomplished *yogis*—the Vedic seers—experienced these impulses of intelligence in their awareness through meditation, and were able to bring them forth to the level of audible speech. They were the "scientists" of consciousness, who discerned the existence of two types of sound: expressed sound—the struck sound of the senses—which they called *shruti*, and the soundless sounds, which they called Veda.

The flow of Veda is the precise, orderly sequence of primordial sounds. From this perfect soundless "score," all the laws of nature unfold to structure and direct the development of matter and energy from consciousness. Maharishi Mahesh Yogi describes primordial sound as "the hum of the vibration produced by the self-interacting dynamics of pure consciousness as it transforms itself from one mode to another"—that is, from unmanifest Being to manifest creation.

This process by which the unstruck sounds of nature become the struck sounds of speech is described in a Vedic treatise on sound, language, and grammar called the *Panini Shiksha*. The mind, united with the Self in pure consciousness, perceives an aspect of Self, and this idea, this impulse, stimulates the mind to expression. Excited by the desire to speak, the mind stirs the force of agni—the force of transformation—which gives rise to the breath. Breath moving through the body creates a vibration in the organs of speech. Connected with the memory of experience, this vibration this—"soft humming tone"—flows out in the specific rhythms of speech, the most fundamental of which is known in Veda as the *Gayatri meter*. These rhythms reflect the patterns of intelligence in consciousness that we call natural law.

The transformation of primordial sound into speech is the basis for the Vedic sounds called mantras. The science of mantras is the science of the first sounds of creation based upon the rishis' direct experiences in consciousness. The mantras are specific sounds or series of sounds known to resonate with natural vibrational energies of the body—that is, the body's intelligence—to produce specific effects. The sequence of sounds of the Gayatri meter, for example, forms a mantra that you will learn in Chapter 13.

The basis of all disease and disorder is the disruption of these innate patterns of intelligence in the psychophysiology. Used correctly in meditation, mantras realign the disrupted vibrational patterns with the perfect sequence established in consciousness by means of resonance, and thereby restore balance to the constitution. In *The Physiology of Consciousness*, Robert Keith Wallace likens this correction process to a musical conductor rehearsing an orchestra. The conductor knows in his mind how the piece should sound and how each instrument relates to the whole; he understands the proper pitch, sequence, timing, and intensity of the notes—the perfect intelligence of the musical score—and compares what is being played to his perfect knowledge of it. If a mistake is made, Wallace writes, the conductor interrupts the rehearsal and reconnects "the performance of the orchestra with the patterns of musical intelligence in his own awareness.... This process resets the perfect sequential unfoldment of the music."

The chanted sounds of the Veda, which are the source of the mantras, are not only the language of creation; they are the music of creation as well. According to the Vedic masters, we can improve mind-body coherence by *listening* to the chanting of primordial sounds as well as by meditating with them. Again, knowledge of their meaning is unnecessary since it is the sound value of the Vedic chants that creates balance. Traditionally, Vedic chanting is performed by *pundits*, who spend their lifetime learning these sounds and their music, which must be sung precisely to be effective. Today we are fortunate to have the benefit of recorded Vedic chants so that we can enjoy their balancing influence without a trip to India.

Recently, Western scientists have started to look at these claims experimentally. Preliminary studies conducted at Ohio State University have found evidence that the primordial sounds of the Veda decreased growth of cancer cells in rats. Other research suggests a possible theory to explain how Vedic chanting and music in general could reverse cancer growth. In the 1980s, the geneticist Dr. Susumu Ohno discovered that the sequence of nucleotides in the DNA molecule formed patterns similar to those found in music. These nucleotide molecules carry the genetic codes that create life. When these patterns of biological intelligence in the DNA are damaged and the natural repair mechanisms are unsuccessful, cancer results. By converting some of these genetic patterns into musical notations, Ohno found that different genes created very different sounds, some reminiscent of Bach and the styles of other known composers. He also reversed the process, turning musical scores into DNA patterns, and discovered that a Chopin piece, for example, contained the genetic code for a human cancer. This likeness between patterns of intelligence in human life and music suggests the possibility of using these DNA-linked "melodies" to enliven cellular intelligence and perhaps repair genetic damage. Indeed, this is essentially the principle by which Ayurveda explains the balancing influence of sound.

Of course, this is such a new field of investigation that any findings necessarily break new ground and require corroboration. Nevertheless, the implications are provocative.

FROM VEDIC MANTRAS TO VEDIC MUSIC

The ancient sage Bharata, India's first and foremost authority on aesthetics and consciousness, taught that since human beings are capable of producing the primordial sounds, and since the human voice is the first "instrument," therefore all other instruments and arts should emulate the human voice. As Reginald and Jamila Massey explain in *The Music of India*, Bharata illustrated this idea in a story of a king who wants to make sculptures of deities and asks him for instruction. "To understand the laws of sculpture," Bharata tells the king, "you will have to learn the laws of painting. To understand the laws of painting, you will have to learn the art of dance. And this," he adds, "is difficult if you do not know the laws of instrumental music." Now the king becomes impatient. "Well, teach me the laws of instrumental music then!" he demands. And again Bharata answers, "To fully understand instrumental music, you must study vocal music. This is the source of all arts." The king bows to the sage and begs instruction in this sublime art.

Bharata's idea—that the human voice is the essential instrument and that all art must follow its forms—underlies the development of the *raga*, the classical form of Indian music. Although we can help to balance the doshas by listening to the music of many cultures, as you will see, we cannot do justice to the notion of Ayurvedic sound therapies without considering the nature of the music that grew directly from the Vedic tradition. We have already seen how the primordial sounds of Vedic mantras and chants are powerful tools to create health and beauty. However, all classical Indian music reflects to some degree the Ayurvedic principle that sound is the primary, and therefore most potent, expression of consciousness. The unique sound of the raga derives from this unified view of nature described by Veda.

Unlike classical Western music, which has a linear structure and development, with a beginning, middle, and end, ragas are constructed around a *swara*, a sustained expression of a single sound or note, which gives this music its characteristic repetitive quality and drone. Its one-note-at-a-time principle, which disallows the rich harmonies of Western music, mimics the capacity of the single voice. To the unaccustomed Western ear, this sound initially may seem boring, but in Vedic culture it is the music of the spheres. In contrast to the narrative aim of Western melodies, the goal of the raga is not to tell a story or to display diversity, but to express the underlying unity of all things in creation. Thus, the raga is more circular in structure, exploring and re-exploring the same note from every possible direction in order to evoke a particular emotion or mood. As Dr. Frawley suggests, the raga's lack of a rigid pattern has a "deconditioning" effect on the mind, and helps to align individual awareness with the universal consciousness.

At the same time, the notes of the raga individually and in various sequential combinations produce specific effects on the body. The seven notes of the Indian scale, called the *saptak*, correspond to the seven chakras, the seven dhatus, or body tissues, the seven aspects of existence (the five senses, mind, and soul), as well as to the seven colors of the spectrum, for example. In Ayurvedic terms, this means that these forms— these aspects of existence—share a common energetic frequency, although each form represents a different multiple, or a different "octave," so to speak, of that frequency. (This is akin to the type of correspondence we see in the Western musical scale where the frequencies of the various "A" notes, for example, are 110, 220, 440, 880, and so on up and down the octaves.) Understanding this correspondence of energies, you can see why Ayurveda states that a natural harmony exists among the many forms of creation. Indeed, all phenomena in the universe are nothing but energy in motion—that is, sound expressed in varying sequences and transformations. Thus, ragas use sound frequencies directly to balance the body and harmonize energies throughout nature.

Gandharva Veda, which is an aspect of Veda, describes how to harness music's harmonizing power. In Vedic literature, the Gandharvas were the singers and musicians who entertained at the heavenly banquets. Using the principles of this text, Vedic musicians have developed specific ragas to balance the daily and seasonal cycles of nature, as well as to enliven or calm natural phenomena such as the weather conditions. The sequence of sounds that make up the raga's melody captures the rhythms of nature. These rhythms, or natural laws, are obviously different at dawn, for example, than they are at noon or sundown. The same is true for our biological rhythms, which are also tied to the cosmic pulse. In Gandharva Veda, the day is broken down into eight three-hour segments. By performing or listening to ragas at the appropriate times of the day, we match our internal energies with nature's changing rhythms to bring balance to ourselves and simultaneously restore harmony to the environment.

A story of some court musicians and a noted vocalist illustrates the power of the ragas. Jealous of the vocalist's fame, the other musicians cajole their emperor to request a song called the "Melody of Lights," knowing that this raga creates such an increase of heat that anyone who sings it is quickly consumed in flames. One evening, the emperor requests the song. Soon, the courtyard lamps begin to burn brilliantly and the singer himself begins to heat up almost to the point of burning. A friend, realizing what is about to happen, quickly finds the vocalist's lover, who is a musician herself. Immediately, she sings a raga to bring the rains, and in this way saves her beloved.

Although of course we cannot prove the veracity of this ancient story, modern research indicates that ragas do have beneficial effects on the body and mind of the listener. In a recent study, subjects experienced a noticeable reduction in breathing accompanied by the subjective experience of inner clarity and bliss while listening to Gandharva music.

FROM RAGA TO ROCK: MUSIC TO BALANCE YOUR DOSHAS

Because they enliven all the laws of nature, both Vedic chants and the ragas of Gandharva Veda are beneficial to everyone, regardless of constitutional type. Nevertheless, different types of musical sounds do have different effects on the doshas, and can be used as specific balancing therapies, or simply to create a desired mood.

If you have dry skin, or high Vata, like my client Karen, you are prone to be irregular in your routine and also tend to experience frequent mood swings, especially when you are imbalanced. Physically, you are thin-skinned, quick, and easily fatigued. Consequently, you need slow music to balance your hyperactivity, soft sounds to counter your sensitivity, low-pitched tones to harmonize your high-strung temperament, and easy rhythms to counter your erratic nature. You will enjoy the deeper tones of the bass, cello, viola, bassoon, and saxophone, and the steady beat of soft drums over instruments with higher registers. The classical Indian forms with their melodic, repetitive quality are a good all-around choice for you, as are Gregorian chants and Baroque and early Classical forms of Bach and Haydn in the Western tradition, for example. Folk music, waltzes, and many types of standard ballroom music, all of which are highly rhythmic, also soothe Vata. In general, you will feel better tuned to the "light sounds" and "easy listening" bands on your radio than to the driving beat of rap or the harsh electric quality of hard rock and heavy metal.

If you have sensitive skin, or high Pitta, you are naturally a fiery, passionate, and active individual with lots of energy and a moderate amount of stamina. When you are imbalanced, you become irritable, angry, and sharp. You need sounds that are soothing, somewhat mellow, and fluid, with a medium pitch and moderate tempo—just slow enough to slow you down a bit but fast enough not to spark your impatience. You will enjoy the melodious sounds of the piano, the flute, the clarinet, the oboe, and horn. The music of Paul Horn or James Galway, some Benny Goodman, plenty of cool jazz, Mozart, Beethoven, and Italian opera, and moderate doses of New Age sounds will temper Pitta without tedium.

If you have oily skin, or high Kapha, your slow, heavy nature and solid build always need some stimulation and leavening. Your naturally sweet and loving temperament rarely needs mellow, romantic music to set the mood; however, you can always use some extra spice in your life. The upper registers of the piccolo and violin, large-voiced sopranos, and even the shriller sounds of rock 'n' roll are right for you. The harder the beat and the higher the volume, the better for staid Kapha. You have the stamina to dance until the cows come home, but it takes a strong, quick rhythm to get you going. You enjoy getting stirred up by the passions of the opera, Handel, Beethoven, and any loud electric music. You can handle large doses of Latin rhythms, R&B, rock, and rap when the beat is fast enough. The light sounds of jazz and New Age music have an airy quality that bouys your spirit in small doses, but their irregular tempos and soft edge do not have the oomph you need to go the distance.

Keep in mind that these are general recommendations for music listening when you are feeling imbalanced, not absolute rules. Moods change, and so do our tastes in music. As in all things Ayurvedic, you are the ultimate authority and best judge of what sounds best suit you. Use these guidelines to experiment. Choose your music carefully according to these Ayurvedic principles and see if you notice a difference. The results may surprise you.

CLEANSING THE VOICE
FOR PLEASING SPEECH

Perhaps the most powerful sound effect in life is the one we produce by our own speech. What we say and how we speak not only sends a potent influence into the environment, but also affects our own body and mind. According to Western science, the voice is the most capacious active-sense organ, spewing data into the world at the rate of ten thousand bits per second.

Vocal sound is produced by contact of the breath with various parts of the body, including the throat, windpipe, tongue, teeth, and lips, all of which subtly vibrate in order to produce sound. According to Ayurveda, these vibrations spread throughout the nervous system, affecting every cell of the body as well as all the chakras and other subtle energy centers. Of course, the throat itself is the fifth chakra, which governs expression. The quality of our expression literally shapes the quality of our life, bringing balance and bliss or discord and unhappiness. Nothing casts a shadow on our radiance more immediately than unpleasant conversation or ugly speech.

In Sanskrit, the word for sound is *nada*, which derives from *na* meaning breath and *da* meaning fire. Fire is the force of transformation, and in its primordial expression, rajas, it is also the spur to create. In other words, vocal sound is created in consciousness by the interaction of our individual will and breath acting upon the body. It is breath—prana—transformed by the urge to create. In this sense, every word we speak recapitulates the first creation from unmanifest pure Being to the manifest universe.

As the organ of speech—and the organ of action for the sound sense—the throat receives some special attention in Ayurveda. To purify the sound of the voice, the vaidyas recommend that we gargle daily upon awakening and before meditating. The box below includes the recipe for an Ayurvedic mouthwash to use for gargling. If you prefer, you can also gargle with warm saltwater. Brand-name mouthwash usually contains alcohol, which dries the mouth, as well as chemicals and dyes, and is not recommended.

AYURVEDIC MOUTHWASH

Add 1–2 drops of mint, clove, or cinnamon essential oil to ½ cup warm distilled water. Gargle.

This is by no means Ayurveda's last word on the subject of words. In the final chapters of the book, we will look more closely at the creative power of thought and speech, which, as types of sound, have the most immediate impact of any action on our consciousness. As we have said, sound underlies all things, and indeed, every form in the universe gives forth a unique sound, which is its natural frequency. Together these myriad "notes"—the movement of the wind, the waves on the ocean, the soil and bedrock under our feet, the solar system, the galaxies—create the celestial music. They are the cosmic hum—*Aum*. Likewise, the beating heart, the flow of blood, the play of muscles, the rhythm of breath all create the unique symphony that is a human life. Go with some friends to the seaside, the mountaintop, or the forest's depth. Join hands, close your eyes, and in unison sound a sustained "Aum." As your voices naturally align and your notes become one, you will experience the bliss of being an instrument tuned to the cosmic orchestra. When all sounds, outer and inner, are in harmony, life is in perfect balance.

With this taste of bliss fresh in your awareness, we invite you now to go beyond the limits of sensory life to explore the breath, mind, and consciousness.

PART IV

BEYOND THE SENSES TO THE SOUL: BREATH, MIND, AND SOUL PURIFICATION

PRANA, MANAS, AND ATMA SHUDHI

It is a beauty wrought out from within upon the flesh, the deposit, little cell by cell, of strange thoughts and fantastic reveries and exquisite passions.

WALTER PATER

CHAPTER 12

BREATHING THERAPY

Breath is the bridge which connects life to conscious-
ness, which unites your body to your thoughts.

THICH NAHT HANH

Even if my clients don't know themselves what they think or feel or desire, their skin tells me exactly what is happening in their body and mind. Then, using the five senses, I treat them externally with natural herbs and oils, and internally with diet, massage, aroma, color, and music. Sometimes this flurry of activity, this fuss of foods and herbs and *things*, inadvertently reinforces the notion that sickness and health are essentially physical phenomena. We do indeed inhabit physical bodies that require upkeep and care of a physical kind, and all the Ayurvedic therapies we have so far described do work to restore vigor and balance. However, the path to absolute beauty does not end here.

We began this book with a story of a deer who never finds the source of the heavenly scent he smells because he seeks it everywhere but where it is—within. We are just like that deer, looking in vain with our eyes for a perfection that is ultimately found *beyond* the senses, within our own nature, our Self, our soul. In Ayurveda, the way to the soul ultimately leads us through the nature of thought and perception, and then beyond mind to the experience of consciousness itself, as you will see. But this inner journey begins with the vital essence of our breath that sustains both body and mind.

Indeed, no aspect of physical existence is more central to our notion of life than breath. Without the oxygen that breath provides, the brain stops functioning within minutes, and the rest of the body, quickly after. The need for air differentiates living things from nonliving things. A single breath is the difference between life and death—we speak of "baby's first cry" and "his last words" or "her final breath" as the bookends of an existence. Indeed, life is often described as a series of breaths. We are born with a certain number in us, and when we breathe out the last one, we *expire*—even at the roots of our language, breath and life are intertwined.

PRANA: THE SOURCE OF VITALITY AND BALANCE

In Ayurvedic thought, breath is not just the air and the action of the lungs. It is the physical form of *pranamaya kosha*, one of the subtle layers of existence, and the vehicle by which prana, the life-giving energy, enters and animates the body. Prana itself is the essence of the body's intelligence and vitality; the exchange of this vital force between lovers is one reason that a kiss is such a powerful, intimate expression.

Pranamaya kosha is also the link between annamaya kosha and manomaya kosha—between body and mind. We experience this connection directly every time our breathing or moods change. The two are so bound together in our psychophysiology that we cannot alter one without affecting the other. When we are happy, the breath is spontaneously steady and full; when we are depressed, it becomes spasmodic; when we are angry, it becomes panting; when we are afraid, it becomes rapid, shallow, and irregular. Indeed, breathing is one of the first physiological functions to change when we are under stress.

Respiratory changes affect the skin and body as well as our moods: shallow breathing pales the complexion, for example; heavy breathing makes it flush. If the natural breathing pattern is continuously disrupted as a result of stress, prana becomes depleted. Without sufficient prana, which is the moving life force that along with tejas engenders ojas, the skin loses its vitality and glow. In fact, improper breathing affects immune functioning. As Weil explains in *Spontaneous Healing*, "It is the force of the breath—the rhythmic pressure changes in the chest—that pumps the lymphatic circulation," which is necessary to detoxify the blood.

Other factors besides stress that deplete prana are incorrect breathing habits; "dead" food and improper nutrition; extended periods of speaking, arguing, screaming, crying, or shouting; problems of digestion and elimination, including diarrhea, dysentery, constipation, excessive urination, and excessive sweating; and overuse of sexual energy. A blockage of the nadis, the seventy-two thousand subtle energy channels, also impedes pranic movement. When the flow of prana is disrupted over a period of time for any reason, the doshas become imbalanced, and disorder and disease inevitably ensue.

BALANCED BREATHING FOR A BALANCED LIFE

Thought and breath always work together. When the mind struggles, the breath struggles; when the mind is steady, the breath flows. But the reverse is also true: When the *breath* is steady, the mind and emotions spontaneously settle down and the physiology becomes balanced.

This correspondence is at the root of Ayurvedic breathing techniques. In the midst of emotional crisis, when the breath is halting and the mind is racing, we often try to "talk" ourselves into a calmer state. This mental technique may work temporarily, but anyone who has ever made the conscious effort knows it is hard to do—thought is ever-changing by nature, and trying to control it is like trying to stop a rushing river.

We consciously control the flow of breath much more easily than the stream of thought. "The urge to breathe is the most imperative of all man's basic urges, and the most automatic," says Montagu. Nevertheless, we can hold the breath or inhale and exhale *at will* for brief intervals, and we exercise this prerogative over and over in the course of the day, as when we blow out a candle, take an underwater dive, or simply swallow or speak.

Ayurveda's "neurorespiratory" exercises, described below, harness prana's unique integrating power not only to restore normal breathing patterns but simultaneously to soothe the emotions and harmonize the doshas.

THE NATURAL
BREATHING TECHNIQUE

What is a *natural breath*? The term seems redundant. After all, what is more natural than breathing? It is the most primal instinct. Slap the back of a healthy newborn baby and the process begins automatically and continues for life, no rewinding necessary. Right?

Well, not completely. There is a natural breath—a *right* way to breathe for maximum health and beauty—and babies instinctively do it. Just watch one as it sleeps: The upper body fills with air like a balloon, expanding and contracting in a gentle, even rhythm. It is a three-directional motion in which the chest and back rise and fall, the sides move out and in, and the torso lengthens and shortens in unison. In infancy, this natural breath is a completely effortless activity that uses the full capacity of the lungs. However, by adulthood, many of us have lost the knack because of stress, ill health, poor posture, or bad habit. We unconsciously hold the breath or constrict the body in ways that disturb the flow of air. That interruption is literally killing, because it deprives the body of the prana that is essential to all life.

A normal breath fills the lungs with half a liter of air. A deep breath draws in about seven times as much. On page 308, you will find a simple technique to restore correct, rhythmic diaphragmatic breathing. No matter what type of skin you have, you can do this exercise as often as you like, since there is never a wrong time to take a full, natural breath. However, we recommend doing it at least once a day in the morning, just to start off with an easy breath and a good supply of prana.

As you do this breathing exercise, or any of the exercises in this chapter, you may find that you start to feel anxious, or even start to laugh or cry. This is a natural response, since every breath holds anxiety. Ashley Montagu has called it "a faint phobic stir" that is reminiscent of the anxious moment before life's first breath. Ketul Arnold, who teaches yogic breathing at the Rasa Yoga Studio in New York City, points out that in addition to this primal fear, many people also unconsciously constrict the vocal cords when they speak or contract the muscles of the chest and diaphragm as a way of holding in uncomfortable feelings. As a result, the rib cage loses its flexibility and their breathing becomes restricted. When they finally take a full breath again, emotions that have been unexpressed for years start to surface—and in some cases, as he has witnessed many times among his students, they come in a flood of hysterical laughter or tears.

This is a *good* sign. It means you are breathing correctly. Try to breathe through the feelings. They will subside naturally. In the meantime, you are emptying yourself of toxins—emotional stresses that have been blocking the flow of intelligence in the body like undigested food. In fact, the breath, along with the pores of the skin, the sweat glands, urine, and feces, is one of the five natural means of detoxification. With each inhalation, oxygen-rich blood cleanses the body, and with each exhalation, carbon dioxide and other toxic wastes pass out.

PRANAYAMA:
HEAD-TO-TOE BREATHING

Pranayama is the Ayurvedic term for neurorespiratory exercise. It literally means "breathing from head to toe." The various techniques balance specific emotions and the doshas associated with them.

Some of the exercises involve alternate nostril breathing. Each nostril is linked physically to the opposite half of the brain: The right nostril connects to the left hemisphere, and the left one to the right. This connection is significant because each hemisphere is associated with different types of mental activity. In recent decades, neuroscientists have discovered that, generally speaking, the left brain regulates analytical or "linear" functions such as logic and language, and the right brain, conceptual or "spatial" functions such as intuition and imagination. Research indicates that alternate nostril breathing, by systematically stimulating both halves of the brain, promotes more integrated brain functioning overall. It also improves lung capacity, which typically declines with age, and reduces the pulse rate.

Ancient scientists also recognized the "split" functions of the cerebral hemispheres, but they used different terminology to describe them. According to Ayurveda, the left brain is the center of "male," or solar energy, which is active, calculating, and heating; the right brain is the center of "female," or lunar energy,

NATURAL BREATHING

- Sit comfortably or stand with your feet slightly apart, and breathe normally through your nose.

- Place your palms on the abdomen. Feel how your stomach expands and contracts as you inhale and exhale normally. If you do not feel the stomach moving, keep your hands where they are, gently arch backwards at the head, neck, and shoulders, and then breathe. Repeat the original instructions.

- Place your palms over the breasts with fingertips touching over the sternum. Feel how the hands naturally move apart and together as you breathe normally.

- Place the backs of your hands on either side of the rib cage with the fingers pointing upward. Feel the ribs expand sideways and contract as you breathe normally. Notice if one side is expanding more than the other. If so, lift up the arm on the stronger side, reach it overhead, bend towards the weaker side and breathe. Repeat the original instructions and see if the movement of the ribs feels more equal.

• Keeping the
hands in the same
position, move
them down a few
inches to your waist-
line. Again, feel your middle
blow up like a balloon as you
breathe normally. Notice if one side is expanding more than the other. If so, lift up the arm
on the weaker side, reach it overhead, and bend
sideways toward the stronger side. Repeat the
original instructions to see if the movement of your
waistline is equal on both sides.

• Place the backs of your hands on your midback with
fingertips touching. Feel how the hands naturally
spread apart and come together as you inhale and
exhale normally. If
the back does
not expand,
take another
breath, and as
you exhale, gently
lower your head and
bend forward at the waist until your
fingers touch the toes (or as far as you
can reach without straining). Inhale as
you slowly stand up straight again. Repeat
the original instructions to see if the back
expands.

which is creative, calming, and cooling. Consequently, left nostril (right brain) breathing cools excess fire and calms anger and frustration with its lunar energy. It is good for sensitive skin types, Pitta conditions, or simply to cool down the physiology on a hot day. Right nostril (left brain) breathing, on the other hand, stimulates the mind with its solar energy and warms up the body on a cold day. Alternate nostril breathing relieves Vata conditions such as stress, worry, and fear.

ALTERNATE NOSTRIL BREATHING
(Balances Vata)

- Sit up straight with eyes closed. Using the right hand, close off the left nostril with your ring finger. Slowly inhale through the right nostril, gently expanding the abdomen and then the lungs. Release the left nostril, and with your thumb, close off the right. Slowly exhale through the left nostril as you release the lungs and then the abdomen. Reverse the procedure, inhaling on the left nostril and exhaling on the right. Do nine full breaths: in right, out left; in left, out right equals one full breath cycle.

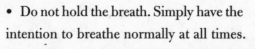

- Do not hold the breath. Simply have the intention to breathe normally at all times. Do not force the breath, but gently push the air in and out so that a dry leaf held directly in front of the nose would not move whether you are inhaling or exhaling. Do not breathe to any particular count, but let each breath continue as long as it needs to be without forcing. The torso will pulsate slowly and gently with your breath. Otherwise, the body should be still as the breath finds its own easy, natural rhythm.

- This exercise is excellent for relieving dry skin conditions and soothing anxiety and fear.

SHITALI BREATH
(Balances Pitta)

• Sit up straight with eyes closed. Curl your tongue into a tube shape and place it with the tip sticking out of the mouth. (If you cannot curl your tongue, touch the tongue tip to the roof of the mouth behind closed teeth.) Inhale through the mouth. With your lungs full, close the mouth, relax the tongue, and slowly and easily exhale through the nose. Do not hold the breath. Repeat for 1–2 minutes. Depending upon your natural rhythm, the number of breaths will vary. As a general rule, take about two times longer to exhale than to inhale; however, do not count the time.

• *Shital* means "cool" in Sanskrit. When you inhale through the nose, the body heat warms the breath before it gets to the lungs. When you inhale through the mouth, however, you can feel the coolness of the air as it enters the mouth and lungs. This cooling breath provides relief from hot tempers, hot temperatures, and all sensitive skin conditions, including rash, acne, eczema, and psoriasis.

KAPALBHATI BREATH
(Balances Kapha)

• Sit up straight with eyes and mouth closed. Place your palms on your abdomen and slowly inhale through the nose, while relaxing the diaphragm and filling up the lower lungs with air. With the rest of the body relaxed, quickly draw the stomach in tightly so that the breath is automatically expelled through the nose. Release the stomach muscles, rounding out the abdominal wall, so that breath is automatically drawn in through the nose in a reflex reaction.

Breathe in this manner for one minute only. Start out slowly, but as you get comfortable with the technique, aim to increase the number of repetitions per minute.

- With this technique, the sensation of exhaling is similar to getting socked in the stomach and having the wind knocked out. In this case, however, no pain is involved. In fact, this breath is extremely exhilarating when done correctly. Kapalbhati literally means "polishing the crown," and you may notice a tingling sensation around your ears, neck, and scalp, or a vibration on the inside of the skull as you practice the breath. The action of drawing the abdomen in tightly has the effect of massaging the spine and circulating the spinal fluid. At the same time, it gives an inner massage to the stomach and guts. Arnold notes that once you get up to speed, kapalbhati breath has a swish-swish sound like an automatic lawn sprinkler.

- This invigorating exercise is good for Kapha conditions. It will lift you out of a slump, alleviate depression, and help you to let go of emotional problems in general. It also improves digestion and provides relief from gas, so it is good for Vata as well.

CARE OF THE NOSE AND MOUTH

Pranayama strengthens the lungs and improves the supply of oxygen to the blood. However, the nose and mouth, which are closely associated with the breath, also require care if you want to keep the breath fresh and prana flowing.

Instructions for nasya, or nasal cleansing, appear on page 250. The instructions for oral cleansing follow here. Since bacteria accumulate on the tongue as well as between the teeth and gums, it is equally important to clean all parts of them if you want to avoid bad breath.

The four steps to a clean, healthy mouth are:

- *Step one.* Scrape the tongue in the morning. For those of you who have never used one, a tongue scraper is a flat metal strip about one-fourth inch wide, which is bent into a U-shape. In India, four types are commonly available:

gold scrapers for Vata types, silver for Pitta, copper for Kapha, and stainless steel for any type. If you don't have a tongue scraper, you can use the edge of a spoon. Starting at the back of the mouth, gently scrape the top of the tongue to remove the accumulated coating.

- *Step two.* Massage your gums with a mixture of 1 drop camphor, clove, or cardamom essential oil + 1 tsp sesame oil. For bleeding gums, rinse with a mixture of fennel tea + 1 pinch tumeric powder.

- *Step three.* Brush your teeth. Brand name toothpastes often have chemicals and artificial sweeteners. To make Ayurvedic toothpaste, mix 1 pinch *each* ground rock salt, cinnamon, and clove + ½ tsp triphala. Add a few drops of water to make a paste, and brush as usual.

- *Step four.* Gargle with saltwater or with Ayurvedic mouth wash (see page 299 for instructions).

The breath is the one aspect of physical life that endures beyond the body. Writer Guy Murchie has compared the number of atoms of air in an individual breath to the total number in the atmosphere, and has concluded mathematically that each breath we take contains "more than a million atoms breathed personally sometime by each and any person on Earth." We literally share our vital essence with every human being who has ever lived, and even after we die, the atoms of our breath continue to sustain other lives.

At the same time, each breath we take rises and falls out of the eternal present. Unlike our thoughts, which flow between past, present, and future, our breath occurs only in the here and now. You cannot breathe tomorrow's breath today, nor can you hold on to the present breath for tomorrow. Each breath comes in its allotted sequence, and cannot be recalled, and thereby binds us to time. Yet, each breath *in itself* is a taste of the timeless. Let the mind be always where the breath is—then you cannot help but live each moment fully.

CHAPTER 13

AYURVEDIC MEDITATION

*The Yogis, the Great Spirits, projected their minds by
an effort of the mind into this unstruck sound . . .
attaining liberation.*

THE SANGEET-MAKARANDA

Like the term "consciousness," "meditation" is a word that means many things to many people, and undoubtedly there are at least as many forms of meditation in the world as there are cultures. Meditation refers to the thinking process, and all meditations are in fact "mental" practices. However, all meditation practices are not the same either in method or result.

Westerners are most familiar with concentration and contemplation techniques. In general, concentration involves the active and disciplined focus of the attention on some particular object, idea, or goal. In Ayurvedic terms, concentration is the same as mindfulness, which we have discussed earlier and will consider further in the final chapter. Contemplation involves thinking *about* a particular idea, such as God or love, in order to uplift the mood and investigate the meaning of life. By definition, it is concerned with thought *content* and therefore engages the mind *in* thinking, since one thought naturally leads to another. As you will read in Chapter 14, both of these mental techniques are valuable in developing conscious *will* and in helping us to change the perceptions that create emotional stress. However, they do not automatically create the one effect that is the necessary condition to achieve balance, bliss, and absolute beauty—that is, the regular direct experience of pure consciousness.

In contrast to contemplation and concentration, the Ayurvedic meditations we describe in this chapter bring the mind *beyond* the process of thinking to the source of thought in pure consciousness. These techniques involve the mechanisms of thinking and perception, as you will see, but they are *not* concerned with ideas per se or their meaning, and they do not require controlling the direction of our thoughts. Rather, these ancient techniques work spontaneously to disengage the awareness altogether from the boundaries of thought, so that we can experience the absolute silence and unbounded bliss of pure Being. This direct temporary experience of pure consciousness is known in Ayurveda as *transcendental consciousness*. It is complete peace of mind.

You can find descriptions of this experience of *transcending*—of going beyond thinking mind to silent mind—in the poetry and philosophy of every civilization. It is an utterly natural, spontaneous, and therefore universal experience, characterized by a sense of inner expansion and inner calm, heightened awareness and perception beyond the senses, deepening silence, clarity, and certitude. In the poem called "Inspiration," Henry David Thoreau wrote, for example:

> *I hear beyond the range of sound,*
> *I see beyond the range of sight,*
> *New earths and skies and seas around,*
> *And in my day the sun doth pale his light.*
>
> *A clear and ancient harmony*
> *Pierces my soul through all its din,*
> *As through its utmost melody,—*
> *Farther behind than they—farther within.*
>
> *More swift its bolt than lightning is,*
> *Its voice than thunder is more loud,*
> *It doth expand my privacies*
> *To all, and leave me single in the crowd.*
>
> *It speaks with such authority,*
> *With so serene and lofty tone,*
> *That idle Time runs gadding by,*
> *And leaves me with Eternity alone.*

Such experiences of transcendental consciousness are not limited to intellectual endeavors, however, and do not require any particular level of education. It is an experience that we can have at any moment during any activity. Athletes, for example, often describe it in terms of effortless "flow" and a sense of perfect mind-body harmony. Roger Bannister, the first runner to break the four-minute mile, wrote of his winning moments: "The earth seemed to move with me . . . a fresh rhythm entered my body. No longer conscious of my movement I discovered a new unity with nature . . . a new source of power and beauty, a source I never knew existed."

Without a specific method of transcending, however, we can only hope for the chance encounter with unbounded bliss—the lucky "slip through the crack" between the boundaries of our thoughts—such as Thoreau and Bannister described. With a technique for transcending, we not only enjoy the regular experience of bliss, but also, as you will see, develop the capacity to have "bliss consciousness" at all times. Below, we will explore the practical value of this experience and then introduce you to two meditation practices that enable the mind to transcend.

THE VALUE OF TRANSCENDING

Of all the Ayurvedic therapies, the techniques for transcending are the most powerful and all-encompassing because they work directly on the level of consciousness—*anandamaya kosha*—which is the subtlest level of bodymind intelligence. By experiencing wholeness in consciousness, we spontaneously bring balance to all grosser fields of existence—ego, mind, breath, and senses. Thus, regular meditation is beneficial to mind and body alike. The mental benefits are rooted in the nature of consciousness itself; and the physical benefits, in the psychoneuroimmunological connection.

We have already described consciousness as ever-present, infinite, silent, and nonchanging. Even without having a clear direct experience of consciousness, we can infer its other qualities from the nature of our thoughts, which are born of it. All thought—no matter what its content—has the qualities of energy, creativity, and intelligence. Thought is a form of energy because it moves—we do not have one unending thought but rather a stream of thoughts that come and go—and all movement displays energy. Thought is creative because thoughts change. Although we can and do have the same thoughts repeatedly, a significant percentage of the thoughts we have each day are brand new. One thought leads to a new thought; so we say thought is creative. Finally, thought is intelligent because its changing nature takes a direction. That is, thought does not change randomly, but builds upon itself in an orderly manner. So, for example, we say to ourselves: "I want to go to the movie. What movie can I see? Let me look in the paper." If thought did not have the inherent ability to take a direction—if thought were not intelligent—we could never take action to fulfill desire. Instead, we might think: "I want to go to the movie. Have a banana. The moon is full." Of course, we sometimes do have thoughts that are "irrational," and occasionally we have thoughts that seem to be non sequiturs. Nevertheless, in the larger picture of human evolution, the thinking brain—the cognitive, discriminative, problem-solving mind—is an adaptive trait. Its very existence indicates a "higher" development. In the sense that all thinking is evolutionary—that is, it takes the direction of survival (life) and growth—it necessarily expresses intelligence.

As the source of thought, consciousness is also the source of the qualities that thought contains. We therefore can infer from the nature of thought that consciousness itself has energy, creativity, and intelligence. Indeed, since our own supply of ideas is endless—no human being ever ran out of thoughts before dying—we also can conclude that the field of consciousness is a boundless source of these qualities. On the level of consciousness, the potential of the mind is infinite.

To experience this unbounded nature of the mind is to experience a state of utter fullness. This experience is pure bliss, as we said. When we regularly and systematically give the mind this deep experience of its full potential through meditation, we eventually develop the capacity to maintain this bliss consciousness at all times. In this state, which Ayurveda calls *cosmic consciousness,* the full resources of the infinite field—that is, infinite energy, creativity, and intelligence—are at our disposal always. When we act from this "quantum" level of awareness, it is not possible to make a mistake because we spontaneously act in accord with natural laws. Action in harmony with our own nature and with our environment always brings success, and this brings joy to life. This is the fulfillment of Ayurveda and the reality of absolute beauty.

At the same time that the experience of transcendental consciousness creates bliss on the subjective level, it also produces profound effects on the body. This is implied by Ayurvedic principles, but now it is confirmed by modern mind-body medicine. As we have seen, all changes in mental experience produce corresponding physical changes, and vice versa. Transcending is a mental process in which the thinking mind becomes less and less active. As the mind goes to quieter levels of awareness in meditation, thoughts themselves become subtler and less distinct until they fade away completely and we experience the silent value of pure consciousness. In this process, as mental activity decreases, naturally so does overall biochemical activity, which is another way of saying that the body gets rest.

Scientific research on the effects of meditation—beginning in 1968 with the pioneering work of Robert Keith Wallace at UCLA and Harvard University on practitioners of the TM technique—has consistently demonstrated the improved psychophysiological functioning that results from transcending. In fact, the initial experimental findings were so unprecedented in terms of known human psychophysiological states that, as a founder of the noted Mind/Body Medical Institute at the New England Deaconess Hospital and the Harvard Medical School has said, they virtually "launched" the field of modern mind-body research. These findings included changes in brain wave activity that indicated increased rest and relaxation and dramatic reductions in oxygen consumption, breath rate, and blood lactate levels. High lactate levels are associated with uneasiness and anxiety; and low levels, with peace and calm. The lactate levels of the meditators *were among the lowest ever measured in human beings.*

All together, these results indicated that transcending produced a state of rest and relaxation that was unknown except under the conditions of sleep or hibernation. However, when further research revealed that meditation was neither of these states, Wallace recognized that Western science had recorded a uniquely different and highly significant mode of mind-body functioning. Its uniqueness rests in the fact that the deep relaxation of meditation, unlike sleep, is accompanied by wakefulness and clarity of mind. Wallace called this state of "restful alertness" the "fourth state of consciousness" after waking, dreaming, and sleep. Of course, this is the perfect description of the experience of transcendental consciousness, which Ayurveda described thousands of years before. Its practical significance rests in the fact that Western medicine now had a proven noninvasive, nonpharmaceutical, natural, easy, cost-effective method to reduce stress levels in the body to an unprecedented degree within just a few minutes of starting the meditation practice.

Although not every form of meditation produces such immediate or profound results, daily meditation can produce long-term benefits, such as reduced anxiety, reduced cigarette smoking and substance abuse, reductions in cardiovascular disease and in the factors that cause it, lower free radical activity, improved reaction time, improved memory and intelligence, improved perceptual acuity, and increased self-esteem and self-confidence. Apparently, you can gain these benefits at any age. In a three-year Harvard study of nursing home residents—average age: eighty-one— meditators showed a higher reduction in stress and blood pressure than subjects who practiced only a relaxation technique.

As mentioned, the benefits come from the regular practice of meditation. As with all the other techniques of Ayurveda, you cannot expect to achieve balance after a few meditations, or if you do it only once in a while on your bad days. Although a single session of meditation will help to relieve the immediate effects of stress and increase your energy and clarity, only consistent daily practice produces the long-term health benefits and the lasting experiences of wholeness and bliss which we have described.

The meditation techniques below involve transcending with sound, which is the most direct path to consciousness, since it is the first manifestation of it. We described these techniques in Chapter 11 as mantra meditations, which use the unique vibrational quality of mantras to take awareness beyond sensory perception to the inner experience of "soundless" or "unstruck" primordial sound. It is the combined effects of the mantra's subtle vibrations on the mind and body, plus the experience of deep inner silence when we transcend, that produce the holistic results of meditation.

SOUND:
THE MEDIUM
FOR TRANSCENDING

Silence is the basis of sound. Indeed, sound is simply the silence of pure consciousness when it is no longer silent. It is the first manifestation of Being and therefore leads back to Being most quickly.

In its subtlest aspect, sound is the primordial vibration from which all creation evolves. Actually, all forms first exist as sound, or vibration, and all sound produces form on some level. In other words, every *thing* is nothing but primordial sound, or vibratory energy, appearing in progressively denser, more manifest forms. Different "octaves of vibration" produce the different kinds of sensory experience. The seven notes of the musical scale and the seven colors of the light spectrum, for example, represent two different vibratory octaves: the first note, "do" or "C," has the same frequency as the first color, red, but in a subtler octave of manifestation. The senses are essentially "instruments" that are tuned to these different vibratory levels. Thought, which is a form of sound, is the subtlest vibratory level.

The language of Sanskrit is the pure language of Veda, and therefore the language of creation. According to Ayurveda, the Sanskrit word for an object—that is, its sound and its "thought"—and the object itself are one and the same. When you perceive the world from the subtlest level of individual consciousness—which is just at the point where vibration manifests and the first faint stirrings of thought arise in the mind—you can have the idea of a rose, for example, and actually experience the form of the rose in consciousness, complete with its sweet essence, soft petals, vibrant color, and cooling taste. At this level of existence, the thing itself, and therefore all knowledge of it, is contained within its name. Ayurveda calls this the *principle of name and form*, and it is the theoretical basis for the legendary ability of saints to manifest objects seemingly out of thin air.

Quantum mechanics describes essentially the same phenomenon in terms of wave/particle duality: the theory that waves sometime act like particles; and particles, like waves. In other words, at the quantum level there is no distinction between energy and matter—between name (vibration) and form. Ayurveda simply adds to this scientific understanding the knowledge that thought is energy, and that pure consciousness, which is the basis of thought, is the same as the quantum field. Nevertheless, we do not have to look further than ordinary life to observe, at least in some degree, the cause-and-effect relationship of name and form, or thought and matter. For instance, the mere thought of a lemon can activate the salivary glands just as if we have tasted lemon juice.

The meditation techniques below use mantras, or sounds, to transcend. Sound is the most subtle and powerful of the senses, as we said, and in its form as thought, it is the most direct avenue to retrace the mind's path back to its source in the silent field of pure awareness. In Chapter 2 we explained that in daily life we do not transcend and experience the unity of the mind all the time, because we ordinarily direct our attention outward through the senses towards activity and diversity. A key function of the mantra in meditation is that it turns the attention inward due to its primordial nature and thereby facilitates the process of transcending. Other meditation techniques (such as those involving visual images called *mandalas*) use senses other than sound to transcend, but they generally do not produce such immediate results. As the primordial vibration from which all other forms derive, sound has the power to influence every other layer of creation. Consequently, it is the most subtle and most potent of healing energies. In India, even the giant cobra can be lured to sleep with a sweet flute. Closer to home, everyone has felt sound's healing power when relaxing to music at day's end.

In the process of going beyond "struck" sound to "unstruck" sound in meditation, we experience four strata: sound we speak aloud; sound heard by the sense organ; sound heard in the mind; and finally the "root" sound, or *para*, which is expressed only on the level of prana, the life force itself. As we said, the mantra is the particular sound that takes awareness inward from the level of articulate expression to subtler and subtler levels of mental experience. When the mind transcends the subtlest sound and goes to the level of para, or the soundless sound, we enter *samadhi*, or the state of bliss. Then, the skin radiates with the light of inner peace.

GETTING READY TO MEDITATE

Although the meditation practices are different, you prepare for them in the same way. Here are the general guidelines:

Choose one time of day and one place. It is preferable to meditate each day at the same hour and in the same place. The best time is in the morning, soon after awakening and before eating. If necessary, have a light snack so you will not feel hungry. The best place to meditate is a quiet and comfortable room where you will not be disturbed. You can meditate sitting on your bed. However, your body and mind are already habituated to sleeping in that spot, so we recommend that you sit on a mat on the floor or in a chair instead, to avoid the inclination to doze. If you have the space, set up an area just for meditating. In my own home, my family has a spare room not much larger than a walk-in closet where we have a little altar, as is traditional in my culture. You may want to do something similar but in a way that honors your own traditions, or you may simply prefer to surround yourself with your favorite objects and furnishings. The purpose here is not to replace your religious practices or to impose new ones, but to create a personal "sacred" space—a place devoted to the activity of reconnecting to the Self, just as other places in the home are devoted to specific activities. Although many religions teach meditation, meditation is not a religious practice per se, but a spiritual one. Meditation and prayer are distinct activities with distinct purposes. As one astute person once described the difference: Prayer is when we talk to God. Meditation is when God talks to us.

This regularity of time and space helps create the habit of meditation so that it becomes a natural part of your routine. Paradoxically, it also lifts the experience above the mundane. When made into a daily ritual—a sadhana—meditation is a means to honor your unique existence. When you meditate, you give yourself the gift of your own time and attention, the two most precious commodities you have, and in so doing, you nurture the body, mind, and spirit. Just as regular exercise builds up the body, regular meditation also structures the experience of silence in your awareness. As you continue to meditate regularly over months and years, you will notice that the benefits of calm, clarity, and energy that you enjoy during and immediately following meditation begin to be there no matter what you are doing. This, of course, is the ultimate purpose of the practice.

Finally, regular meditation creates silence, not just in your inner life but also in your surroundings. By meditating in the same place every day, you actually change the subtle energy of that room so that you will feel a settled, peaceful quality as soon as you enter it. Amid the stress of the workaday world, your meditation space becomes an oasis of calm and light that literally rejuvenates the parched spirit.

Purify the body and throat. Cleansing the body helps to purify the mind, so do your morning beauty routine before you meditate. This should include emptying your bowels, brushing your teeth, scraping the tongue, bathing, and gargling to clear the throat, which is the organ of sound.

One of the many benefits of meditation is that it increases the production of ojas by the skin, which is why people tend to glow after they meditate; so you do not want to wash away this subtle effect by showering when you finish. Instead, try to bathe prior to meditating in the morning. If you have time, do a full body massage and then cleanse and moisturize as usual. At the very least, do your daily beauty regimen, including a shower and a light application of essential oil to the skin.

Wear clean, comfortable clothing. Tight clothes that restrict breathing, cut off circulation, or do not "breathe" with the skin can be a distraction during meditation and prevent you from relaxing deeply. When possible, meditate in loose-fitted clothing made of cotton, silk, or other natural fibers.

Create a conducive atmosphere. If you meditate at home or in a private space, make the room as quiet and comfortable as possible. Turn off the telephone and do whatever else you can to avoid interruptions. Turn off the television, the radio, and the stereo. Although music has its own beneficial effects, as you have seen, they are not the same as meditation. Listening to music keeps the mind active and the attention outward; the purpose of meditation is to allow the attention to go inward, so you can experience the silent depths of consciousness. If you prefer, use fresh flowers or incense, particularly calming scents like sandalwood and jasmine, to purify and settle the atmosphere of the room.

Do asanas and pranayama. You can enhance the effects of meditation by taking a few moments before your practice *first* to do some simple bending and stretching exercises, such as the yoga postures described in Chapter 8; and *then* to do one of the breathing exercises described in Chapter 12. These help to integrate mind-body functioning and produce deeper experiences of pure consciousness.

Sit with the eyes closed. Practice these techniques sitting up straight and with eyes closed. Since mind and body are closely related in their functioning, it is difficult to quiet one while the other is active. However, do not recline or lie down because you are apt to fall asleep. Also, do not try to meditate while you are exercising, driving, walking, working, watching TV, or engaged in any kind of activity. If you do, you will only end up creating mental strain, and you may even feel slightly disoriented. We meditate with the eyes closed for similar reasons. A large portion of sensory input to the brain is visual, so closing the eyes automatically cuts down the stimuli and helps to settle the mind and body.

Be flexible. Notwithstanding all we have said about the importance of regularity, cleansing, proper atmosphere, and breathing, *if you do not have time to do all parts of the preparation routine or if you cannot meditate every day at the same time and place, you are better off to meditate under any conditions than not to meditate at all.* Just try to find a quiet, comfortable spot, and avoid doing it right after a large meal or before bedtime. Although sitting in the "lotus" posture with the legs crossed over each other is beneficial to the clear experience of consciousness, you can meditate successfully in an ordinary sitting position with the eyes closed, as you will see, so there is no reason you cannot do it even while sitting on a bus or train, in a park, a museum, a building lobby, or in any public place. To the onlooker, you simply appear to be napping. Even if the environment is noisy, you can still meditate—noise is just one other form of sound, or thought, that the process will transcend.

MANTRA MEDITATION

Mantra means "the thought that liberates and protects." It liberates in the sense that the mantra is the thought that takes the attention inward and enables the mind to transcend. By transcending, we go beyond the boundaries of thought, which fracture the wholeness of Being, to experience the freedom of our own unbounded nature. The mantra protects in the sense that the proper mantra used correctly always creates healing and balance in the psychophysiology. In fact, mantras are Sanskrit syllables used in meditation for their *known* sound value—that is, their vibratory quality. As we said, every sound produces a specific effect on the physical level, so it is very important to use a sound that is right for your constitution and lifestyle. The wrong sound does not protect, and it may even create imbalance.

Indeed, mantras have their greatest effect on the mind and body at the most subtle levels of mental experience where awareness itself is most powerful. All sound activates each cell of the body, and different vibratory patterns create different effects. Consequently, you want to be sure that you are using a sound that has been proven to have a soothing, purifying, and balancing effect at the deepest levels of thinking. Although Western science has been able to demonstrate the health benefits of transcending, it does not have the knowledge or technology to fully understand the process by which meditation accomplishes this goal—which does not occur on the level of the body, but on the level of intelligence where vibratory energy has its greatest effect. Proponents of relaxation techniques will argue that it does not matter what sound you use to meditate—some even recommend counting numbers—but these are statements based on ignorance of the primordial nature of sound. Two of the most important contributions of the Vedic tradition to meditation are the knowledge of the proper selection of mantras, based upon the authority of the ancient rishis—who understood from firsthand experience the relationship of name and form—and its thousands of years of proven experience in their use.

On the next page you will find two mantra meditations. As you will see, the instructions are simple. However, meditation itself is a rich, changing, and evolving experience that varies from individual to individual and from meditation to meditation, because of our different constitutions and the state of the psychophysiology at the time of practice. This is why in India we learn meditation from *gurus*, or spiritual teachers. Experienced in the practice of meditation, they are expert guides to the unfamiliar territory of consciousness. Without someone to answer questions about experiences and to provide new information based on those changing experiences, many Westerners find it difficult to keep up their meditation practice, since they do not fully understand the process. Unfortunately, in a book, we cannot address the almost infinite variety of thoughts, feelings, images, and sensations that may arise during meditation. Therefore, we have kept the instructions basic and the guidelines general. If you are getting full benefit from your practice, by all means find an experienced teacher. You may also find that you enjoy the increased silence that naturally occurs when you meditate in a group.

THE BIJ MANTRAS

The *bij* mantras are one-syllable sounds that we use to balance the five elements and the five energy centers, or chakras, that are associated with them. These "seed" sounds are derived from more complex combinations of primordial sounds, and they have no meaning. The bij mantras and their effects appear in the following table:

MANTRA	ELEMENT	CHAKRA	EFFECTS
Lum (or Lam)*	Earth	1st	Creates stability, groundedness, joy, life. Governs reproductive system.
Vam	Water	2nd	Governs lower abdomen, kidneys, the elimination process. Reduces swelling due to water retention.
Ram	Fire	3rd	Stimulates solar plexus, the power center. Increases agni (digestive power), willpower, perception, and orderliness.
Yam	Air	4th	Stimulates heart chakra; increases love and compassion.
Ham	Space	5th	Stimulates throat chakra. Improves speech, communication, success.

Pronounce the letter "a" like the vowel sound in the word "the." Pronounce "u" as in "put."

TO USE THE MANTRAS
- *If you have no particular imbalance at this time,* use all five mantras in order, beginning with the first chakra sound. Meditate for about 15 minutes once a day, spending about 2–3 minutes on each mantra. The exact timing is not

important, in the sense that you do not have to worry about counting the minutes. Do not use an alarm. However, you can time yourself by occasionally glancing at your watch or clock.

- *If you have a particular imbalance*, spend the full time of meditation on one or two mantras, depending upon your imbalance. Do this for 3–4 days in a row; then go back to using all the mantras, as above.
- Get ready to meditate as instructed above.
- With eyes closed, let your attention go easily to the chakra, and after a few seconds begin to repeat the mantra aloud *softly* and at a *natural pace*—not too fast or slow. Say the sound, then wait for it to settle and come to mind again. As you repeat it, let your voice get gradually softer and softer until the sound becomes a voiceless whisper and then just an unspoken thought. This process may happen slowly or very quickly. Do not force it in either case, and do not try to keep up with any particular rhythm. There is no one "right way" it is supposed to happen.
- Any time you forget the mantra or notice that you are distracted by noise, thoughts, feelings, or sensations, just *gently* reintroduce it in the mind (you do *not* have to start again *aloud*). This may never happen, it may happen once or twice, or it may happen countless times. Do not strain to focus on the sound and do not try to control your thought process. Such behavior by definition occurs in the field of *doing*, while consciousness is the field of *nondoing*—the field of Being. Therefore, strain and effort only take the mind away from the fulfillment of unbounded silence. Take the attitude that whatever happens, happens.
- When you finish your meditation, keep your eyes closed, and sit or lie down for two minutes before resuming regular activity. Just as getting up too quickly from deep sleep can feel jarring, so too can getting up quickly from meditation.

THE GAYATRI MANTRA

The Gayatri mantra is a twenty-four-syllable mantra that contains within the sequence of its sounds the entire spectrum of human knowledge—that is, Veda. When we repeat this mantra, the sound of each syllable by means of resonance enlivens one of twenty-four corresponding energy centers in the body. Although the words of the Gayatri mantra have a meaning, as you will see, we do not repeat them in order to contemplate any idea. Intellectual understanding has nothing to do with a mantra's effectiveness. In fact, for the purpose of transcending, we do not want to engage the mind in the content of thought at all. Rather, we are only concerned with the vibratory quality of the sounds themselves and their balancing effect upon consciousness.

The sounds of the Gayatri mantra are written in Sanskrit—the language of Veda. Their transliteration and translation appear below:

Tat savitur varenyam bhargo devasya Dheemahi dhiyo yo nah prachodayat. *

"O god, Thou art the Giver of Life, the Remover of pains and sorrows, the Bestower of Happiness; O Creator of the Universe, may we receive Thy supreme, sin-destroying light; may Thou guide our intellect in the right direction."

* Pronounce the letter(s): "a" as in "father;" "e" as in "heyday;" "ee" as in "deep;" "i" (appearing in the middle of a word) as in "it;" "i" (appearing at the end of a word) as in "bite;" "o" as in "no;" "oo" as in "poor;" "u" as in "put."

To meditate using the Gayatri mantra, repeat the complete Sanskrit mantra for about 5–10 minutes using the same method given for the bij mantras—with one exception. Do *not* begin the Gayatri mantra with your attention on the chakras but simply on the sound itself. You can also say the English version as a prayer or use it as a contemplative meditation.

As you continue to practice meditation regularly over a period of time, you will notice a deepening of your experiences both in and out of your practice. That is, inner silence will continue to increase when you are meditating, and the sense of peacefulness and inner strength that you feel in your practice will be carried forth more and more into daily activity. Ayurveda teaches that pure consciousness is infinite and eternal. Therefore, there is no limit to the growth you can experience as a result of meditation. As your experiences develop, you will want to speak to teachers and other students of meditation, to understand the many levels of development of consciousness—from transcendental consciousness, to cosmic consciousness, to the sublime state of unity consciousness, in which all knowledge—Veda—is there for you to know.

CHAPTER 14

SOUL PURIFICATION
AND THE PRINCIPLES
OF ACTION

*I become what I see in myself. All that thought suggests
to me, I can do. All that thought reveals in me, I can
become. This should be man's unshakable faith in himself
because God dwells in him.*

SRI AUROBINDO

When we know our purpose in life, thought is clear and action is clear—body, breath, mind, and spirit work in unison. This is what we mean by balance, which is necessary for healthy, glowing skin. When we are uncertain of what we want or where we are going, the mind, hobbled by doubts, becomes distracted and confused. This mental "noise" is actually a stress that divides the mind: Physically we are here, but our thoughts are elsewhere. As a result, our whole physiology becomes upset. The breath becomes uneven, the heart pounds, the biochemistry changes. We lose the quiet value of consciousness, and become caught up in a vicious cycle of fear, strain, and imbalance. With our energies diffused and our mind unfocused, action becomes less efficient and therefore less likely to achieve its objective. Without success in action, we can never attain the inner fulfillment that is the very essence of absolute beauty.

Along with daily meditation, having a clear purpose in life is essential to the process of soul purification—the ultimate step on Ayurveda's fourfold approach to balance. If worry and confusion lead to imbalance, then clarity of purpose, which harmonizes thought and action, is the secret to health and radiance. We have seen already how meditation is the most powerful and direct means to release deep-rooted stress and unify mind and body. By taking our awareness beyond the thinking process, meditation creates the experience of wholeness, and thus automatically frees the mind from fears and worries—it literally cleanses the spirit. From this state of equanimity, we naturally cope with problems more effectively because

/ 330

we have greater clarity, energy, creativity, and focus. At the same time, we make fewer mistakes, so we automatically create less stress for ourselves. Nevertheless, until the unbounded nature of the Self is established permanently in our awareness—that is, until we ourselves are rishis—we are apt at times to feel overwhelmed by life's problems. In these moments, we need additional knowledge and skills to handle the challenges and to change patterns of thinking and behavior that impede success. On the surface, daily existence has its ups and downs, even for the enlightened. If the commuter train is delayed, everyone on it will be late for work, no matter the person's state of consciousness. Mundane events do not necessarily change as the qualities of consciousness grow in our awareness. What changes is our perception of events, and in that change of perception lies freedom from stress and disease, and our only chance at unshakable happiness.

In this final chapter we will look at the role of perception, conditioning, and fear in the stress response, and at the beliefs that breed fear. We will also describe eight principles of action for cultivating a strong sense of purpose and a totally new and liberating way of viewing the events of your life. With these basic tools for changing self-defeating habits and attitudes that induce stress, you will be able to structure a fearless, wise, and powerful way of life—the way of Chymunda and the path to absolute beauty.

STRESS IS A POINT OF VIEW

"One of the greatest discoveries of my generation was that human beings can alter their lives by altering their attitudes of mind."
WILLIAM JAMES

Stress is the leading cause of problem skin, aging, and disease. But what is it? As we said in Chapter 3, no event in life is intrinsically stressful; rather, it becomes a stressor because we assign certain meaning to it based on our knowledge and conditioning. To a New Yorker lost in the woods, an encounter with a snake can be mortally frightening; to a snake charmer, it's all in a day's work. The stress response is just that: a specific physiological reaction precipitated by a thought. What triggers it is not the objective circumstance, but our subjective interpretation. If we can change our thought about the circumstance, we change the body's response.

Scientists estimate that the average person has about sixty thousand thoughts a day. Of these, ninety percent are the same thoughts as yesterday's. When we "change" our minds—that is, when we break out of old patterns of thinking, or transcend thought altogether through meditation—two things happen to free us from disease. One, our conditioned responses get short-circuited, and the same old stressors no longer cause us to react in the same old way. Two, our neurochemistry spontaneously changes. The fearful New Yorker and the fearless snake charmer send very different chemical messages to their cells based on the differences in their points of view. One message can create poisons in the body, the other bliss.

The matter that composes the human body is recycled through the universe constantly. The dust from which we are made is not the same exact dust we leave behind when we die, but is continually exchanged during the course of our lifetime via biochemical reactions for other cosmic dust—the subatomic particles that dart unseen through the air. At the same time, ninety-eight percent of our cells die and regenerate annually; skin tissue itself regenerates every month. This means that on the material level, each of us is brand new from year to year (although our replacement particles are as old as the universe). The reason we continue to look the same is that we continue to think the same. Until we change the memory of the cells—the intelligence that orchestrates growth—we cannot change its manifestation.

This understanding is consistent with the Vedic teaching: What we think, we become. The lesson of both ancient and modern mind-body science is that mind and emotions—that is, perception and meaning—are as important to health as proper diet and water. Once we have understood this truth, we ultimately cannot avoid asking ourselves the basic existential questions: What *do* I think and feel? Why do certain events upset me? What would make my life less stressful? What would make me happy? What is my purpose? Who am I? As you will see, in the answers lie freedom from the effects of stress.

The stress response is the body's preprogrammed reaction to fear. In prehistoric times, when clubs and cunning were our only weapons, the natural world presented a real and constant threat to human life. To survive attack by man or beast required great physical strength, either to ward off the assailant or run to safety. Fear was the brain's signal to produce the adrenaline rush and other heightened psychophysiological responses that enabled our forebears to prevail. Our problem in modern times is that this primitive mechanism is like a well-trained fireman: It responds with the same efficiency to every alarm, even if it is a false one. Today few of us are called upon to defend ourselves against actual physical attack, yet many of us exist in a perpetually reactive state because of the pace and pressures of contemporary life. Since the bodymind cannot distinguish between real and imagined fears (if it could, we would not wake up from a nightmare with our heart pounding) we have to find new ways to adapt to this overload of demands if we want to avoid its damaging effects on the skin and body. One of the most effective ways to do this is to develop new habits of thinking and action.

"[A] perception," says Ashley Montagu in *Growing Young*, "is a sensation which has been endowed with a meaning." As we said in the very first chapter, we are born with the capacity to recognize two basic sensations: pleasure and pain. One we instinctively react to with contentment, the other with aversion—that is, with balance, or with stress. Later we learn to call these opposing feelings by many names: calm and anxiety; satisfaction and frustration; happiness and unhappiness. But all of them are simply subtle variations of the primal human emotions: love and fear. Each of us tends to convert pain and fear into specific emotional responses depending upon our innate constitution, as explained earlier. In fact, all negative mental traits—depression, rage, dishonesty, doubt, envy, possessiveness, and so forth—arise out of fear, or lack of the experience of wholeness, which is essentially the same thing.

FROM FEAR TO FREEDOM

"The attention of most people is fixed solely on the cure of bodily inharmony, because it is so tangible and obvious. They do not realize that their mental disturbances of worry, egotism, and so on, and their spiritual blindness to the driving meaning of life are the real causes of all human misery."

PARAMAHANSA YOGANANDA

What do you fear? The loss of youth? The loss of a loved one? The loss of money? The loss of esteem? The loss of control? Or the loss of life itself? Whatever your secret terror, know that all fears arise from the same source: the mistaken perception that we have something to lose at all.

All emotional stress is rooted in this mistake of the intellect. The true nature of the Self is infinite and nonchanging consciousness, and being wholeness itself, nothing can be added to or subtracted from its existence. This Self is what is born into the world singular and complete, and leaves the same way. "Our materialistic bias forces us to keep looking at molecules as the source of life," writes Deepak Chopra in *Ageless Body, Timeless Mind*, "disregarding the fact that a newly deceased body contains precisely the same molecules as it did before it died, including a full complement of DNA." Indeed, when I die, my body will be right here, but the doctor will pronounce: "She's gone." *Who* is gone? Something does leave the body, and what leaves *is me*. *I* did not bring anything into the world and there is nothing I take when I leave except *who I am*. Everything else I think I own is rented property, including the body. When you give up your "materialistic bias," you start to appreciate the wondrous unity and fluidity of existence. You are not the body but the bodymind, not cells but cell memories, not molecular structures but the intelligence behind structures, not fleeting quantum particles but indestructible vibratory consciousness that assumes infinite forms. All things do pass, but *you* are eternal. Therefore, what is there to fear?

The nature of consciousness is *ananda*—pure bliss. The nature of fear is *maya*—illusion. Fear is a misperception of who you really are and what your purpose in life is. When your perception and purpose are clear, the light of consciousness dispels fear, just as light dispels dark. Without fear, there is no stress arousal, and without constant stress arousal, the major cause of aging and disease is removed.

Clearly, Ayurveda offers a perspective that is as different from the materialistic view of modern Western culture as quantum mechanics is from classical physics. These most ancient and most modern of sciences agree on a common idea: Human consciousness influences the state of the physical universe—that is, our thoughts create our reality. In Veda, a corollary principle states: Whatever we put our attention on, grows. These ideas are the basis of a different kind of approach to stress reduction and beauty, using techniques of mindfulness—the active value of consciousness and the conscious value of action. Mindfulness is intentional attention—a moment-by-moment choice of mental focus.

Human beings are distinguished by our capacity for free will. We tend to think of this as the ability to *do* as we want. This is not the case at all. We do not have unlimited freedom of action in the sense that all ordinary human actions (miracles are a special case) are subject to and limited by physical laws. We are free to act, but only within the constraints of those laws; for example, we cannot walk from the earth to the moon. Fundamentally, free will is not the the capacity to act as we choose but to *mind* as we choose—that is, to apply the power of our attention—our consciousness—however we desire. This capacity is truly without limit. We can on the level of consciousness fly without machines, travel faster than light's speed, and even contemplate the mind of God, if we desire. We can imagine the worst evil or the highest good. We can conceive the inconceivable. In fact, the capacity to choose a different thought or a different purpose at every moment is the one absolute freedom we have. Our own conscious attention is also the one aspect of our existence over which we exert complete and final authority. Yet we constantly relinquish this power by acting without mindfulness, by allowing destructive thoughts and feelings to run our lives, and by passively accepting given circumstances and learned perceptions or meanings as ultimate truths. We stop questioning, we stop looking, we stop wondering, we stop listening, we stop trying, and we settle for a very narrow scope of possibilities. Moreover, we cling fearfully to the small vision we have, believing falsely this is all we can hope to be.

We do this not only as individuals, but as communities and societies as well. The authors of *Thinking Body, Dancing Mind* tell the story of Roger Bannister's record-breaking run as a stunning example of the limiting power of collective beliefs. Before Bannister broke the four-minute-mile barrier in 1954, fifty medical journals had published research "proving" that humans were constitutionally incapable of running this fast; and in the sports world, this was accepted fact. Yet, within eighteen months after Bannister achieved the "impossible," forty-five more athletes did the same thing. Did all of these runners just happen to better their performance in that brief span of time? "A more likely explanation," write Huang and Lynch, "is that once the four-minute barrier was broken, they all believed it could be broken again."

Ayurveda offers an alternative life strategy based not upon the knowledge of molecules, but the knowledge of consciousness, which is the field of all possibilities. This vision teaches us to make maximum use of our freedom to choose our thoughts, our beliefs, and the focus of our attention, with the understanding that what we put our attention on grows. If we let the mind dwell in misery, life will be misery. If we exercise our power of choice—the power of all possibilities—we can transform even misery into miracles.

THE STEPS OF SOUL PURIFICATION

"Soul doesn't pour into life automatically. It requires our skill and attention."

THOMAS MOORE

No one has ever said that it is easy to change lifelong perceptions and conditioning. Some experts suggest that you have to replace an old behavior with a new one consistently for at least twenty-eight consecutive days in order to break a habit. Nevertheless, for many people, it is not lack of desire or discipline that stands in the way of change, but lack of skills, knowledge, and understanding. It is one thing to recognize that your current coping strategies do not make you happy or healthy. It is quite another thing to figure out a new and improved approach. If a better alternative were that obvious to you before now, chances are you would have tried it already.

The principles of action below are an approach to experience based upon mindfulness and the exercise of our free will. If you apply these principles to your daily activities, self-defeating thoughts and habits will fall to the wayside with greater ease; self-confidence and self-expression will flourish; body, breath, mind, and spirit will be in perfect accord; everything you do will become an opportunity for growth and bliss; and the abundance of the universe will flow your way. A life lived on this basis is truly one of grace and beauty.

KNOWLEDGE IN ACTION

"Truly there is in this world nothing so purifying as knowledge. . . . "

BHAGAVAD GITA

"To cultivate a beautiful soul is to live in defiance of all that is drab in life. It involves the need to know oneself and one's surroundings."

MULK RAJ ANAND AND KRISHNA NEHRU HUTHEESING

Knowledge, the aspect of Chymunda represented by the goddess Saraswati, is the first and most important principle of action, because any action performed without it is doomed, at best, to fail, and at worst, to bring harm. Knowledge, in the Vedic sense, has two aspects: Self-realization, which is knowledge of the unchanging value of consciousness achieved through regular meditation, and self-reflection, which is knowledge of our changing beliefs, desires, and behaviors, achieved through the intentional examination of our thoughts, feelings, and actions. Both are necessary to achieve the goal of soul purification.

Self-knowledge is the basis and goal of all human experience. In the *Bhagavad Gita*, Lord Krishna tells the great warrior Arjuna, "Established in Being, perform action." When we act not from a state of disquiet and distraction, but from the silent level of pure consciousness, then we do not make mistakes, which are the source of human misery. We cultivate this balanced awareness during the practice of meditation and help to maintain it by balancing the body and emotions through the many other tools of Ayurveda.

It is important to understand that the inner state of equanimity is not a mood you create intellectually. Rather, it is the result of a particular state of bodymind functioning that has its own unique set of biochemical and neurophysiological markers, as the first research on transcendental consciousness showed almost thirty years ago. Techniques such as affirmations and visualizations are useful to help alter perception, but alone they do not create this holistic experience of transcendence, which is the prerequisite of bliss. If you do not take practical steps to balance the body and ease the emotions, the constant effort of trying to think positively when you feel terrible can cause stress and strain in itself. You can reduce stress levels through positive thinking, but you heal the bodymind only by going beyond thinking in meditation. This direct experience of inner wholeness in meditation is the knowledge that purifies. You cannot think your way into enlightenment, which is beyond thought. Nevertheless, as you will see, you can think your way away from stress, which impedes the highest enjoyment of life.

The aspect of knowledge that we call self-reflection is necessary to overcome the fears and conditioning that make us vulnerable to stress. Before we can change our emotional responses, we first must know what they are. Self-reflection enables us to bring unconscious feelings into consciousness to increase our capacity for free choice. Conditioned responses have the immutable pull of gravity on the body, holding us to one view of the world and one set of behaviors. For example, the person who has learned to believe she is unlovable will hear every criticism as proof of her unworthiness. This belief precludes various other possibilities: one, that mistakes have no bearing on individual worth, and are in fact a natural part of learning and growing—the person who never stubs her toe is the person who never takes a step forward; or two, that something positive could be gained from hearing about a different approach to a problem, and that the person who has taken the time to point it out may have done so because he or she loves you, not because you are unlovable; or three, that a criticism is just another person's opinion, which may or may not be an accurate view, and does not determine who you are in any case. Any of these three alternative beliefs would render the situation stress-free. On the other hand, the belief of unworthiness is self-perpetuating—indeed, it is self-fulfilling, as is all belief—because it is an attitude that always fosters stress. In turn, stress lowers immunity and leads to sickness, which then becomes proof positive that our self-concept is right: we are weak and unworthy.

When we begin to pay attention to and question our deep motivations and unspoken attitudes, we break the cycle of negativity. Self-reflection involves three processes: attunement to our inner voice, which is the intelligence or knowingness inherent in individual consciousness; openmindedness, which is the willingness to observe ourselves honestly, to see other points of view, and ultimately to try new behaviors; and, most importantly, forgiveness, which is giving unconditional permission to ourselves and others to be imperfect—that is, the willingness to overlook mistakes without judgment.

By developing mindfulness, we expand possibilities. In *Growing Young*, Montagu defines intelligence (as opposed to instinct) as "the most appropriately successful response to the challenge of a particular situation." Conditioned responses, like instinctual responses, temporarily usurp our powers of heart and mind—our capacity to discern and elect the response that would create the greatest benefit to ourselves and others in every circumstance. To work towards becoming mindful in all situations is to work towards living the full potential of human existence. An unexamined life is not necessarily unworthy, but it is necessarily less than everything it could be.

Self-reflection creates true personal power—that is, power independent of position or money. Whether we are conscious or unconscious of our behavior, we are always responsible for its effects. That is the law of *karma*, or action. When we behave unconsciously and the results of our actions are unpleasant, we tend to feel like victims of circumstance—being oblivious to our own motivations, we do not see the central part we have in the outcome of events. Even when the outcome of unconscious behavior is favorable, we tend to feel we were just lucky rather than deserving—and we remain utterly fearful because, not perceiving ourselves as the creators of our luck, we believe it can be taken from us as magically as it came. The choice to live mindfully is a choice to give up victimhood and to accept full responsibility for all outcomes in our life. That is a very powerful place to stand (the creative power of human consciousness *is* great), and great power entails great responsibility. Many individuals would rather sleepwalk through existence—they would rather remain unconscious—than take responsibility for themselves and their happiness. With the power of the king comes the responsibility for the kingdom. Out of fear, many people remain lifelong subjects. Nevertheless, we have the capacity to choose personal power and lose fear at any moment.

By definition, the path of self-reflection is a uniquely personal journey; yet it is very hard to travel completely on our own. Indeed, it is often impossible to gain a clear picture of ourselves without the mirror that others provide. Counseling, support groups, workshops, spiritual teachers, loving family and friends, and at times even strangers are all sources of information and guidance on this path. Scripture, mythology, philosophy, history, stories, art, science, theater, film, books, or any medium that explores human nature is also a place to find a mirror if you are willing to look deeply and reflect.

Observation of the world itself is a valuable tool of self-examination, since everything in nature is an expression of the same consciousness. In a Vedic story, a king learns this lesson when he asks a great saint how he achieved his sublime state of life. The saint says that he has studied at the feet of twenty-four gurus, or masters. In Vedic tradition, saints typically have but one master, so the king is greatly surprised. The saint explains: "The universe is full of teachers if you are mindful, reflect on what you see, and apply the knowledge to your life." Then he introduces the king to his "gurus"—earth, air, sky, water, fire, sun, moon, dove, python, ocean, river, moth, honeybee, honey gatherer, elephant, deer, fish, maiden, courtesan, osprey, arrowsmith, snake, wasp, and spider—and tells the lesson he has learned from each.

Of course, the skin, which is the mirror of the soul, is the first "guru" on the path of self-knowledge. As you have seen, you can discover the inner forces that drive you, as well as the aspects of your emotions that need to be balanced, simply by observing the subtle changes on the complexion. Indeed, Ayurveda offers the most fundamental lesson in self-reflection and self-knowledge, in teaching us how to recognize our basic nature and innate tendencies, and how to live in harmony with them—that is, how to take care of the Self and minimize suffering. In this regard, Ayurveda teaches us to balance all levels of life—that is, to attack problems from all angles. This implies that we should take advantage of all stress-reduction tools that are available, including medical interventions. (The vaidyas invented surgery, after all.) Drugs are not cures, but used wisely, they can be effective tools to reduce symptoms of pain, depression, and anxiety. In some instances, physical discomfort or mental instability can render holistic health strategies virtually useless, because our attention is so distracted—and bodymind healing requires conscious attention. In the long term, drugs have toxic effects, but in the short term, they can relieve physical and mental suffering. Over time, the pain and fear created by unrelieved suffering are just as poisonous to the body as medication, so you gain nothing by eschewing drugs that could ease your distress. Once the acute symptoms have diminished, you can work towards balancing the mind and body. Rigidity of any kind—including rigid ideals of purity—has no place in Ayurveda because it is the antithesis of balance and wisdom. Fanaticism, whether materialistic or Ayurvedic, always blinds, and can cause us to overlook the very truth we think we have found. Indeed, to be wise is to know the right questions, not the right answers.

This idea is illustrated by the tale of a self-styled holy man who refuses to leave his house when a flood threatens his town. As the waters rise in the street, the police drive up and urge the man to ride with them to safety, but he sends them away, saying, "I have no fear. God will take care of me." When the water is up to the second floor, rescuers appear in a rowboat. Again they plead with the man to leave, and again he refuses, saying, "God will take care of me." When the water is at the roof, rescuers come in a helicopter. For the third time, they beg the man to escape to safety, and for the third time he waves them off. Finally the floodwaters peak, and sure enough, the holy man is swept off the roof and drowns. When he gets to the heavenly gates, he is furious. "How could you let me drown!" he cries to God. "I believed you would save me!" "What are you talking about?" God replies. "I tried to save you three times, but when you refused the car, the boat, and the helicopter, I thought you wanted to die."

We can miss even a miracle if we do not keep our minds open to see the wisdom sent in all things. If the divine is omnipresent, then even a speck of dust carries a holy message. In *Song of Myself*, the poet Walt Whitman wrote of this complete willingness to pay attention and see every gift in the present:

> *Why should I wish to see God better than this day?*
> *I see something of God each hour of the twenty-four,*
>   *and each moment then,*
> *In the faces of men and women I see God, and in my*
>   *own face in the glass,*
> *I find letters from God dropt in the street, and every*
>   *one is sign'd by God's name,*
> *And I leave them where they are, for I know that*
>   *wheresoe'er I go,*
> *Others will punctually come for ever and ever.*

Inward reflection and the knowledge it brings are preparation for action. It is the necessary pulling back of the arrow on the bow in order to hit the target—the retirement within ourselves—that is the foundation for dynamic action in the world. Both knowledge and experience are necessary for progress, however, so it is important not to get lost in the inner search. The test of what you know and who you are occurs in the world of action, and in that sense action itself is the ultimate teacher.

PURPOSE IN ACTION

"If we think of defeat, that is what we get.
If we are undecided, nothing will happen for us.
We must just pick something great to do and do it.
Never think of failure at all,
for as we think now, that is what we will get."

MAHARISHI MAHESH YOGI

"Yea, all those that have ears to hear, listen. For I tell you this: at the critical juncture in all human relationships, there is only one question: What would love do now? No other question is relevant, no other question is meaningful, no other question has any importance to your soul."

NEALE DONALD WALSCH

Clarity of purpose, as we have seen, is essential to soul purification because it frees us from the inner turmoil that divides the body and mind. If knowledge is the ocean that upholds action and gives us clear vision, intention is the rudder that gives direction and keeps us steady on the course. Without it, we are at the mercy of the tides. Purpose is the aspect of Chymunda symbolized by the goddess Lakshmi, who represents individual will, desire, expansion, dreaming, wishing, imagination, and abundance.

In the larger sense, clarity of purpose refers to the spiritual meaning of our life—the answers to the questions: Why am I here? Who am I? Such answers are rarely easy or simple. They may take a lifetime to find, or they may change many times within a lifetime. We cannot advise what unique purpose you will find, but only how to approach the search—and how important it is to your health and beauty to be mindful of the questions.

Such questions naturally arise from the process of self-reflection, which often reveals not only our unconscious beliefs but also our deepest desires. In many cases, the goals we pursue in life or the reasons we pursue them have less to do with our own choice or sense of self than with the early conditioning we receive from our family and culture. Of course, the dissatisfaction many people feel stems from an inner sense that their lives do not express or nurture their deepest selves. By reflecting upon our thoughts and actions, our purpose usually starts to become clear. This revelation does not necessarily require a change of direction, lifestyle, or career. Wherever we are, we can find a means to fulfill our purpose without suddenly abandoning other responsibilities. However, living according to our purpose usually requires that we reorder priorities, make attitudinal shifts, or redefine our reasons for doing what we do. By making these adjustments, we bring thought and action into accord, and thereby save ourselves from stress.

Finding and living your purpose is a goal and process unto itself. In the culture of instant gratification and quick fixes, however, few of us have the patience to let this process unfold naturally. We think we have to have all the answers and proofs now. We cannot tolerate the state of not knowing. Few people are willing to admit, even to themselves, "I don't know" or "I feel lost." It is hard to be the one standing still when everyone around you seems to be rushing by so purposefully. Yet sometimes this is exactly what is required if you are to discover what you want to know. In a Hasidic tale, a young man hurrying down the street is stopped by a sage who wants to know what he is chasing. "I'm rushing after my livelihood," the man answers. "And how do you know that your livelihood is running on before you so that you have to rush after it?" asks the sage. "Perhaps it is behind you, and all you need to do is stand still."

In fact, the state of not knowing can be a very interesting place to stop for a while, because it is a state of all possibilities. If you "still" the mind in meditation, and then reflect on life as you go about your day, you may discover a purpose you never could have imagined and never would have seen had you only kept rushing about thoughtlessly. In the end, this "hurry sickness" only creates more stress anyway. As Larry Dossey writes in *Space, Time and Medicine*, "Just as Pavlov's dogs learned to salivate inappropriately, we have learned to *hurry* inappropriately. . . . Our 'bells' have become the watch, the alarm clock, the morning coffee, and the hundreds of self-inflicted expectations we build into our daily routine. The subliminal message . . . is: time is running out; life is winding down; please hurry." As a result of these perceptions, body rhythms such as breath rate and heart rate speed up, too, resulting in the stress syndrome that causes aging and disease.

Nevertheless, not knowing your purpose is not an excuse for inaction. The *Bhagavad Gita* says, "No effort in the world is lost or wasted; a fragment of sacred duty saves you from great fear." Purpose in action applies in the broad sense to the meaning of life, as we have said, but it also applies to every smaller objective on the way to discovering and fulfilling our highest goals. Indeed, we can live very satisfying and successful lives without having a clear vision of our ultimate destination, as long as we at least take action with a clear vision of our immediate purpose, which may be as mundane as earning money for the rent. To name a few: being responsible for ourselves, being kind to ourselves and others, striving to do the best we can in all circumstances, and enjoying our blessings—all are worthy purposes for the present. From our clear purpose comes our commitment to the task, and from our commitment comes complete absorption in the action. As you will see, this capacity to be fully present at each moment is the source of fulfillment and poise.

Obviously, purpose in action requires commitment to a goal, not just knowledge of it. Unfortunately, this idea causes as much stress for some people as it relieves in others. If this fear keeps you from choosing a direction, keep in mind that not all commitments are for a lifetime. Purpose represents the principle of stability in behavior—it keeps us on course and ensures progress. But, as you will see, flexibility has an equally important role in the field of action, which is by its very nature the field of constant change. How do we reconcile these seemingly contradictory principles? Again, we simply understand that we have the ability at all times to change our minds with a single thought. This may be obvious to many people, but others become so paralyzed in the face of a choice or commitment that they never decide on a purpose and never take action. Such stagnation is a sure path to stress and bitter old age.

One technique for moving beyond fear and lack of purpose and towards success and radiance is to commit yourself to a direction, any direction that is even remotely appealing, for a week or a day or even fifteen minutes. Do the job, and if you can find nothing else likable about it, at least use it as an opportunity for self-reflection. For example, observe all the messages, negative and positive, that you give yourself in a challenging or undesirable situation. See if you can find some humor in the situation. (You can also try this technique in social situations.) If you are willing to learn something new in every circumstance, no effort is ever wasted. Once your commitment is up, there is no rule that says you cannot make a new decision and find a new occupation if you are unhappy with this one. (In fact, job dissatisfaction increases the risk of disease—more fatal heart attacks occur at 9 A.M. on Monday, the start of the work week, than at any other time.) For the time your purpose is set, however, put a hundred percent of your energy and attention into the job. It's like a coin toss: If you commit yourself to "heads" and "heads" turns up, your heart will tell you instantly if you really prefer "tails." Unless you commit totally to the action, however, you will never be certain that it is *not* the right choice for you. Keep looking, keep asking questions, keep taking action. Do not give up the search. As the Buddhists teach: The arrow that hits the bull's-eye is the result of a hundred misses.

The search for a purpose is not exclusively an inner search. There are many practical steps you can take to help clarify your goals. Of course, Ayurveda can help illuminate your natural strengths and weaknesses, both physical and mental, which are useful to know in deciding what you want to do in life. By the same token, career counseling, psychological testing, job research, networking, and education are just a few of many tools available to help you choose a worthy endeavor and develop a plan to achieve it.

Each human life is a unique expression of consciousness, and in that regard has a unique contribution to make to the world. Yet, collectively we have a purpose, too. That universal purpose is helpful to keep in mind always, as it is a great comfort when the meaning of our own individual life seems to escape us. In the words of a spiritual teacher: "The first purpose of earth is to love. When you have begun to learn this very deeply, then the purpose of your activity, your relationships, your creativity does become quite clear as an inner feeling. Therefore, then, work to the first purpose if you would know the many other purposes in life."

COURAGE IN ACTION

"Whether you think you can, or think you can't,
you're probably right."

HENRY FORD

The principle of courage in action, symbolized by Chymunda's appearance on the back of a tiger, teaches the importance of affirming our purpose through thought and action. Even though the vision of our purpose may be quite clear, sometimes we fail to act because of fear and doubt. Ayurveda teaches: "As you think, you become." Consequently, it is important to train the mind not to dwell on worry and negativity, but to choose positive thought as well as constructive action. These two functions must go together, because even if we take a step forward but do not let go of doubt, then the action will be weak and will not achieve the highest results.

Doubt, not fearfulness, is the opposite of courage. Indeed, we can take bold steps and succeed even when we are scared—it *is* possible to "feel the fear, and do it anyway." Firemen and others who perform heroic deeds often balk at the notion that they are brave, and they do not deny the fearfulness of their job. Yet they never seem to voice *doubt*. They do not hesitate or second-guess themselves. To the contrary, such heroes often speak of their single-minded purpose—"I didn't think about the danger, I just knew I had to help" is the common refrain. Fear is necessary for acts of courage; it gives us the heightened responses that we need to overcome a challenge. In day-to-day life, a little "edge" promotes eustress—"good" stress—or the mental and physical arousal necessary for top performance. But doubts never heighten performance. They only drain energy and distract attention from the goal. They are the "traitors," as Shakespeare said, that "make us lose the good we oft might win by fearing to attempt."

To build courage, therefore, we need to remove doubt from our thinking. Doubt takes many forms, but all express a sense of lack, a feeling of incompleteness. We doubt our abilities, we doubt our motives, we doubt we deserve success, we doubt we are good enough, we doubt others appreciate us, we doubt we are lovable, and so forth. Much self-doubt is the result of early conditioning by parents, teachers, and whole communities, who may have sent repeated messages that we are unworthy, untalented, or even unwanted (all messages that reflect the fears and doubts of the messenger, by the way, *not* the truth of who we are). Nevertheless, when, where, or how we acquired these doubts—and even their particular form—is not important to our progress. What is important is that when we notice self-doubt, we take immediate steps to correct it. There are five simple steps to turn self-doubt into courage: acknowledgement and correction; self-acceptance; self-assessment and education; affirmation, visualization, and playing "as if"; and success-building.

Acknowledgement and correction. The first thing many of us do when we recognize self-doubt is to heap more negativity upon ourselves in the form of judgments and self-criticism. That is, we make the stress worse by punishing ourselves for our "bad" thoughts. All negative thoughts are just the mistake of the intellect; they are the symptoms of a loss of wholeness—of fear. They arise from misperceptions due to our conditioning, and in the case of overwhelming or obsessive negative thoughts, from psychophysiological imbalance. Whatever the case, Ayurveda teaches us to deal with all negative feelings, including doubt, in the same easy way: acknowledge the feeling for what it is (a learned response or a sign of stress and imbalance); feel it in the body (take a full breath, relax, and let your attention go to any physical sensations); gently release the thought from your awareness (some techniques below will help you let go); and correct the imbalance through diet, massage, and meditation to ease the mind.

In this regard, balance does not mean an end to emotions. According to Ayurveda, there are nine emotions, or *rasas*—the Sanskrit word which also means "taste"—that are natural to life and necessary for full experience. These are: romantic and erotic love, which is a source of rejuvenation, vigor, and beauty; the qualities of humor, playfulness, and jest (laughter is especially important to health and happiness); compassion, which includes mercy, charity, and helpfulness; fearlessness or courage, which includes a charismatic or heroic quality; contentment or tranquility, which includes qualities of tolerance and sociability; wonder, surprise and awe; natural fury or passion, or the emotional readiness to confront danger (a good description of the fight-or-flight response); spontaneity, impulsiveness, and shock; and fierceness or forcefulness, which is a special use of emotions reserved for warriors going into battle. Flexibility is built into the human physiology. We have the innate capacity to feel the full range of emotions, and we should enjoy them. But when the same negative emotion is triggered over and over, when it becomes dominant in our mood, it indicates imbalance.

The equivalence between tastes and emotions—both called rasas—provides a useful insight into the nature of feelings and how to release our doubt and other negative emotions. When we taste food, we have the sensation, sweet or bitter, and whatever it is, it passes. We savor it in the moment and let it go, knowing new and different tastes will follow. Emotions are the flavors of our experience, and like tastes on the tongue, they naturally dissolve in our awareness. However, if we resist or deny the full feelings of experience, the emotions become like undigested food, creating toxins in the body. If we hold on to them, they become like cravings that imprison our attention. Thus we say, recognize doubt, or negative emotion, for what it is; taste it fully; and let it go. In this way, like a well-savored taste, the experience burns in the strong fires of agni and then naturally dissolves, clearing the palate for new sensations.

Self-acceptance. Doubt always arises when we compare ourselves to others, a common habit in the competitive culture that conceived of the phrase "keeping up with the Joneses." Such comparisons indicate a lack of self-acceptance, a lack of a clear sense of one's own purpose, and a lack of trust in life (a concept we discuss below). The *Bhagavad Gita* contains a very clear lesson on the idea of making such comparisons. The Lord Krishna tells his student, Arjuna, that it is literally better to die performing one's own life purpose, or dharma, even if it seems lesser in merit, than it is to strive to be like another. Why is it better? Because one *can* perform it, the text says—that is, because we naturally have the capacity for our dharma alone, or it would not be ours to perform. As a modern thinker has stated the idea: It is always best to be yourself, because you only can be the second-best somebody else.

Ayurveda reiterates this idea in the teaching that we cannot change our innate nature—our innate constitutional balance. For example, a steady, easygoing Kapha type cannot sustain the natural aggressiveness of a Pitta personality without creating imbalance in the physiology. By the same token, a Pitta type does not have the natural stamina to continually work the long, concentrated hours of a Kapha type. The secret to absolute beauty and to success in general is to recognize and accept your innate strengths, as well as your vulnerabilities and shortcomings, and to work with them, not against them. If you push yourself to succeed in ways that are unsuited to your nature (or if you push yourself to perfection, which is impossible to achieve in relative existence), you not only put strain on the body but also cause doubt in the mind. Of course, this is not an excuse to be complacent and lazy. As you will see below, if something is difficult for you, it does not necessarily mean you are unsuited to the task; it can also mean that you have to learn specific skills or develop your innate tendencies.

Self-assessment and education. Self-doubt is *not* the same thing as honest self-questioning. As we said, self-reflection is a necessary part of successful action. You may have very sound reasons for doubting your ability to accomplish a goal because you have taken an inventory of your knowledge and skills and realistically concluded that you come up short.

Take a long, hard, honest look at yourself to assess your level of abilities as well as your natural inclinations and true passions. The right training and disciplined practice may be all that it takes to permanently remove your doubts. Again, do not make this an occasion for beating yourself up for what you may have failed to do in the past. Wherever you are now is where you are meant to be, and it is never too late to learn new skills. Ashley Montagu has noted that while development becomes arrested in all other species, humans are "capable of growing behaviorally and spiritually to the end of life." Many renowned artists and thinkers, including Leonardo DaVinci and Thomas Jefferson, accomplished their greatest works late in life, so age is not a legitimate excuse to stop trying or doing. If you think it is too late to accomplish your goals, consider this story of a woman who had always wanted to play the piano but had never had a chance to learn. At fifty-five, she decided to take lessons. Although she clearly had musical talent, she became frustrated with the slow learning process. Telling herself that it was just too late to become a pianist, she gave up after a year. Of course, her deep love of music never left, and when she was seventy-three she could no longer resist her desire. She resumed piano lessons and made great progress quickly, yet she had lost twenty years of study and enjoyment for the very reason that she *thought* she had been too old.

Affirmations, visualizations, and acting "as if." Many times we think negatively because we have developed a habit of self-defeating thought due to old conditioning. As we said, a simple technique to handle this kind of doubt and negativity is to recognize the particular thought and then release it gently, without self-judgment. Once you address your emotions in this way, the techniques of affirmation, visualization, and acting "as if" are useful to restructure the thinking patterns that repeatedly create these negative feelings.

Affirmation is a method of positive thinking in which you introduce a constructive thought in place of a destructive thought. For example, you can replace a self-doubt with an affirmation of an accomplishment: "I have created many wonderful friendships in my life" or "I won the tennis championship in high school" or "I am a great cook" or "I contribute excellent ideas to my job." The accomplishment can be big or small, old or new. Write the affirmation on a piece of paper (write a few accomplishments if they come to mind), take a full, easy breath, and then read the affirmation aloud a few times. Take a couple of more breaths and repeat it silently as you do. Do not concentrate on the idea; *contemplate* it—that is, do not fix your thoughts on the words per se, but let yourself think about the concept and its significance to you. The purpose is to stimulate a genuine memory of success, because by that new thought we induce a completely different biochemical reaction in the body. Remember, the mind does not distinguish between the real taste of a lemon or the memory of the lemon taste—both create the response of salivation. In the same way, affirmations can alter the psychophysiology, releasing us from stress. A good way to strengthen the affirmation is to accompany the thought with a constructive action. Rather than sitting with your worry or doubt and then sitting with the new idea, change your activity as you change your thought. Do any simple, quick task such as reorganizing your desk or washing your hair, so that your action reinforces your sense of purpose and accomplishment as you repeat and think about the affirmation.

This is a natural process. It is simply the conscious decision to change the object of our attention in the present moment. It is not necessary to force the mind to push away doubt and concentrate on a positive idea. Rather, whenever you notice doubts and then release them, your very shift of awareness is a sign that the emotion has begun to loosen its grip, and then you can interject a new thought easily. Keep in mind that it is the nature of thought to come and go. In fact, you can see for yourself how impossible it is to purposefully cling to one idea, positive or negative, by shutting your eyes and silently repeating the word "elephant." Within a few seconds, you will notice that the thought moves on spontaneously; for example, "elephant" may give way to "Republicans . . . politics" and so forth. The ultimate value of affirmations is that they develop a new habit of thinking. The recognition of a negative thought eventually becomes the mental bell, so to speak, that stimulates a conscious choice to redirect your attention to a positive thought and action. You will be surprised to see after a period of regular practice how the conscious effort of positive thinking becomes an effortless habit.

Visualization and "as if" techniques are ways to preempt doubt, so to speak, by creating positive expectations. Just as affirmations use memories of past accomplishments to induce a balancing biochemical response, visualizations and playing "as if" use imagination—the thought of future events and possibilities—to create a sense of well-being in the moment. Visualization involves creating a positive mental picture of your desired action and outcome—for example, imagining yourself successfully going through each step of an upcoming job interview, giving your best answer to each question, receiving a positive response, and ending with the image of yourself getting the new job. Acting "as if" is the adult version of the childhood game of make-believe, and is a powerful method for changing limiting beliefs about ourselves. It is an application of the lesson of the twenty-four gurus. You not only reflect upon the qualities of another person or creature to see what they can teach you, but also become that individual or animal in your imagination, and then behave as if you actually possess those qualities to experience how it feels.

In *Thinking Body, Dancing Mind,* Huang and Lynch offer some examples of the effectiveness of these techniques in sports training. They tell of an offensive guard who helped to bring his football team to victory by playing like a "mean, wild badger" and of a team of cross-country runners that produced a number of All-American distance winners by approaching their daily training process as if they were already champions. To demonstrate the power of visual imagery, they describe an experiment in which two groups of basketball players worked to improve their free-throw percentage: "One group shot one hundred free throws every day for three weeks; the other group simply visualized doing the same." The results? The "visualizing group showed significant improvement over those who actually shot the ball."

One psychologist has suggested a visualization that is particularly meaningful for anyone plagued by self-doubt and self-criticism. Say to yourself, "I love you. I think you're absolutely beautiful," whenever you look in the mirror, advises Dr. Susan Olson. "That affirmation may seem ridiculous, but it's important, because you're not going to believe those words coming from anyone else until you believe them from yourself."

Success-building. The English statesman Lloyd George once advised, "Don't be afraid to take a big step if one is indicated. You can't cross a chasm in two small steps." This is inspiring advice on the subject of courage. However, if you have never taken even a little leap before, you may want practice on something smaller than a chasm that perhaps has a safety net as well. After all, in order to visualize success, it helps to have had the experience at least once.

You can build mental fortitude and a "winning" psychology in the same way that you build muscle—through regular progressive exercise. Start with a "lightweight" challenge, some simple task that you have never tried to do—learn to play poker or fix a flat tire; bake a cake or knit a scarf. If you are *not* athletic, take up a simple sport—start walking every day, building up the distance you cover in an allotted time. (Building physical endurance helps to build mental stamina, a lesson which athletes know well.) The point is to pick an activity totally unrelated to your work life or personal responsibilities—in other words, an activity where there is absolutely nothing at stake except a new feather for your cap if you succeed. Stay with the activity until you complete it, accepting no thought of failure. When you master the first challenge, go on to one a step more difficult, and then another. With each success, your self-control, self-confidence, and fearlessness will grow. Eventually, your mind will develop a habit of success, so that your natural response to any new challenge will no longer be hesitation or doubt, but rather optimism and clear, focused action. In no time, you will be leaping chasms boldly.

FOCUS IN ACTION

*"If we take care of the minutes, the years
will take care of themselves."*
BEN FRANKLIN

"Work is love made visible."
KHALIL GIBRAN

A teacher is giving a lesson in archery to five princes. Pointing to a sparrow on the branch of a distant tree, he instructs his young students to shoot for the eye. As they aim their bows and arrows, the guru asks them what they see. The first brother describes in detail the tree trunk, the branches, and the bird. The second brother describes the branches and bird only. The third describes the single branch and the bird. The fourth describes the bird alone. But Arjuna, the great warrior prince, says, "Master, I can see nothing, except a bird's eye!"

This Vedic story illuminates the principle of focus in action. It is the single-pointed attention to the task at hand, which is the product of a disciplined will. When we are completely immersed in action, the mind is not distracted by any other thoughts. Yesterday and tomorrow have no existence; we are literally one with the present—timeless. This total absorption in action is an experience of unboundedness. The mind is still, yet dynamic; the body is restful, but energized. Actions performed from this quiet awareness have the effortless precision and grace of Arjuna's arrow. It is this bodymind unity in motion that makes Olympic athletes and great dancers so mesmerizing to watch.

We have all experienced this "flow" in our lives even for just a few moments. It carries with it a sense of inner expansion and bliss, and appears to others as an energetic yet serene quality. This skill of focus is symbolized by Chymunda's trident, the three-pronged spear with which the goddess Kali destroys physical, emotional, and spiritual pain. These three pains are also associated with the three gunas—the primordial forces of consciousness—sattva, rajas, and tamas, which give rise to all the activity of creation. Absolute focus occurs when we transcend the field of the gunas in meditation—that is, when we vanquish the activity of the mind—and experience the total concentration of awareness in the field of Being. This lesson is contained in the *Bhagavad Gita* in Lord Krishna's teaching: "Be without the three gunas, O Arjuna, freed from duality . . . possessed of the Self." However, as long as we remain in the field of duality—the field of action—the technique of focus, or mindfulness, does save us from stress, or pain, as well. The Vedic teacher Maharishi Mahesh Yogi states this principle of focused action with the simple advice: "See the job, do the job, avoid misery."

Focusing on a task *results* in concentration, but focusing itself is not concentration. Ideally, it is a relaxed state of mind in which the attention, free of all distraction, is unified on the single activity to which we direct it. This relaxed state is not attained by mind control, which by definition requires effort and concentration, but by yielding or surrendering the mind to the task. The question is, how do we "surrender" the mind? What exactly is it that we are letting go? According to Veda, we let go our attachment to action's *result*. In effect, we give up our concern for the outcome so we can focus on the *process* alone.

When we worry about the results of action, a part of our attention is always on the future. As a result, the mind is divided and the action itself is weakened. On the other hand, when we mentally let go of expectations, we free our attention to be fully on the task at hand, and this is what guarantees greater success. Letting go of *concern* for a result is not the same thing as letting go of *desire* for it, however. Desire is the spur to all creation. Why would we act at all if not to fulfill some desire (even one so basic as the need for food)? As we said before, a vision of the goal is a prerequisite to action and achievement, but *attachment* to the goal actually holds us back from reaching it. According to Veda, the focused mind is *impartial* to success or failure, but not indifferent, and in that impartiality lies the key to a blissful life.

This idea of impartiality to success or failure is a radical departure from the success-driven approach of modern Western culture. In a society that is hungry for power, we perceive control—meaning control of outcomes—to be the ultimate weapon in the game of life. We believe that control gets us what we want, and in that sense we equate it with happiness. Indeed, even mind-body research indicates that having a *sense* of control is a crucial factor in minimizing the effects of stress on health. So why would we advise giving it up? Because, as Arjuna learns in the *Bhagavad Gita*, "You have control over action alone, never over its fruits." That is, we have ultimate power only over our thoughts, our purpose, and attention, as we said, and over our choice of how we put them into use. We also can control to a degree the actual physical activities we perform, and we exercise that control through total mindfulness to the task. But after that, any notion of control in relative existence is purely illusory. Indeed, we think stress arises from the sense that we lack control over getting what we want, but stress actually comes from the misguided belief that we ever have such control in the first place. We do not control the forces of nature, which function according to their own laws, and we simply do not have the omniscience that is necessary to foresee how all events in the universe will conspire to aid or impede our efforts. We *do* have control over our actions, as Veda says, and it is our responsibility to exercise that control. However, when we realize that we simply do not possess power over ultimate outcomes (nor can we change past results), we actually liberate the mind to be totally focused on the present, and the present moment is the *only* place where we do control action and where we can know bliss. "Live not for the fruits of action, nor attach yourself to inaction," Lord Krishna teaches Arjuna. By all means, *act*—act with passion, and with compassion—but surrender totally to the joy of *doing*, let go *wanting* (have *preferences* rather than expectations; be prepared for everything and expect nothing, as Huang and Lynch suggest), and you will experience fulfillment in each and every moment, no matter the final result of what you do. This is absolute success and the secret of absolute beauty.

Living by this principle of surrender requires a big leap of faith in much the same way that jumping a chasm requires a big leap of courage. Below we will talk about trusting the abundance of creation to send us what we need. However, unlike chasm-jumping, which cannot be accomplished in two small steps, focus in action actually is best accomplished by breaking down large goals into many little tasks. Again, much of our worry about outcomes, and therefore much of our distraction from the present, arises from the fear that we do not have what it takes to achieve what we want, or that we do not deserve to enjoy the fruits. This doubt and fear of failure is quickly and easily overcome by following the lead of a sixty-four-year-old ultra-distance runner who said, "I don't run a hundred miles; I run one mile—a hundred times."

In this manner, no task is ever too great for anyone to accomplish. No matter where you are in relation to your goals, you can always find one small job that you can do successfully to move yourself forward, even in the face of great stress or loss of self-confidence. For example, getting in shape is a challenge for many people because it requires daily discipline without the satisfaction of immediate results. Especially at the beginning, when your stamina and self-esteem are low, the goal can seem unattainable. Yet there is always a way to take a step forward, even if it is just eating one less bite of food at your next meal, or starting a new hobby as an alternative source of gratification, or taking a short walk. In fact, research from the University of Pittsburgh shows that women who worked out for ten minutes, four times a day actually did more exercise and lost more weight in the same time period than those who exercised forty minutes, once a day—so you can break up your exercise program into small steps and still get the results you want. Build endurance and confidence one day, or even one hour, at a time. Whatever you do, do not give up. Heed the advice of the runner, and cheer yourself on with the words of this African proverb: "No matter how slowly you go, you can still get there." Then focus on the task in front of you—avoid the misery, gain success.

FLEXIBILITY IN ACTION

"Biologically, your body is perfectly set up to live in the present and acquires its greatest joy and satisfaction there. Your body never knows what its blood pressure will be the next second, so it has a built-in flexibility to allow a wide range of pressures; the same flexibility is built in to every other involuntary response. This is the wisdom of uncertainty, which permits the unknown to take place and welcomes it as a source of growth and understanding. We see this widsom expressed in the spontaneity of every cell and organ."

DEEPAK CHOPRA

Even when our purpose is clear, our action bold, and our mind focused, we still do not exercise absolute control over events in our lives. Sometimes circumstances present themselves that seem to defy our best efforts to grow and succeed. Of course, these are the times that *flexibility* as well as focus in action are particularly handy skills. A simple story illustrates the point:

One day a poor, hardworking farmer and his son found a stray stallion on their land. News spread fast through the small village that the old man had finally gotten a beast to ease his burden, and neighbors came to congratulate him on his good luck. "Good luck, bad luck! Who can say?" responded the old farmer. The next day, he and his son set about to hitch the animal to their plow. Unused to domesticity, the horse reared up and ran away, crushing the son's leg as it went. Soon the neighbors came to console the man on his bad luck. "Bad luck, good luck! Who can say?" responded the old farmer. The next day, war was declared in the province and word went out that all able-bodied young men must report to fight. The villagers wept as they sent their sons to battle. Only the farmer's boy, handicapped by the stallion, was allowed to stay behind. Again the neighbors came to make a point of the farmer's good fortune. Again he told them, "Good luck, bad luck! Who can say?" The next day, the fighting spread to the outskirts of the village not far from the farm. Eventually the enemy retreated, but not before they burned the surrounding hillside and destroyed the farmer's old stone fence. The next day, their sons returned to safety and their own homes unscathed, the villagers came once more to cluck at the farmer's ill fate. And once more he replied, "Bad luck, good luck! Who can say?" Days went by. The farmer worked from dawn to dusk to restore his fence. One afternoon, the tired old man fell fast asleep, forgetting to rope off the unfinished part. As he napped, the stray stallion found his way back to the farm, having abandoned the scorched hills in search of greener pastures. Seeing the downed fence, he whinnied for his herd to follow. When the farmer awoke, a hundred horses were grazing happily on his land.

It takes a flexible as well as a focused mind to be one with the present, because the present is moving ever onward. As a child, I remember my mother saying that nothing in life remains constant. "Good or bad," she would tell me in my moments of distress, "all things change in time." My mother lived as she taught. I know she did not have everything she wanted in life, and I know there were things she had that she did not like, but she accepted her circumstances without complaint or unhappiness because she understood that everything occurs for a reason. This does not mean that she passively resigned herself to life or behaved like a martyr. She took charge of her life where she could. But she also understood that the rudder which gives us fixity of purpose must itself be movable. Flexibility is what enables us to navigate the shifting waters without losing sight of the goal.

When our preferences, if not our expectations, are unmet, the best strategy is just to continue about our business, like the impartial farmer, and not get caught up in the momentary circumstance. Life always moves on, and it always progresses in cycles. We create much unnecessary stress and strain for ourselves by resisting the natural ups and downs rather than learning to go along with them. Martial artists have mastered this secret of success well. Instead of countering attack, they align themselves with its movement and force, and thereby defuse its power. Huang and Lynch suggest that we can approach personal challenges in the same way: "If you see your setback as an opposing force, you can still accept and blend with it and internalize its lessons to your advantage. By doing this, the power of the opposing force no longer exists. You redirect the force and forge ahead."

Of course, even the most adaptable people occasionally find themselves knocked down by a crisis. At such times, when the stress or pain is extreme, it is difficult to take things as they come and remain philosophical. Again, this is a time to stay on a balanced diet and routine and to try additional therapies such as massage, aroma therapy, and pranayama to soothe the emotions. Writing about difficult experiences is also an effective tool for minimizing the effects of stress. In one study, subjects showed a significant boost in immune activity after spending twenty minutes a day for four days writing about a traumatic event in their lives and the negative feelings it produced. I have suggested this technique to many clients, and have seen improvements in their skin once they confronted and expressed their painful emotions. Resisting our feelings is what makes us ill; *feeling* and *releasing* them—in effect, aligning in the moment with their force—enables us to heal. Healing also is achieved by openness, patience, reflection, prayer, happiness for others, and the language of silence—meditation.

Once the grip of emotions is loosened in these ways, we can begin to regain some perspective on our circumstances and examine the subtle messages we are giving ourselves about the meaning of the situation. Crises often bring out the best in us, but they frequently stir up our worst fears and doubts about ourselves as well, sending us into a state of depression. Although we may feel at such moments that the forces of the universe are against us, these are the times when we most need to *align* ourselves *with* them. Without the cosmic vision to comprehend the ultimate purpose and meaning of each life event, we have to trust that we will come to see the rightness of all our experiences in time, no matter how difficult and painful they are in the moment. Consider this true story of a radio traffic reporter who suddenly lost her job because of corporate politics that had nothing to do with the quality of her performance. She was so distraught over her unjust treatment that she went into a deep depression. A few weeks later, the station's traffic helicopter crashed during a routine broadcast and killed the announcer who had taken her place.

Of course, we do not always get to see with such immediate clarity the blessings hidden in life's seemingly "bad" events. However, if we keep an open mind—and a questioning mind—and regard every turning point as an opportunity rather than a crisis, and every setback as a learning experience rather than a failure, we eventually see that we are indeed always at the place we need to be. Trust, surrender, and forge ahead. As one woman remarked looking back on the winding course of her own fate, "I have lived to thank God that all my prayers have not been answered."

BALANCE IN ACTION

> *"Don't mentally review any problem constantly. Let it rest at times and it may work itself out; but see that you do not rest so long that your discrimination is lost. Rather, use these rest periods to go deep within the calm region of your inner Self. Attuned with your soul, you will be able to think correctly regarding everything you do; and if your thoughts or actions have gone astray they can be realigned."*
>
> PARAMAHANSA YOGANANDA

This principle reiterates the basic Ayurvedic strategy for achieving timeless beauty. The secret of Chymunda's serenity atop a tiger is found in the balance of the five senses, mind, ego, and spirit, which is symolized by the presence of Kali, Lakshmi, and Saraswati.

If stress is defined as any overload on the mind, body, or emotions, then the obvious way to avoid stress is to avoid overloading yourself through a balanced lifestyle in accord with your nature. Deepak Chopra defines the four components of balance as moderation, regularity, rest, and activity. We have already seen that fulfillment depends upon clear thinking and successful action. These, in turn, depend upon regular periods of deep rest. According to Ayurveda, sleep is rest for the body, but meditation is rest for the mind and rejuvenation for the spirit; both types of rest are needed daily for optimum health and absolute beauty. Where all other animals maintain a cycle of rest and activity by instinct, however, human beings must make a mindful choice everyday how to live.

In this sense, balance also refers to the understanding of opposites. As you can see by these eight principles, life progresses through the alternation of opposites—stability and adaptability, action and inaction, single-pointedness and open-mindedness, wisdom and innocence, creation and destruction. Ayurvedic balance is always a conscious, dynamic process, like captaining a ship in rough waters. Automatic pilot only works on calm seas, and even then, you have to keep an eye out for changing conditions. When we align with the cycles of change, rather than resisting them, we are never out of balance and the journey is much less of a struggle.

There is another aspect of the principle of balance, which is commonly understood as the law of karma. In the language of modern physics, this universal law states that every action has an equal and opposite reaction. According to the rishis, this fundamental principle of nature applies as much to human actions as it does to the movements of planets and billiard balls. Every expression of ourselves, from our outward behavior to our innermost thoughts, sends a ripple of influence through the universe and binds us forever to its results. Energy cannot be destroyed; this is also a universal law. Our thoughts are a form of energy, and whether expressed silently or through words or action, they have an infinite life whose ultimate effects are impossible to compute on the level of the intellect. Harold Kushner, author of *When Bad Things Happen to Good People*, has described the human soul as anything in individual life that is *not physical*, including our personality, our values, and our ideas. Because they are not physical, he reasons, they cannot die. Just as a joke cannot die, and an idea cannot die, and the soul cannot die, words and actions also have an unending existence, carried forth by the unbroken chain of reactions that each word and act creates through its effects on others. As the one who sets the ripple in motion, each of us is responsible *for all time* for the outcomes of what we create.

For this reason, according to Ayurveda, it is important not only to be moderate in action, but also to be wise in action, so that we do not bring harm to ourselves or others through anything we think or do. These are the lessons of *yama*, or ethics, and *niyama*, or self-restraint, the first two steps of what is known as Patanjali's eightfold path of *yoga*, or union with God. Yama teaches:

- Do not hurt anyone through thought or action. This, of course, includes the principle of nonviolence.

- Always be truthful to yourself and others—that is, tell the "sweet" truth; do not attack with the truth.

- Do not steal. This includes the act of envy. To want what someone else has—even another's charm or grace—is a form of stealing. In accord with the idea that life unfolds as it should for each person, Veda teaches that we each have what is necessary and due us in the moment, whether or not we immediately perceive its true value. Moreover, we are each created *unique*—we are not meant to be exactly like anyone else, but we are meant to find our own talents and beauty. Therefore, there is nothing to want that we do not already have. Again, this does not mean we passively accept our circumstances no matter what they are. Rather, it means to accept the present situation as it is, look to understand the meaning of it, then let that moment go in order to be completely present in the next moment, where all possibilities exist.

- Do not judge others since you are not in their shoes. We each make choices based upon our previous experience, and not having had the totality of anyone else's experiences, we are not in a position to judge anyone's decisions but our own.

- Be balanced in all activity, including the actions of the five senses. In other words, do not overindulge any sensual appetite. Attune yourself to the body's natural intelligence, and you can easily recognize the signs of imbalance that signal excess.

The lessons of niyama include the cultivation of purity of mind and body—that is, the cultivation of balance and health—of contentment, and of surrender and devotion to the divine. This includes prayer as well as action. There is no greater act of devotion to God than to honor the gifts you possess by always making the best use of them in the most loving and creative way you can. Offer all your actions, and the fruits of your actions, to the Divine—to whom all creation belongs anyway. This is the humbleness of prayer, to let the ego melt, and to think and act with God as partners in creation.

Indeed, aligning in prayer and meditation with the cosmic mind—with universal law—is the only way to ensure right action, because we as individuals simply do not have the prescience to know the ultimate outcomes of our undertakings—"good" or "bad." A story about the prophet Moses conveys this idea. God sends a teacher to prepare Moses for his great work. For his first lesson, the teacher instructs him that whatever happens, he is only to observe, but never to act or even speak. In the course of their travels, they come upon a river where a small boy is drowning while his mother stands on the shore helpless and crying. Moses asks if they cannot do something to help, but the teacher merely answers, "Silence!" Moses obeys, although he is shocked by this hard-heartedness. They continue their travels and soon they arrive at the seashore, only to see a great ship sinking in the ocean with its entire crew aboard. Again Moses implores his master to help, and again he is silenced. Now Moses is deeply troubled, and when he returns home he takes up his question with God. "Your teacher was right," the Lord tells him. "Had you saved the drowning child, he would have lived to start a terrible war in which multitudes would have perished. As for the ship, it was manned by pirates who were setting sail to plunder and destroy an innocent fishing village."

Ayurveda is the practical science of how to live according to your own nature—which on the level of consciousness is aligned absolutely with natural law—so that you spontaneously live and act in harmony with the evolutionary principles of creation. Short of achieving enlightenment or direct divine guidance as Moses received, the best course to follow is daily prayer and meditation to develop the value of unbounded awareness. The intellect makes decisions on the basis of previous experience—that is, on the basis of a finite amount of information. In meditation, we go beyond the intellect directly to the source of experience. When knowledge comes directly from this infinite source of intelligence, there is no confusion. We act without mistakes. Until we are living this level of pure knowledge daily, however, we must rely on the principle of right action given by a great biblical scholar, who two thousand years ago summarized the message of the scripture in this way: "What is hateful to you, do not do to your neighbor. That is the whole law. The rest is commentary."

Implicit in this message, of course, is the teaching to "love thyself." If we do not know how to do this for ourselves, we cannot know how to extend love to another. If we are always attacking ourselves with negative thoughts, we are not likely to hesitate to attack someone else. In this sense, all love truly begins and ends with the Self. What we do to another, we do to the Self, by virtue of the fundamental law of action. Therefore, the path to absolute beauty—the path to bliss and wholeness—is not a path of vanity but of great compassion. To learn to love ourselves truly is to learn to love the whole of God's creation.

EFFORTLESSNESS IN ACTION

"Concerning all acts of initiative (and creation), there is one elementary truth, the ignorance of which kills countless ideas and splendid plans: That the moment one definitely commits oneself, Providence moves too. All sorts of things occur to help one, that would never otherwise have occurred. A whole stream of events issues from the decision raising in one's favor all manner of unforseen incidents and meetings and material assistance which no man could have dreamed would have come his way.

JOHANN WOLFGANG VON GOETHE

"Relax. God's in charge."

T-SHIRT SLOGAN

Effortlessness is a spontaneous effect of living in harmony with natural law. It is the economy of energy—the grace—that we automatically experience due to the frictionless flow of right action. The law of karma is only in effect as long as there are opposing forces at work. When our nature is totally aligned *with* nature, however, there is no opposition to our action and therefore no opposing reaction. Instead, the ripple of influence we create by right action resonates with the infinite waves of bliss that compose the primordial energy of creation. In this sense, *our* creation actually amplifies *the* creation. If we can conceive of a way to increase what is already eternal and bigger than the biggest—if there is a way to thrill God's Being—surely this is it.

How do we achieve effortless right action? It is first and foremost the effect of silence structured in consciousness through regular meditation. The experience of transcendental consciousness is a state of yoga—union of individual mind with cosmic mind. When we can maintain that unity of awareness at all times—in meditation and in action—we live in the total effortlessness of Being.

We also develop effortlessness in action by cultivating trust and surrender in the mind and heart. If you follow the principles of knowledge, purpose, courage, focus, flexibility, and balance in action, then yielding to the moment becomes not only a natural habit but also a joyful one because it automatically defuses stress. When the attention is totally absorbed in the process of doing, the effort of doing disappears. As we've said, this process of surrender is an exercise of will, a choice of where to place your attention. It is neither miraculous nor arduous. It does not require divine intervention or ascetic austerities. It is, indeed, child's play. Watch a child playing, and you will see the simplicity and joy of a will that has given itself over a hundred percent to the task at hand. Surrender *creates* miracles. It transforms all tasks into pleasure.

A few years ago, I decided to begin an exercise program using a "skiing" machine. Unused to either the equipment or the hard work, I found myself getting tired after only a few minutes of effort. I thought about quitting. And then I thought about surrendering. I looked for a new way to regard what I was doing, and began to put my attention on the motion itself. I changed my breath, inhaling through the nose, exhaling through the throat, and I began to *be* in the present instead of thinking about the number of minutes left in my workout, or about the long labor of getting into shape, my recriminations about being out of shape, or my fantasies of being in shape. In the present, I discovered the essence of the experience—the beautiful coordination of body, mind, and spirit that exercise is. It was exhilarating. I do not remember what happened to the rest of the time; I only remember that thirty minutes later my workout was done, and I was not at all fatigued. It was a simple lesson in yielding to the instant—a lesson that we can practice at any moment of the day.

We learn to build faith and trust through these moment-to-moment acts of surrender, just as we learn to build courage through small steps of success. Even more so than the idea of jumping chasms, however, the notion of *total* surrender tends to arouse our deepest fears, and of course, our prejudices and doubts about what we would be surrendering to.

We began this chapter with the premise that stress is a result of our perceptions, and that by examining our self-limiting thoughts and beliefs and consciously altering our perception, we can free ourselves from stress and achieve absolute beauty. In the spirit of this inquiry into what we believe is possible, we want to go one step further: to consider the "inconceivable"—that is, the Creator and the creation, and more specifically, what you believe about your relationship to them. Our purpose is not to debate or decide the existence of God. Rather, in order to get "out of the box" of habitual thinking that results in stress, we want to consider an alternative perspective that may increase your experience of effortlessness in life. Try this as a simple "as if" exercise: *If* you believed there was an all-loving, almighty Creator, how would it change your experience?

We suggest that your experience would not be much different unless you also believed with all your heart, mind, and soul that *you* are the Almighty's perfect and beloved creation to whom every desire has already been granted. Indeed, the essential question to ask is not whether you believe in God at all, but how much—or how little—you believe that you are worthy to receive the greatest abundance and highest love the universe can provide. If you do not believe you *can* have such fulfillment, you will not even seek it, and may even overlook the significance of blessings that are given to you. After all, it is not the giver who determines the true value of a gift, but the one who accepts it.

In the literature of modern mind-body medicine, there are numerous examples of the power of beliefs to completely alter, in some cases instantaneously, physical life. Studies of psychosocial dwarfism demonstrate clearly the creative power of our perception of self-worth. Young children who do not receive adequate nurturing contact often stop producing growth hormones, resulting in arrested physical development. In such cases, injected growth hormones tend to be ineffective, although hormone replacement frequently stimulates normal growth in other types of developmental problems. Yet once they are placed in loving homes, these same abused children often will respond to treatment or even start producing enough natural hormones again to completely reverse their condition.

Again we are simply asking, "What if?" All these principles of action are exercises in changing perception, based upon the quantum understanding that the world appears to us as it does because of how we think about it. Our perception determines our reality. What universe would appear if we asked ourselves, "What kind of universe do I think I am worthy of receiving?" instead of, "Is this all I get?" Does your belief in yourself and your expectation of reality require that miracles appear as chariots come from the sky, instead of as rowboats or helicopters? Indeed, what if everything, every event and every aspect of your life—large or small, significant or seemingly inconsequential—contained within it a letter of love to you "sign'd by God's name"? How would this perception alter your approach to life? How would it alter your experience?

A Vedic story tells of an infant in her high chair who sees a beautiful shiny object just beyond her reach. She cries for her mother, who stands nearby, to give her the precious toy, but the mother refuses. Of course, this makes the child cry all the more, so desperately has she set her heart on the thing. Yet no matter how long the baby wails, the mother will not give in, because she knows what the child does not—the brilliant object is not a toy, but a bejeweled and golden knife. Though her own heart breaks for them, the loving mother will never give her children what can cause them harm.

Total surrender is a clear, unquestioning, and deeply felt commitment to an idea—in this case, the possibility of a universe created out of absolute, unbounded love. In effect, we do not surrender to something outside us, but rather to a thought—an impulse of intelligence, an impulse of the *Self*, which is united at all times (whether or not we know it in conscious awareness) with the all-encompassing intelligence of creation. Therefore, surrendering is not relinquishing control of anything. It is not giving up anything except fear. It is *taking* our attention, which is all we do control, and giving it *over* wholeheartedly to the eternal *now*, wherein lie our own wholeness and highest wisdom—and all possibilities. *If* you were to surrender in this way, with total belief in your worthiness, to the perfection in the present, all action would become effortless. In one sense, we surrender ourselves—our Self—all the time to our thoughts. What we do not do all the time is consciously choose the thoughts to which we surrender. All we are proposing, and all Ayurveda proposes, is that we surrender to the highest thought: the perfect love of the Creator for the creation—that is, for us. That thought renders all action effortless, because it renders all outcomes the perfect outcome at each moment in time.

JOY IN ACTION

"I do not call one greater and one smaller,
That which fills its period and place is equal to any."
WALT WHITMAN

"Spiritual success comes by understanding the mystery of life;
and by looking on all things cheerfully and courageously, realiz-
ing that events proceed according to a beautiful divine plan."
PARAMAHANSA YOGANANDA

At the peak of the baby boomers' career-climbing frenzy in the mid-eighties, a now-famous T-shirt slogan facetiously pronounced what many seemed to believe were the stakes in the game of life. It said: "The one who has the most toys when he dies, wins." My own son, a successful engineer who is now thirty and has lived most of his life in the United States, once asked me to explain the Vedic philosophy that he has inherited by birth, but that he does not see expressed too often in the play-to-win world: "If we shouldn't expect any fruits of our action, why should we do anything at all?"

Life *is* a game of sorts. It takes place on God's playing field, and we are all in it by virtue of our birth. Whether we like it or not, *we have to play*. We do not get to make up the rules, which were set in motion at the dawn of creation, but we do get to choose *how* we play. Although we have the option to sit around and do nothing, eventually the ball will come into our corner of the field—and if we do not at least get up and out of the way, we are sure to be hit or knocked over by the players.

To do nothing is to stand in the way of the flow, although our inaction does not stop the course of evolution any more than standing still in the water stops a river. So we might as well play, if only to keep up with the action and avoid the pain. And once we join in, we might as well play our best game possible—and we might as well enjoy ourselves, and even learn to laugh at ourselves as we do. This means not that we play to win, but that we play to *play*.

This is the principle of joy in action. It is symbolized by the radiance of the goddess Chymunda. Like effortlessness, joy in action is a spontaneous result when we have mastered the other principles. When we give ourselves over to the process of what we are doing, the action itself is its own reward. No matter what the ultimate outcome is, internally we have already won! These internal victories, as I call them, are what make us truly beautiful.

The ultimate expression of joy in action is the bliss of unity consciousness. This is the state of attainment of the great seers who declared, "I am That, thou art That, all this is That." A vision of this sublime life is contained in a beautiful story about a gifted young student of Veda who eventually rose to become one of the great spiritual leaders of India. When he was not much older than a boy, this seeker left home to find his guru, his spiritual teacher, who sent him into the Himalayas to meditate, as is customary among monks. After many months alone in a cave, the boy received a visitor, another disciple, with a message that their teacher wished to come and check on his progess. When the messenger asked for his reply, the student answered, "Please tell Teacher, there is no empty room."

The messenger was dismayed. The truth was, he and the other older disciples were jealous that such a beginner was receiving so much attention from their master. Now he took the chance to put the boy in his place. "Here you are alone in this cavernous mountain in the midst of the vast Himalayas, and you dare to say you have no place for our honored teacher, who has given you so much! Have you learned nothing?" he asked angrily.

"I meant no dishonor," said the student. "I have followed Teacher's instructions and meditated faithfully every day. At first, my mind grew quiet until all thought ceased. In its absence, I experienced infinite Being. I realized: I *am* this unbounded consciousness—and I felt the joy of unshakable wholeness. In time, I began to notice this vibrant force of life not just within me during meditation, but all around me throughout the day. Objects began to appear less concrete than the energy permeating them. Finally the world of boundaries became like a transparent veil through which I could perceive the ever-present light of consciousness. Then, all difference dissolved between inner and outer awareness: I saw my Self in all things, and all things in my Self. The love and bliss that filled my heart embraced the fullness of the universe.

"Now, wherever I am, I am this fullness. Whatever I do *is* fullness. Truly, I have searched everywhere, and everywhere is filled with bliss. So please tell Teacher," the boy repeated, "I have no empty room."

When my mother first gave me Chymunda's picture, I did not understand who she was. I thought to myself, "So, she is just a good-looking goddess." I was much too young to comprehend the symbol of her radiance. Now that I have traveled her path for many years, I have a glimpse of the infinite layers of experience—the fullness of life and utter self-sufficiency—that her beauty represents. This wholeness is where Chymunda's path ends and the fulfillment of absolute beauty begins. Yet there is much beauty, much wholeness, much bliss along the way. This has been the gift of Ayurveda, which has taught me the language of the skin, the secrets of balance, and the gentle wisdom of the purification of the soul—the same lessons you have found here. As they have done for me, I hope they will help to make your own journey to absolute beauty an effortless, joyful, successful adventure. If at times the way seems difficult to see, just remember that the essence of this knowledge lies in three words of Veda:

Sattyam. Shivam. Sundaram.

"Wherever pure knowledge and happiness is, there lies beauty."

DHATUS AND THE SKIN
(Sanskrit names appear in italics)

TISSUE *(Dhatu)*	RELATED SKIN LAYER	FUNCTIONS GOVERNED BY DHATU
Plasma *Rasa* (Lymph, blood plasma)	Stratum Corneum *Avabhasini* (External horny, keratinized layer)	Menstruation, lactation, circulates nutrients, makes skin smooth, soft.
Blood *Rakta* (Red blood cells)	Stratum Lucidum *Lohita* (Contains blood capillaries)	Blood vessels, muscle tendons, sensory-motor functions, oxygenation, maintains color of skin & lips, governs liver function.
Muscle *Mamsa* (Skeletal muscle)	Stratum Granulosum *Sweta*	Flat muscle, skin, covers skeleton, provides protection, movement.
Adipose *Meda* (Fat in limbs & torso)	Stratum Spinosum *Tamra* (Pigmentation layer)	Joint tendons, subcutaneous fat, lubricates body, maintains water balance, sebaceous secretion.
Bone *Asthi*	Stratum Germinosum *Vedini*	Teeth, gives support, firmness.
Nerves *Majja*	Dermis Papillary *Rohini*	Brain functions, nerve impulses, lubricates eyes and skin, fills bony cavities, provides sensations of touch, heat, cold, pain & love.
Reproductive *Shukra* (Male and female repro-ductive fluids)	Reticular *Mamsadhara* (Contains ojas)	Produces ojas, governs or provides immunity, stamina, reproduction, hormonal balance.

WASTE *(Malas)*	PHYSICAL MANIFESTATIONS WHEN IMBALANCED
Mucus, tears, saliva, menstrual blood.	Dry skin, wrinkes, dermatosis, loss of glow, and skin sensation, pain in joints, restlessness, Kapha imbalance.
Bile.	Dull complexion, redness or rash, acne rosacea, itching, burning, discoloration, psoriasis, moles, birthmarks, scleroderma, alopecia, bleeding tendencies, blood too hot or cold, weak eyesight, broken capillaries, allergic reactions, Pitta imbalance
Ear wax, teeth tartar, navel lint, smegma, sebaceous secretions.	All of the above plus warts, basal cell carcinoma, benign moles, cystic acne, muscle inflammation, or increased sebaceous secretion, teeth tartar, and ear wax; glandular swelling, Kapha & Pitta imbalance.
Perspiration.	Dry, dehydrated skin, leucoderma, eczema, Pitta enters in tambra level causing acne rosacea, loss of shine on hair & nails, weight gain, tumors, Kapha imbalance.
Nails, body hair, beard.	Ericiples, discoloration or breakage of teeth and nails, osteoporosis, Vata & Kapha imbalances.
Breast milk.	Unusual, hard-to-cure skin rash at the joints, dryness, enlarged lymph glands, elephantiasis due to blockage of lymph channels, joint pains, osteoporosis, multiple sclerosis, imbalance of Kapha and air element.
None.	Fissures, piles, abscesses, loss of skin vitality and luster, slow healing, cysts in breasts, ulcers, spontaneous abortions, sterility, birth defects in children, impotency, loss of ojas, weakened immune and reproductive systems, slow cellular rejuvenation premature aging, Pitta & Kapha imbalance.

MAKING AND STORING
HERBAL PREPARATIONS

GHEE (CLARIFIED BUTTER)

(The following process takes about 10–15 minutes.)

• Melt one pound unsalted butter in a saucepan.

• Continue to heat over low flame until it boils gently and a foam rises to the surface. Do not remove foam.

• Continue cooking gently until the foam thickens and then settles to the bottom of the pot. When the remaining liquid turns a golden brown color and starts to boil silently with only a trace of small bubbles on surface, the ghee is ready. When it starts to cool and before it starts to harden, pour the liquid ghee into a glass jar.

MOISTURIZING AND MASSAGE OILS

For face oil, add 20–25 drops pure essential oil into 1 oz base oil. To use, mix 2–3 drops of face oil with 4–6 drops of water, and apply to wet skin as directed.

For body oil, add 10–15 drops pure essential oil into 1 oz base oil. To use, apply to wet skin as directed.

DECOCTIONS

For a mild decoction (teas), mix 1 Tbsp selected herbs + 1 cup water. Bring to a boil and then simmer on low flame for 7–10 minutes, or until liquid has partly evaporated. Strain and use remaining liquid as directed.

For a strong decoction (baths), mix 1 part selected herbs + 4 parts water, bring to a boil and then simmer on a low flame until half the liquid has evaporated. Strain and use remaining liquid as directed.

HERBAL OILS

- Make a strong decoction of the selected medicinal herb.
- Mix 1 cup of the decoction + 4 cups selected base oil. Bring mixture to a boil, then immediately simmer on a very low flame until the water from the decoction evaporates. (This could take up to 2–3 hours, depending upon the original amount.) When the decoction has boiled off, what remains is the medicinal herbal oil. To use, add a few drops of selected pure essential oil.

HERBAL INFUSIONS

For a water-based infusion, pour into a jar 2 cups boiling water over 2 Tbsp herbs. Let stand for a couple of days, stirring occasionally. Strain and use liquid as directed.

For an oil-based infusion, mix 1 Tbsp herbs + 1 cup base oil. Do not boil. Let stand for 2–3 days, stirring occasionally. Strain and use liquid as directed.

HERBAL BATHS

To use essential oils, add a few drops directly to the bathwater to make a dilution.

To use dried herbs, wrap about a handful of the selected herb(s) in a cheese-cloth, tie it closed, and let it soak in the bathwater.

STORING THE PREPARATIONS

- Keep all liquid formulations in sterilized, airtight, dark amber glass containers, and store in a cool place to protect them from heat, light, and air. The essential oils in these formulations act as a natural preservative. Stored properly, they have a natural shelf life of several months.
- Ghee does not need refrigeration. It has an unlimited shelf life—the older the ghee, the more medicinal value it has.
- Ghee-based products and some oil-based products will harden or liquefy according to the temperature. To melt, run the tightly closed bottle under hot water and shake lightly to mix the ingredients.
- Store herbal powders in spice jars, just as you do in the kitchen.

AYURVEDIC RESOURCES

To order Ayurvedic oils and products created by Pratima Raichur, N.D., please contact:

Bindi Facial Skin Care
A Division of Pratima Inc.
P.O. Box 750–250
Forest Hills, NY 11375
(800) 952-4634 (To order products)
(718) 268–7347
(Bindi Skin Care Products are also available in many natural food stores throughout the United States.)
or
Tej Ayurvedic Skin Care, Inc.
162 West 56th Street, Room 204
New York, NY 10019
(800) 310–0179 (To order products)
(212) 581–8136

Additional suppliers of Ayurvedic herbal products and essential oils

Auromere Inc.
1291 Weber Street
Pomona, CA 91768
(800) 925–1370

Auroma International
P.O. Box 1008, Dept. ABC
Silver Lake, WI 53170
(414) 889–8569

Ayurveda Center of Santa Fe
1807 Second St., Suite 20
Santa Fe, NM 87505
(505) 983–8898

Ayurvedic Concept
6950 Portwest Drive, Suite 170
Houston, TX 77024
(713) 863–1622

Ayurvedic Institute & Wellness Center
11311 Menaul N.E., Suite A
Albuquerque, NM 87112
(505) 291–9698

Ayush Herbs, Inc.
10025 N.E. 4th Street
Bellevue, WA 98004
(800) 925–1371

Banyon Trading Co.
P.O. Box 13002
Albuquerque, NM 87192–3002
(505) 275–2469

Bazaar of India Imports, Inc.
1810 University Avenue
Berkeley, CA 94703
(800) 261–7662

Frontier Herbs
P.O. Box 229
Norway, IA 52318
(800) 669–3275

HerbalVedic Products
Ayur Herbal Corporation
P.O. Box 6054
Santa Fe, NM 87502
(414) 889–8569

The Heritage Store Inc.
*(for castor oil, organic ghee, and
essential oils)*
P.O. Box 444
Virginia Beach, VA 23458

Janca's Jojoba Oil and Seed Company
*(for naturally pressed vegetable and
nut oils and essential oils)*
456 E. Juanita #7
Mesa, AZ 85204
(602) 497–9494

Kanak
P.O. Box 13653
Albuquerque, NM 87192–3653
(505) 275–2469

Lotus Herbs
1505 42nd Avenue, Suite 19
Capitola, CA 95010
(408) 479–1667

Maharishi Ayur-veda Products
International Inc.
417 Bolton Road
P.O. Box 541
Lancaster, MA 01523
1–800–ALL-VEDA

Raven's Nest
4539 Iroquois Trail
Duluth, GA 30136
(404) 242–3901

Ayurvedic studies
American Institute of Vedic Studies
(Dr. David Frawley)
P.O. Box 8357
Santa Fe, NM 87501
(505) 983–9385

Ayurvedic Healing Arts Center
16508 Pine Knoll Rd.
Grass Valley, CA 95945
(916) 274–9000

Ayurvedic Institute & Wellness Center
(Dr. Vasant Lad)
P.O. Box 23445
Albuquerque, NM 87192–1445
(505) 291–9698

Blue Sky Foundation
220 Oak Street
Grafton, WI 53024
(414) 376–1011

AYURVEDIC RESOURCES (CONTINUED)

The Chopra Center for Well Being
7630 Fay Avenue
La Jolla, CA 92037
(619) 551-7788
Fax: (619) 551-9570

Himalayan Institute
RR1, Box 400
Honesdale, PA 18431
(800) 822-4547

Institute for Wholistic Education
33719 116th Street, Box SH
Twin Lakes, WI 53181
(414) 889-8501

Lotus Ayurvedic Center
4145 Clares Street, Suite D
Capitola, CA 95010
(408) 479-1667

New England School of Ayurvedic
Medicine
1815 Massachusetts Avenue
Cambridge, MA 02140
(617) 876-2401

Wise Earth Institute
(Bri Maya Tawari)
Route 1, Box 484
Chandler, NC 28715
(704) 258-9999

MEDICINAL HERBS AND FOODS

FOR DRY SKIN (VATA)

USES	HERBS AND FOODS
Alterative (Purifies blood)	Heating—bayberry, black pepper, cayenne, cinnamon, garlic, myrrh, prickly ash, sassafras.
Antiparasitical (Reduces yeast, candida)	Ajwan, aseofida, cayenne pepper, garlic.
Astringent (Firming, drying, healing, rejuvenating, stops bleeding)	Black pepper, buttermilk, cinnamon, ginger, haritiki, nutmeg, triphala, yoghurt.
Carminitive (Digestive)	All hot spices such as ajwan, basil, bay leaves, calamus, cardamom, cinnamon, orange peel, valerian.
Diaphoretic (Induces perspiration, reduces edema, detoxifies)	Basil, cardamom, eucalyptus, ginger.
Diuretic (Increases urination, reduces edema)	Does not need diuretics.
Expectorant & Demulcent (Hydrating, soothing)	Bamboo, chickweed, comfrey, flaxseed, Irish moss, licorice, milk, rowsugar, slippery elm.
Laxative (Induces bowel movement)	Caster oil, Epson salt, figs, psyllium seeds, raisins, warm milk + 1 tsp ghee.
Nervine (Strengthens mind, nervous system, relieves spasms)	Basil, camphor, eucalyptus, garlic, hing, neroli, nutmeg, sandalwood, valerian, vanilla.
Nutritive & Tonic (Grounding, builds muscle & fat)	Almonds, bala, ghee, ginseng, kanda, milk, raisins, sesame seeds, vidari.
Rejuvenating	Aswagandha, calamus, garlic, ginseng, guggul, haritaki.

MEDICINAL HERBS (CONTINUED)

FOR SENSITIVE SKIN (PITTA)

USES	HERBS AND FOODS
Alterative (Purifies blood)	Cooling—Aloe vera, blue flag, burdock, chaparral, dandelion, echinacea, goldenseal, manjista, neem, plantain, red clover, sandalwood, yellow dock.
Antiparasitical (Reduces yeast, candida)	Goldenseal.
Astringent (Firming, drying, healing, rejuvenating, stops bleeding)	Stops bleeding: Agromony, goldenseal, hibiscus, marshmallow, nettle, saffron, self heal, turmeric, white oak, yarrow. Heals wounds: Aloe, amalki, comfrey, honey, slippery elm.
Carminitive (Digestive)	Catnip, chamomile, chrysanthemum, coriander, cumin, dill, fennel, lime, peppermint, spearmint, wintergreen.
Diaphoretic (Induces perspiration, reduces edema, detoxifies)	Boneset, burdock, catnip, chamomile, chrysanthemum, percoriander, elderflowers, horehound, horsetail, mint, pepper, spearmint, yarrow.
Diuretic (Increases urination, reduces edema)	Asparagus, barley, buchu, burdock, cleavers, coriander, cornsilk, dandelion, fennel, gokshura, grass, horsetail, lemon, punarnava, uvaursi.
Expectorant & Demulcent (Hydrating, soothing)	Same as Vata.
Laxative (Induces bowel movement)	Aloe vera powder, blue flag, cascara sagrada, rhubarb, senna, yellow dock.
Nervine (Strengthens mind, nervous system, relieves spasms)	Bhringraj, chamomile, gotu kola, lavender, mullein, passion flower, sandalwood, St. John's wort, vervain.
Nutritive & Tonic (Grounding, builds muscle & fat)	Amalki, bala, coconut, comfrey, ghee, licorice, shatavari, wild yam.
Rejuvenating	Aloe vera, amalki, brahmi, comfrey, saffron, shatavari.

FOR OILY SKIN (KAPHA)

USES	HERBS & FOODS
Alterative (Purifies blood)	Needs both heating & cooling herbs (see Vata & Pitta).
Antiparasitical (Reduces yeast, candida)	Same as Vata.
Astringent (Firming, drying, healing, rejuvenating, stops bleeding)	Same as Vata. Also, bibhitaki.
Carminitive (Digestive)	Basil, clove, ginger, thyme, turmeric.
Diaphoretic (Induces perspiration, reduces edema, detoxifies)	Angelica, basil, bayberry, camphor, cardamom, cinnamon, clove, eucalyptus, ginger, juniper, sage, wild ginger.
Diuretic (Increases urination, reduces edema)	Ajwan, cinnamon, garlic, juniper berries, mustard, parsley, wild carrot.
Expectorant & Demulcent (Hydrating, soothing)	Calamus, cardamom, cinnamon, cloves, dry ginger, hyssop, mustard seeds, pippali, sage, wild ginger.
Laxative (Induces bowel movement)	Ginger, fennel.
Nervine (Strengthens mind, nervous system, relieves spasms)	Basil, bergamot, pennyroyal, sage, skullcap.
Nutritive & Tonic (Grounding, builds muscle & fat)	Does not need tonic herbs.
Rejuvenating	Bibhitaki, guggul.

MEDICINAL HERBS (CONTINUED)

FOR ALL SKIN TYPES

USES	HERBS & FOODS
Aphrodisiac	Increases ojas & immunity: Ashwangandha, ghee, lotus seeds, shatavari. Increases sexual vigor: Cloves, fenugreek, saffron, wild yam.
Antipyretic (Detoxifies, dispels heat)	Aloevera, bayberry, calumba, chaparral, chirata, gentian, goldenseal, gold thread, kutki, neem.
Emmenagogue (Regulates menstruation)	Heating herbs are better for delayed menstruation: Angelica, cinnamon, ginger, hing, mugwort, myrrh, parsley, pennyroyal, turmeric, valerian. Cooling herbs are better for excessive bleeding: Blessed thistle, chamomile, hibiscus, manjista, primrose, rose, yarrow.

ESSENTIAL OILS

FOR DRY SKIN (VATA)

(Needs sweet, sour, salty, calming, warming,
hydrating, alteratives, carminatives, nervines)

ESSENTIAL OIL	ODOR/DESCRIPTION	USES
Aniseed	Sweet, warm, herbal.	Calming.
Bergamot	Fresh, sharp, citrus.	Sedative.
Cajuput	Strong, warm.	Warming.
Chamomile	Fruity, herbal, sweet, astringent.	Sedative, soothing, cooling; relieves insomnia, depression
Cardamom	Sweet, woody, floral.	Hot, warming, calming.
Cinnamon bark	Strong, sweet, warm, spicy.	Sedative, warming.
Cypress	Sweet, rich.	Sedative.
Frankincense	Lemony, spicy.	Sedative; relieves dwelling on past.
Geranium	Strong, rose-like, sweet, slightly astringent.	Relieves anxiety, depression; soothing, slightly warming.
Ginger	Lemony, warm, spicy, woody.	Warming, calming.
Jasmine	Floral, fruity, herbal, sweet, bitter, astringent.	Sedative, relieves fear, apathy, hypersensitivity; cooling, calming.
Lemon	Light, fresh, citrus, sour, bitter.	Refreshing, hot, stimulating; promotes fearlessness, optimism.
Nutmeg	Fresh, spicy, warm, astringent, slightly sweet.	Warming, calming, rejuvenating; stimulates hair growth.
Orange	Sweet, light, citrus, sour.	Calming, warming.
Rose	Sweet, rich, floral, astringent.	Sedative, cooling, calming; relieves anger, jealousy.
Spearmint	Sweet, warm, herbal, minty.	Sedative.
Thyme (White)	Sweet, warm, herbal.	Warming.
Ylang-Ylang	Strong, floral, spicy.	Aphrodisiac; soothing.

VATA HERBAL OILS:

Brahmi, bay, ashwangandha.

ESSENTIAL OILS (CONTINUED)

FOR SENSITIVE SKIN (PITTA)

*(Needs sweet, bitter, astringent, calming,
soothing, cooling, alteratives.)*

ESSENTIAL OIL	ODOR/DESCRIPTION	USES
Calamus	Bitter, astringent.	Warming.
Chamomile	Fruity, herbal, sweet, astringent.	Sedative, soothing, cooling; relieves insomnia, depression.
Cardamom	Sweet, woody, floral, astringent.	Warming, calming.
Coriander	Light, spicy.	Cooling.
Fennel (Sweet)	Sweet, spicy, sour.	Warming, calming.
Frankincense	Lemony, spicy.	Sedative, relieves dwelling on past.
Geranium	Strong, rose-like, sweet, slightly astringent.	Relieves anxiety, tension, depression; soothing, slightly warming.
Jasmine	Floral, fruity, herbal, sweet, bitter, astringent.	Sedative; relieves fear, apathy, hypersensitivity; cooling, calming.
Lavender	Fruity, herbal, woody.	Sedative, soothing, energy balancing.
Orris	Sweet, woody, floral.	Soothing.
Patchouli	Earthy, spicy, woody.	Calming.
Rose	Sweet, rich, floral, astringent.	Sedative, relieves anger, jealousy; cooling, calming.
Sandalwood	Sweet, woody, bitter, astringent.	Sedative, aphrodisiac, cooling, calming.
Spearmint	Sweet, warm, herbal, minty.	Soothing, cooling.
Vetiver	Sweet, woody, earthy, bitter.	Cooling, calming, sedative.
Ylang-Ylang	Strong, floral, spicy.	Aphrodisiac; calms anger, frustration.

PITTA HERBAL OILS:

Neem, brahmi, burdock, licorice, shatavari.

FOR OILY SKIN (KAPHA)

(Needs bitter, pungent, astringent, stimulating,
heating, diuretics, diaphoretics.)

ESSENTIAL OIL	ODOR/DESCRIPTION	USES
Basil	Bitter, astringent, pungent.	Hot, stimulating, uplifting, refreshing.
Bay Leaf	Fresh, spicy.	Depression, confusion.
Cajuput	Strong, warm.	Warming, stimulating.
Calamus	Bitter, astringent.	Warming.
Cinnamon Leaf	Spicy, clove-like, warm.	Stimulating.
Clove Bud	Warm, spicy, woody, bitter.	Hot, stimulating.
Eucalyptus	Strong, warm.	Hot.
Ginger	Lemony, warm, spicy, woody.	Uplifting.
Patchouli	Earthy, spicy, woody.	Refreshing, stimulating.
Pepper (Black)	Hot, spicy.	Stimulating.
Peppermint	Fresh, minty, sweet.	Cooling to touch, but hot, stimulating effect.
Rosemary	Strong, fresh.	Stimulating; relieves mental fatigue, sadness.
Sage	Warm, herbal, spicy.	Activating.
Thyme (Red)	Sharp, warm, herbal, spicy.	Invigorating.

KAPHA HERBAL OILS:

Jatamansi, neem, sage.

SADHANAS

TO BALANCE VATA

*(Needs warm, nourishing, calming,
hydrating, and lubricating qualities.)*

- Take walks by the water.
- Listen to sounds of flowing river or ocean waves.
- Listen to the soft music of a flute, violin, or sitar.
- Look at the rising sun.
- Plant and garden.
- Take warm baths with aromatic oils of neroli, lemon, geranium, honeysuckle, or nightqueen.
- Have gentle massage with sesame oil and sweet, warming herbal oils.
- Read a romantic novel in soft candlelight.
- Adorn your body temple with warm-colored clothes, gold jewelry, and exotic perfume.
- Give and receive hugs from friends and family.
- Embrace small children and cuddle with loved ones.
- Make a warm, loving home by cooking and spending time with family.
- Take a nap in the afternoon.
- Do gentle physical activity or yoga.
- Take time each day for meditation, sitting quietly, and paying attention to the breath.
- Relax in front of a fireplace and watch the burning logs.

TO BALANCE PITTA

(Needs cooling, soothing, and hydrating qualities.)

- Walk in a flower garden.
- Look at nature.
- Listen to birds chirping and singing.
- Swim in a cool river.
- Take a cool bath with aromatic oils of rose or sandalwood.
- Massage your body with cool herbal oils.
- Adorn your body with soft blue, pink, or green clothes and silver jewelry.
- Make a flower arrangement.
- Decorate your house and mow the lawn.
- Laugh.
- Listen to the inner sound.
- Look at beautiful art.
- Walk in the moonlight.
- Bathe in silver water.

TO BALANCE KAPHA

(Needs hot, stimulating, and invigorating qualities.)

- Take a mountain hike.
- Watch the sunrise.
- Put on some loud rock 'n' roll and get up and dance.
- Clean out your closets and throw out things you haven't used in years.
- Rearrange the furniture in your home.
- Cleanse your body with herbs.
- Have a vigorous massage, dry or with safflower oil and a pinch of saffron.
- Cook a spicy meal for friends.
- Enjoy good sex.
- Play games and sports.
- Help others.

DAILY ROUTINE
TO BALANCE DOSHAS

	VATA	PITTA	KAPHA
Wake up at	½ hour before sunrise.	1 hour before sunrise.	1½ hour before sunrise.
Teeth & gums	After brushing teeth, massage gums with triphala, honey & sesame oil.	After brushing, massage gums with cardamom, honey, pinch of rock salt & sesame oil.	After brushing, massage gums with ginger, honey, rock salt & sesame oil.
Gargling	After massaging gums, gargle with triphala tea. Chew 2–4 tsp of black sesame seeds daily in the morning.	Gargle with fennel or licorice tea. Chew 2 tsp of black sesame seeds in the morning.	Gargle with ginger tea.
Eyewash	Triphala eyewash.	Triphala or rosewater eyewash.	Triphala or cranberry juice eyewash.
Drink	a) Warm water with lemon juice & honey, or b) Water kept in gold vessel overnight.	a) ¼ cup aloe vera juice, or b) Water kept in silver vessel overnight.	a) Warm herbal teas such as ginger or triphala, or b) Water kept in copper vessel overnight.
Bowel movement (if constipated)	Massage stomach; take ½ cup warm milk with 1 tsp ghee, or 1 tsp triphala with warm water at night.	Usually Pitta does not have this problem.	Massage stomach & do not drink coffee.

	VATA	PITTA	KAPHA
Nose wash	Wash with water, then put a few drops sesame oil or ghee in both nostrils.		
Exercise— (yoga asanas)	Alternate nostril breathing, sun salute, shoulder stand, corpse pose, cobra, vajrasana, yoga mudra, shavasana, walking, slow stretching.	Shitali breathing, shoulder stand, bow, sun salute, fish pose, shavasana, running, swimming, walking.	Head stand, vajrasana, peacock, lion pose, shoulder stand, palm tree pose, jumping, running, or any aerobic exercise.
Bath or shower	Massage with sesame oil before taking bath. Take bath with few drops of essential oil such as basil, geranium, jasmine, or champa. After shower, massage body with Tej Body Massage Oil on wet skin.	Massage body with coconut oil before bath. Bathe with essential oils such as rose, sandalwood, or vetiver. After shower, use Tej Pitta Body Massage Oil on wet skin.	Massage with dry herbal powder like barley before bath. Bathe with a few drops essential oil such as lavender or sage. After shower use Tej Kapha Body Massage Oil on wet skin.
Face care	Follow cleansing, nourishing & moisturizing routine for your skin type (see Chapter 5).		
Spiritual practice	Spend at least 15-20 minutes in meditation or prayers.		
Breakfast	Cooked cereal.	Cold cereal.	Fresh fruit.
Lunch (12:30– 1:30 p.m.)	Follow the balancing diet for your dosha. Take a few minutes of rest after lunch whenever possible.		

DAILY ROUTINE (CONTINUED)

	VATA	PITTA	KAPHA
Afternoon (3–4 p.m.)	Have a light snack to balance your leading dosha (see page 214).		
Sunset	Spend at least 15-20 minutes in meditation or prayers. Before you begin, wash feet, hands & face; put on clean clothes.		
Dinner (7:30–8:30 p.m.)	Follow the balancing diet of your leading dosha.		
Night	Follow cleansing, nourishing & moisturizing routine for your skin type.		
	Massage scalp and feet with brahmi oil. Use Vata aroma therapy (1 drop) on pillow for restful sleep.	Massage scalp & feet with brahmi oil.	Usually Kapha does not need a massage to settle down before sleep.
	Say prayers and let go of worries, conflicts & frustrations of the day. Lights out between 10:00-11:00 p.m.		

PHYSICAL EFFECTS
OF THE SIX TASTES

TASTE	BENEFITS	PROBLEMS DUE TO EXCESS
Sweet	Helps build tissues, rejuvenating, hydrating, healing, tones muscles.	Clogged pores, congestion, oily skin, blackheads, acne, toxicity, puffiness under eyes, hypoglycemia, diabetes.
Sour	Cleanses the skin, stimulating, carminitive, diaphoretic, refreshing.	Muscle weakness, diarrhea, dark circles under the eyes, hyperacidity, burning sensation, bleeding, broken capillaries.
Salty	Helps in digestion, opens up blocked channels, improves circulation, stimulating, carminitive, awakens mind and senses, strengthens heart.	Contraction, edema, wrinkles, general debility, thirst, hyperacidity, high blood pressure, heartburn, itching, burning, swollen glands, impotency, loose teeth, darkening of skin pigmentation, premature aging, graying, balding, kidney damage.
Pungent	Promotes sweating, improves circulation and digestion, clears channels, relieves nerve pain, gives glow to skin.	Burning sensation, dryness, tissue depletion, broken capillaries, redness of skin & nose, dry lungs, dry cough, dehydration of skin.
Astringent	Antiseptic, shrinks pores, promotes healing, tightens tissues, diuretic, cooling, reduces sensitivity, controls excessive sweating.	Dryness, colon problems, gas, constipation, contraction, tension, nerve pain, irritability.
Bitter	Promotes weight loss, reduces fat, purifies blood, detoxifies, clears mind & skin, anti-inflammatory, antibacterial.	Heart problems, anemia, low blood pressure, insomnia, cold, vertigo, constipation; dehydration of the skin, premature wrinkles.

DIETARY SUGGESTIONS

FOR DRY SKIN (VATA)

These dietary suggestions reflect the pure aspect of one dosha only. Each person is a combination of all three doshas, with a predominant tendency toward one or two. Therefore this chart serves only to draw general awareness to diet, in order to encourage a regimen that will promote good health.

TASTES:
Sweet, sour, salty.

DAILY DIETARY PROPORTIONS:
50% whole grains.
20% proteins.
20–30% fresh vegetables & fruits.

GENERAL:
YES: Warm, heavy, cooked foods. Keep regular routine, drink 6–7 glasses of water daily.
NO: Cold foods, salads, raw foods, soda, ice cream.

VEGETABLES:
YES: Asparagus, beets, cooked cabbage and cauliflower, carrots, cilantro, cucumber, daikon radish, fennel, garlic, green beans, green chilies, horseradish, leeks, mustard greens, okra, black olives, cooked onions, parsnip, cooked peas, sweet potatoes, pumpkin, radishes, rutabaga, cooked spinach, squash, tomatoes, taro root, watercress, zucchini.
NO: Artichoke, beet greens, bitter melon, broccoli, brussels sprouts, burdock root, raw cabbage and cauliflower, celery, fresh corn, dandelion greens, eggplant, kale, kohlrabi, leafy greens, lettuce, mushrooms, green olives, raw onions, parsley, peas, sweet & hot peppers, white potatoes, prickly pear, radish, spaghetti squash, sprouts, raw tomatoes, turnip greens, wheat grass, sprouts.

FRUITS:
YES: Sweet apples, apricots, avocados, berries, cherries, coconut, fresh dates, ripe figs, grapes (red & purple), kiwi, lemons, ripe mangos, melons, sweet oranges, papaya, peaches, pineapple, sweet plums, raisins, rhubarb, strawberries, tamarind.
NO: Dried fruits, raw apples, cranberries, dried dates, dried figs, prunes, pears, persimmons, pomegranates, watermelon.

GRAINS:

YES: Amaranth, durhum flour, oats (cooked), pancakes, quinoa, rice, seitan, sprouted wheat bread (Essene), wheat.

NO: Barley, bread, buckwheat, cereals, corn and corn chips, couscous, crackers, granola, millet, muesli, oat bran, dry oats, pasta, polenta, rice cakes, rye, sago, spelt, tapioca, wheat bran.

LEGUMES:

YES: Lentils (red), Miso, mung beans, mung dal, soy cheese, soy sauce, soy sausages, tur dal, urad dal, tofu.

NO: Aduki, black beans, chickpeas, kidney beans, brown & red lentils, lima beans, navy beans, pinto beans, soya beans, white beans.

DAIRY:

YES: Most dairy is good—butter, buttermilk, cheese, sour cream, cottage cheese, warm goat's milk, yogurt (if diluted & spiced).

NO: Hard cheese, plain yogurt, ice cream.

ANIMAL FOODS:

YES: Beef, buffalo, chicken, duck, egg, freshwater or sea fish, salmon, sardines, seafood, shrimp, tuna fish, turkey (dark), lamb, crab, oyster, trout.

NO: White chicken, pork, rabbit, venison, turkey (white).

CONDIMENTS:

YES: Black pepper, chutney, coriander leaves, dulse, gomasio, hijiki, kelp, ketchup, kombu, lemon, lime, lime & mango pickle, mayonnaise, mustard, pickles, salt, scallions, seaweed, soy sauce, tamari, vinegar.

NO: Chili peppers, chocolate, horseradish, sprouts.

NUTS:

YES: All nuts, but only in small quantities (about 10 nuts).
NO: None.

SEEDS:

YES: Flax, halva, psyllium, pumpkin, sunflower, chia, sesame, tahini.
NO: Popcorn.

OILS:

YES: Sesame, ghee, olive, most other oils.
NO: Flax seed.

FOR VATA (CONTINUED)

SWEETENERS:

YES: Barley malt, fructose, fruit juices, maple syrup, rice syrup, sucanat, honey, jaggary, molasses, turbinado.
NO: White sugar.

BEVERAGES:

YES: Almond milk, aloe vera juice, apple cider, apricot juice, beer (occasionally), berry & carrot juice, chai, cherry juice, grain coffee substitutes, grape juice, lemonade, mango juice, miso broth, orange & papaya juice, peach nectar, pineapple juice, rice milk, sour juices, soy milk, wine (occasionally). Herbal teas: All spicy teas, including ajwan, bancha, basil, chamomile, cinnamon, clove, comfrey, eucalyptus, fennel, fenugreek leaves, fresh ginger, juniper berry, kukicha, lavender, lemongrass, marshmallow, oat straw, orange peel, peppermint, raspberry, rosehips, saffron, sarsparilla, sassafras, spearmint.
NO: Apple juice, black tea, caffeinated beverages, carob, chocolate milk, coffee, cold dairy drinks, cranberry juice, iced tea, icy cold drinks, mixed vegetable juice, pear juice, pomegranate juice, prune juice, soy milk, tomato juice, V8 juice, vegetable bouillon. Herbal teas: Alfalfa, barley, blackberry, borage, burdock, catnip, chicory, chrysanthemum, cornsilk, dandelion, ginseng, hibiscus, hops, hyssop, jasmine, lemon balm, Mormon tea, nettle, passion flower, red clover, sage, strawberry, violet, wintergreen, yarrow, yerba mate.

SPICES:

YES: Basil, black pepper, cardamom, cinnamon, coriander, cumin, dill, fennel, spearmint, turmeric, vanilla, wintergreen, ajwan, allspice, almond extract, ginger, mint, neem leaves, orange peel, parsley, peppermint, saffron, anise, asafoetida (hing), basil, bay leaf, cayenne, cloves, fenugreek, garlic, dry ginger, mace, marjoram, mustard seeds, nutmeg, oregano, paprika, pippali, poppy seeds, rosemary, sage, salt, savory, star anise, tarragon, thyme.
NO: Caraway.

FOOD SUPPLEMENTS:

YES: Aloe vera juice, calcium, magnesium, zinc, spirulina, blue-green algae, amino acids, bee pollen, royal jelly, iron, vitamins A, B, B$_{12}$, C, D & E.
NO: Barley green, brewer's yeast.

DIETARY SUGGESTIONS

FOR SENSITIVE SKIN (PITTA)

These dietary suggestions reflect the pure aspect of one dosha only. Each person is a combination of all three doshas, with a predominant tendency toward one or two. Therefore this chart serves only to draw general awareness to diet, in order to encourage a regimen that will promote good health.

TASTES:
Sweet, bitter, astringent.

DAILY DIETARY PROPORTIONS:
50% whole grains.
20% proteins.
20–30% fresh vegetables and fruits.

GENERAL:
YES: Cool, heavy foods; drink water regularly.
NO: Salty, hot, spicy & oily foods, sour fruit, yogurt, tomatoes, vinegar.

VEGETABLES:
YES: Sprouts, cabbage, carrots (cooked), cauliflower, celery, cilantro (sweet and bitter), artichoke, asparagus, beets, bitter melon, broccoli, brussels sprouts, dandelion greens, fennel, green beans, leafy greens, kale, leeks (cooked), lettuce, mushroom, okra, parsley, parsnips, peas, sweet potatoes, peppers, pumpkin, cooked radish and spinach, sprouts, squash, wheat grass, zucchini.
NO: Pungent—beets (raw), carrots (raw), corn (fresh), radish, eggplant, garlic, green chilies, horseradish, mustard greens, olives (green), onions (raw), hot peppers, spinach, tomatoes.

FRUITS:
YES: Sweet fruits—apples, apricots, avocados, berries, cherries, coconut, dates, figs, grapes (red & purple), ripe mangos, melons, sweet oranges, pears, pineapple, plums, pomegranates, prunes, raisins, watermelons.
NO: Sour fruits—bananas, cranberries, grapefuit, green grapes, lemons, papaya, peaches, strawberries, tamarind.

FOR PITTA (CONTINUED)

GRAINS:
YES: Amaranth, barley, dry cereals, durum flour, granola, oat bran, pasta, rice (Basmati), tapioca, wheat, wheat bran.
NO: Bread with yeast, buckwheat, corn, millet, brown rice, rye.

LEGUMES:
YES: Aduki, black beans, chickpeas, kidney beans, brown & red lentils, lima beans, mung beans, navy beans, pinto beans, soya, tofu, white beans.
NO: Miso, soy sauce, tur dal, urad dal.

DAIRY:
YES: Unsalted butter, cheese, cottage cheese, cow's milk, ghee, soya milk.
NO: Buttermilk, hard cheese, sour cream, frozen yogurt or yogurt with fruit.

ANIMAL FOODS:
YES: White chicken, egg whites, freshwater fish, turkey, shrimp, venison.
NO: Dark chicken, duck, egg yolk, seafood, lamb, pork, salmon, sardines, tuna fish.

CONDIMENTS:
YES: Sweet chutneys, coriander leaves, lime, black pepper.
NO: Chili pepper, chocolate, dulse, kelp, ketchup, lime & mango pickles, mayonnaise, scallions, seaweed, soy sauce, vinegar.

NUTS:
YES: Soaked and peeled almonds, coconut, charole.
NO: Black walnuts, brazil nuts, cashews, hazelnuts, peanuts, pecans, pines, pistachios, walnut.

SEEDS:
YES: Flax, unsalted popcorn, pumpkin, sunflower.
NO: Chia, sesame, tahini.

OILS:
YES: Sunflower, ghee, canola, olive, soy, primrose, walnut.
NO: Almond, apricot, corn, safflower, sesame.

SWEETENERS:

YES: Barley malt, fructose, fruit juices, maple syrup, rice syrup, sucanat.

NO: Honey, jaggary, molasses.

BEVERAGES:

YES: Almond milk, aloe vera juice, apple, apricot & berry juice, beer (occasionally), black tea, carob, chai, cherry juice, cool dairy drinks, grain coffee, grape, mango, mixed vegetable & pear juice, peach nectar, pomegranate & prune juice, rice & soy milk, vegetable bouillon. Herbal teas: All astringent teas, including alfalfa, bancha, barley, blackberry, borage, burdock, catnip, chamomile, chicory, chrysanthemum, comfrey, cornsilk, dandelion, elder flower, fennel, fresh ginger, hibiscus, hops, jasmine, kukicha, lavender, lemon balm, lemongrass, licorice, marshmallow, nettle, oat straw, orange peel, passion flower, peppermint, raspberry, red clover, saffron, sarsparilla, spearmint, strawberry, violet, wintergreen, yarrow.

NO: Hard alcohol or wine, apple cider, berry juice (sour), caffeinated beverages, carbonated drinks, carrot juice, cherry juice (sour), chocolate milk, coffee, cranberry & grapefruit juice, iced tea, icy cold drinks, lemonade, orange juice, miso broth, papaya, pineapple, tomato, V8, & sour juices. Herbal teas: ajwan, basil, cinnamon, clove, eucalyptus, fenugreek, ginger, ginseng, hawthorne, hyssop, juniper berry, Mormon tea, pennyroyal, rosehip, sage, sassafras, yerba mate.

SPICES:

YES: Basil, black pepper, cardamom, cinnamon, coriander, cumin, dill, fennel, ginger, mint, neem leaves, orange peel, parsley, peppermint, saffron, spearmint, turmeric, vanilla, wintergreen.

NO: Ajwan, allspice, almond extract, anise, asafoetida (hing), basil, bay leaf, caraway, cayenne, cloves, fenugreek, garlic, dry ginger, mace, marjoram, mustard seeds, nutmeg, oregano, paprika, pippali, poppy seeds, rosemary, sage, salt, savory, star anise, tarragon, thyme.

FOOD SUPPLEMENTS:

YES: Aloe vera juice, barley green, brewer's yeast, spirulina, blue-green algae, vitamins D & E, minerals—calcium, magnesium, zinc.

NO: Amino acids, bee pollen, royal jelly, vitamins A, B, B_{12}, C, minerals—iron

DIETARY SUGGESTIONS

FOR OILY SKIN (KAPHA)

These dietary suggestions reflect the pure aspect of one dosha only. Each person is a combination of all three doshas, with a predominant tendency toward one or two. Therefore this chart serves only to draw general awareness to diet, in order to encourage a regimen that will promote good health.

TASTES:

Bitter, pungent, astringent.

DAILY DIETARY PROPORTIONS:

30–40% whole grains.

20% proteins.

40–50% fresh vegetables & fruits.

GENERAL:

YES: Warm, light foods.

Keep active & vary routine.

NO: Dairy, heavy foods, fried foods, iced foods & drinks, sweets.

VEGETABLES:

YES: Artichoke, asparagus, beets, bitter melon, broccoli, brussels sprouts, cabbage, carrots, cauliflower, celery, cilantro, burdock root, corn, daikon radish, dandelion greens, eggplant, fennel, garlic, green beans, green chili, horseradish, Jerusalem artichoke, kale, kohlrabi, leafy greens, leeks, lettuce, mushrooms, mustard greens, okra, onions, parsley, peas, peppers (hot & sweet), potatoes, prickly pear, radish, rutabaga, spinach, sprouts, squash, tomatoes, turnip greens, turnip, watercress, wheat grass sprouts.

NO: Cucumber, olives, parsnips, potatoes (sweet), pumpkin, spaghetti squash, taro root, tomatoes (raw), zucchini.

FRUITS:

YES: Astringent fruits—Apples, applesauce, apricots, berries, cherries, cranberries, figs, peaches, pears, persimmons, pomegranates, prunes, raisins, strawberries.

NO: Avocado, bananas, coconut, dates, figs (fresh), grapefruit, grapes, kiwi, lemons, limes, mangos, melons, oranges, papaya, pineapple, plums, rhubarb, tamarind, watermelon.

GRAINS:

YES: Amaranth, barley, buckwheat, cereal (cold or dry), corn, couscous, crackers, durum flour, granola, millet, muesli, oat bran, oats, polenta, Basmati rice, rye, sago, seitan, sprouted wheat bread (Essene), tapioca, wheat bran.

NO: Bread with yeast, cooked oats, pancakes, pasta, quinoa, brown or white rice, rice cakes, spelt, wheat.

LEGUMES:

YES: Urad dal, aduki beans, black beans, black-eyed peas, chickpeas, lentils (brown & red), lima beans, miso, navy beans, peas, pinto beans, soy milk, soy sausage, split peas, tempeh, tofu, tur dal, white beans.

NO: Kidney beans, mung beans, mung dal, soybeans, soy cheese, soy flour, soy powder, soy sauce, tofu (cold).

DAIRY:

YES: Unsalted butter, cottage cheese, ghee, goat's cheese, goat's milk (skim), diluted yogurt.

NO: Salted butter, buttermilk, cheese, cow's milk, ice cream, sour cream, plain or frozen yogurt, or yogurt with fruit.

ANIMAL FOODS:

YES: White chicken, fish, rabbit, shrimp, turkey, venison, freshwater fish in small quantities.

NO: Beef, buffalo, chicken, duck, lamb, pork, salmon, sardines, seafood, tuna fish, turkey (dark).

CONDIMENTS:

YES: Black pepper, chili peppers, chutney (spicy), coriander leaves, horseradish, mustard, scallions, sprouts.

NO: Chocolate, sweet chutney, dulse, gomasio, hijiki, kelp, ketchup, lemon, lime pickle, mango pickle, mayonnaise, pickles, salt, seaweed, soy sauce, tamari, vinegar.

NUTS:

YES: Charole.

NO: Almonds, coconut, black walnuts, brazil nuts, cashews, hazelnuts, peanuts, pecans, pines, pistachios, walnuts.

SEEDS:

YES: Chia, flax, popcorn, psyllium, pumpkin, sunflower.

NO: Halva, sesame, tahini.

FOR KAPHA (CONTINUED)

OILS:

YES: Corn, canola, sunflower, ghee, almond, flax seed.

NO: Avocado, apricot, coconut, olive, primrose, safflower, sesame, soy, walnut.

SWEETENERS:

YES: Fruit juice concentrates, honey (raw).

NO: Barley malt, fructose, jaggery, maple syrup, molasses, rice syrup, sucanat, turbinado, white sugar.

BEVERAGES:

YES: Aloe vera juice, apple cider & juice, apricot & berry juice, black tea, carob, carrot & cherry juice, grain coffee, grape, mango, mixed vegetable juice, peach nectar, pear juice, pomegranate juice, prune juice, soy milk, vegetable bouillon, wine (occasionally). Herbal teas: All spicy teas including ajawan, alfalfa, basil, blackberry, burdock, cinnamon, clove, cornsilk, dandelion, elder flower, eucalyptus, fenugreek, ginger, hops, lavender, lemon balm, nettle, peppermint, raspberry, red clover, saffron, sage, spearmint, strawberry, wintergreen, yarrow.

NO: Alcohol (sweet wine & beer), caffeinated beverages, carbonated drinks, coffee, cold dairy drinks, grapefruit juice, lemonade, orange, papaya, pineapple juice, rice milk, soy milk, miso broth, tomato juice. Herbal teas: Comfrey, red ginger, rose hips.

SPICES:

YES: All spices are good, except salt.

FOOD SUPPLEMENTS:

YES: Aloe vera juice, amino acids, barley green, bee pollen, brewer's yeast, royal jelly, spirulina, blue-green algae, vitamins A, B, B_{12}, C, D, E, minerals—copper, calcium, iron, magnesium, zinc.

NO: Potassium.

SENSE THERAPIES
AT A GLANCE

	VATA	PITTA	KAPHA
Sound	Soft music with low tones; e.g., soft rhythmic drumming.	Soft music with middle tones; e.g., flute.	Loud music with high tones; e.g., drums.
Mantras	Ham, Yam.	Ram.	Vam, Lam.
Touch	Gentle massage with sweet, sour, heating oils in base of sesame oil. Scalp & foot massage.	Moderate massage with sweet, cooling, astringent, soothing oils in base of sunflower oil.	Dry massage, vigorous deep tissue massage, Swedish or Shiatsu massage are good. Use warm, spicy, stimulating, bitter, pungent oils in base of safflower oil.
Color	White, yellow, violet, blue, deep red, warm colors.	Soft pastel colors such as pink, rose, green, blue, cool colors.	Red, orange, hot stimulating colors.
Taste	Sweet, sour, salty.	Sweet, bitter, astringent.	Pungent, bitter, astringent.
Smell	Sweet, mild, warm smells such as geranium, kewada, champa, lemon, neroli, nutmeg, cinnamon.	Fragrant, cool smells such as sandalwood, rose, jasmine, cardamom, fennel, vetiver.	Sharp, penetrating smells such as ginger, eucalyptus, bergamot, camphor, clove, mint.

THE DOSHAS AT A GLANCE

	VATA	PITTA	KAPHA
English name	Air	Fire	Earth
Governing life force	Prana	Tejas (or agni)	Ojas
Associated functions	Movement, breath, expansion.	Metabolism, digestion, intelligence.	Structure, secretions, attraction.
Bodily seats	Colon, nervous system, skin.	Intestines, liver, blood.	Stomach, lungs.
Associated senses	Sound & touch.	Sight.	Taste & smell.
Associated sense organs	Ears, skin.	Eyes.	Tongue, nose.
General appearance	Thin, tall or short, bony; dry, rough, cool skin; scanty, dark, curly, coarse hair; dry, small, dull eyes; thin, dry lips, crooked teeth.	Medium build & weight; warm, reddish, freckled skin; moderate, fine, soft, reddish or premature balding or gray hair; sharp green, gray, or hazel eyes; red or pink moderate lips & teeth.	Short, stout, well developed; oily, soft, smooth, cool skin; abundant, thick, oily, lustrous, wavy hair; wide, prominent, attractive eyes; strong, white teeth, thick firm lips.

	VATA	PITTA	KAPHA
Positive mental traits	Resilient, imaginative, sensitive, spontaneous, positive, helpful, flexible, quick, enthusiastic, energetic, communicative, fast to learn, fast to forget.	Sharp, intelligent, confident, clear, enterprising, joyous, ambitious, perceptive, friendly, courageous, warm, independent, good memory.	Calm, loving, forgiving, serene, wise, compassionate, insightful, resourceful, loyal, persistent, patient, nurturing, content, stable, faithful, receptive, shy, slow to learn, slow to forget.
Negative mental traits due to imbalance	Anxious, empty, agitated, fearful, ungrateful, nervous, indecisive, hyperactive, servile, secretive, dishonest, addicted.	Angry, cruel, irritable, critical, jealous, aggressive, hostile, vain, willful, impulsive, dominating, arrogant, vindictive, psychopathic.	Depressed, possessive, lethargic, attached, greedy, slow to change, controlling, insecure, apathetic, vulgar, dull, slothful, materialistic, thieving.
Physical symptoms of imbalance	Insomnia, low appetite, gas, constipation, cramps, arthritis, muscle spasms, low back pain, varicose veins, popping joints.	Hot flashes, heartburn, acid stomach, ulcers, hemorrhoids, burning eyes & feet, allergies, colitis.	Sinus congestion, cough, cold, high cholesterol, asthma, weight gain, obesity, diabetes, bronchitis, emphysema.
Common skin problems	Dry, dehydrated skin, fine wrinkles, dry eczema.	Inflammation, acne rosacea, moles, burning eczema, rashes, sunburn.	Cystic acne, wet eczema, deep wrinkles.

abhyanga: Head-to-toe oil massage.

adhipati: Vital energy point, or marma, located at the "crown" chakra on top of the scalp midway between the ears.

agni: Biological fire that governs metabolism. Also called the "digestive fire." Corresponds to tejas.

agrapata: Tip of sternum, solar plexus; marma point.

ama: Toxins formed in the body due to improper digestion and imbalance.

amalaki: A sour fruit that acts as a tonic, rejuvenating, and laxative herbal remedy; works on all elements; increases ojas.

ananda: Transcendental bliss; pure joy.

anandamaya kosha: The highest or most subtle of the five body sheaths or "envelopes"; corresponds to the state of bliss.

annamaya kosha: Literally the "food sheath." The lowest of the five body sheaths; corresponds to the physical body.

Arjuna: Warrior prince who is the central character in the Bhagavad Gita.

asana: Means "seat." The physical postures of Hatha-Yoga.

ashwangandha: "Winter cherry"; gives vitality and sexual energy; rejuvenating, tonic, increases ojas.

atma: Soul.

aum: (Also "om.") The root mantra; the "soundless sound" of creation. The universe comes from aum, rests in aum, dissolves in aum.

Ayurveda: The science or knowledge of life; the world's oldest system of health and healing; based upon systematic study of body, mind, and soul.

Bhagavad Gita: "Lord's Song." Considered the "Bible" of Vedic literature and the earliest popular writing on yoga, the path to God-realization.

Bharat: Ancient sage who established study of Indian aesthetics.

Bhringaraj: Ancient sage of India.

bij mantra: "Seed word." Monosyllabic meaningless sounds that are derived from a more complex combination of primordial sounds and that are used to balance the five elements in the body.

Bindi: A product line named after the red dot worn by Indian women over the "third eye."

bindu: "Dot." Point where creation begins and ends.

brahma randhra: Marma point situated at the center of cranium.

chakra: Energy "wheel." One of the seven main centers of consciousness in the body which are situated (starting with the first chakra) at the base of the spine, the navel, the solar plexus, the heart, the throat, the "third eye," and the crown of the head.

Chymunda: The goddess who embodies the three goddesses Kali, Lakshmi, and Saraswati, or the three feminine principles of courage, abundance (wealth and happiness), and wisdom.

dashamula: Ayurvedic herbal medicine made with ten roots.

deha shudhi: Purification of the body.

Dhanurveda: The science of warfare.

dharma: Right purpose or action; duty.

dhatu-agni: The metabolic fire in each bodily tissue.

dhatus: The seven bodily tissues, or structural elements of the body, including: rasa (plasma), rakta (blood), mamsa (muscle), meda (fat), asthi (bone), majja (bone marrow, nerves), shukra (reproductive tissues).

dosha: Literally means "impurity." One of the three metabolic principles (Vata, Pitta, and Kapha) governing mind and body; the intelligence of the body-mind.

Gayatri mantra: A sound made up of the twenty-four primordial sounds in which the knowledge of creation is structured.

Ghandharva Veda: An aspect of Veda concerned with the balancing effects of music

ghee: Clarified butter. Used widely in Ayurveda as an ingredient in skin care preparations as well as in cooking

gunas: "Quality." The three primordial forces within consciousness (sattva, rajas, and tamas) that give rise to creation; also the twenty fundamental physical attributes of the universe made up of pairs of opposites, such as hot and cold, dry and oily, and so forth.

guru: Spiritual teacher. The "remover of darkness" from life.

hanu: Chin.

Hatha Yoga: A system to achieve spiritual union with God based upon the purification of the body and the opening of subtle energy channels through physical postures.

hridaya: Heart.

Jyotish: The Vedic system of astrology and astronomy.

Kali: Goddess who destroys evil and represents courage.

kapalbhati: A type of breathing exercise characterized by continuous rapid expirations.

Kapha: One of the three doshas; the principle of bodily structure; provides lubrication and strength to the body; corresponds to water and earth elements.

karma: "Action." Refers to action in general and also to one's fate as determined by the outcomes of one's past actions.

kashadra: Marma point in center of underarm.

khichadi: A dish made of rice and lentils that is commonly eaten during light fasting.

Krishna: An incarnation of God; the god-man who is Arjuna's teacher in the story of the Bhagavad Gita.

kshipra: "Quick" marma point located on the foot in the groove between the base of big toe and second toe.

kundalini: The spiritual energy in the body. In dormant form it is located in the first chakra at the base of spine, where the energy rests coiled like a serpent. In the process of spiritual awakening the energy moves up the spine.

laja: A sweet rice.

Lakshmi: Goddess who represents wealth and prosperity.

lepas: Herbal pastes for cleansing the skin.

lodhra: Medicinal Indian herb.

lohitaksha: Marma point. "Red eyed": lower frontal insert of shoulder joint and leg joint.

mahamarmas: Literally, the "great marmas"; the main energy points of the body corresponding to the seven chakras.

Maharishi Mahesh Yogi: Contemporary Yoga master and Vedic scholar; founder of the Transcendental Meditation and Maharishi Ayur-Ved programs.

malas: The "unretainable substances" of the body; the bodily wastes, including urine, feces, and sweat.

manas: "Mind." Typically refers to the aspects of mind including the intellect, desire, emotions, and volition.

mandala: "Circle." A circular image sometimes used to focus the mind in meditation.

manjista: "Rubia cordifolia." Best blood-purifying herb, increases blood flow, promotes healing; cleanses and regulates liver, spleen, and kidney.

manomaya kosha: The third subtle body of the Ayurvedic "anatomy," it encompasses the mind and intellect.

mantra: A primordial sound used in meditation for its known balancing effect on the mind and body. It may be a single syllable or a string of syllables that are typically without any cognitive meaning. Also called "the thought that liberates and protects."

manya mula: Marma point situated in carotid triangle in the back of the neck.

marma: One of 107 vital energy points in the body where matter and consciousness converge. Physically located at the sites where nerves, veins, arteries, tendons, bones, and joints meet.

maya: Typically translated as "illusion." Refers to the dualistic nature of the manifest universe as opposed to the unity of unmanifest consciousness.

muladhara: The first of the seven chakras, or energy centers, located at base of spine.

nabhi: Navel.

nada: Sound. *Na* means breath; *da* means fire.

nadi: Subtle energy channel in the body. Ayurvedic name for pulse.

nasya: Treatment for cleansing nasal passages.

nava rasas: The nine fundamental emotions.

neela: Marma point located at the throat.

neem: "Azadiracta indica." Bitter tonic, antiseptic, antifungal, blood purifier, used in all skin diseases internally and externally.

netra basti: Eye cleansing treatment.

niyama: Self-restraint; the second "limb" of Yoga which includes the practices of purity, contentment, nonattachment, study, and devotion.

ojas: The essence of the seven dhatus and the source of physical immunity; one of the three vital forces, along with prana and tejas, that govern life functions.

pancha amrit snan: The "five nectar bath."

pancha karma: The "five actions"; five purification treatments that are part of a medical detoxification therapy.

Panini: Ancient Vedic sage.

para: The root sound existing at the subtlest level of manifestation.

Pitta: One of the three doshas; the transformative or "fire" principle, formed from the fire and water elements.

prajnaparadha: The "mistake of intellect"; the loss of wholeness or natural wisdom.

prakriti: The individual constitution, or mind-body type; literally means "nature."

Paramahansa Yogananda: One of the first modern Yoga masters to come to the West; author of *Autobiography of a Yogi,* and founder of the Self-Realization Fellowship in 1920; lived from 1893–1952.

pichu: Oil compress used on "third eye."

prana: The life-giving force, the life breath.

pranamaya kosha: The pranic sheath; the second subtle body of the Ayurvedic anatomy corresponding to the breath.

pranayama: "Breath control." Yogic breathing exercises.

raga: The classical form of Indian music.

rajas: One of three gunas, or fundamental forces of creation; it is the spur to action and the principle of change.

rasasara: The "essence" of rasadhatu, which is skin. Also described as "the cream that rises to thte top."

rasa: Literally means "essence," "emotion," or "taste." The six basic rasas, or tastes, of foods are sweet, salty, sour, pungent, bitter, and astringent.

Rig Veda: The oldest and foremost of four collections of Vedic hymns that describe the cosmic order.

rishi: Realized soul; seer; yogi.

sadhana: "Means of realization." Activity performed with awareness.

samadhi: State of pure joy.

sangeeta: Indian music.

Sanskrit: The primordial language of Veda; also called the language of creation.

santrasa: Continued discomfort of body and mind; stress.

Saraswati: Goddess of knowledge.

sattva: The pure essence of reality; one of the gunas, the three fundamental forces of consciousness.

sat: Pure knowledge; truth.

shatavari: "Asparagus racemous." Tonic for women. Nourishes ovum and increases fertility. Increases ojas.

shirodhara: Treatment to calm central nervous system, in which warm medicated oil is dripped on head or forehead.

shitali: Means "cooling." It is a breathing exercise to cool the body and balance Pitta, characterized by inhalation through the mouth.

Shiva: Benevolence, bliss. Also the name of a god.

shruti: "That which is heard"; sacred scriptures consisting of the four Vedas and 108 Upanishads.

shudhi: Purification.

snana: Bathing.

snehana: Oil massage. Lubrication therapy.

soma: "Nectar"; the taste of ojas, or bliss, in consciousness.

sthapani: One of main marma points, or chakras, located at the center of eyebrows; also known as the "third eye."

sundar: Beauty

surya namskar: Sun salute; a yogic exercise.

swami: Holy man.

swara: Sustained expression of a single sound or note in raga.

swedana: Sweat-inducing therapy.

Sri Aurobindo: One of India's foremost modern sages; lived from 1872–1950; founder of the "Integral Yoga" path to Self-realization, and author of *The Life Divine* and other books on Indian philosophy.

Taittiriya Upanishad: A Vedic text based on the teachings of the sage Tittiri; introduces the concept of the five bodily sheaths, or koshas.

talahridaya: Marma point in the center of palms and soles.

tamas: One of the three gunas; represents the principle of inertia or destruction.

Tantra: Means "to expand." It is a system of spiritual practices using the kundalini energy. Mantras are gift of Tantra.

tej: Radiance, glow.

tejas: One of the three vital forces; the "fire of the mind"; the transformative principle in the body. Also called agni.

thali: A large steel plate on which Indians traditionally eat their meals.

triphala: An herbal remedy made from a combination of three dried fruits called hirda, behada, and amalki.

upadhatus: The subsidiary dhatus, or bodily tissues.

Vata: One of the three doshas; the principle of movement or "air" in the body; formed by the space and air elements.

Vatsyayana: Author of the ancient writings on eroticism and sexuality known as the Kamasutra.

Veda: Means "pure knowledge"; the cognized knowledge of the principles and structure of creation. Also refers to the sacred texts of Rig Veda, Atharva Veda (which is the source of Ayurveda), Yajur Veda, and Sama

Veda, which were cognized by the
ancient rishis.

vaidya: Ayurvedic physician.

vijanamaya kosha: The fourth sheath of
the Ayurvedic anatomy. Corresponds
to the ego.

vikriti: Imbalanced condition of doshas.

vipak: Post-digestive effect of food.

virya: Heating or cooling actions of food
after digestion.

yama: The "first limb" of Yoga, which
includes five moral or spiritual
disciplines.

Yoga: Union with divine.

Yoga Sutras: The yoga aphorisms of
Patanjali, one of the foremost ancient
Vedic seers.

Agarwal, R. S. *Secrets of Indian Medicine*. Pondicherry: Sri Aurobindo Ashram Publication Department.

Aman, Oscar. *Medicinal Secrets of Your Food*. Mysore: Secretary Indo-American Hospital, 1985.

Anand, Mulk Raj, and Krishna Nehru Hutheesing. *The Book of Indian Beauty*. Rutland, Vermont: Charles E. Tuttle Company, 1981.

Athavale, V. B. *Basic Principles of Ayurveda*. Bombay: V. B. Athavale, 1980.

Athavale, V. B. *Health and Vigour Forever*. 1977.

Boone, Sylvia. *Radiance from the Waters: Ideals of Feminine Beauty in Mende Art*. New Haven: Yale University Press, 1986.

Chopra, Deepak. *Ageless Body, Timeless Mind*. New York: Crown, 1993.

Chopra, Deepak. *Creating Health*. Boston: Houghton Mifflin Company, 1987.

Chopra, Deepak. *Perfect Health*. New York: Crown, 1991.

Chopra, Deepak. *Quantum Healing*. New York: Bantam Books, 1989.

Damian, Peter, and Kate Damian. *Aromatherapy: Scent and Psyche*. Rochester: Inner Tradition, 1995.

Dash, Bhagwan. *Diagnosis and Treatment of Diseases in Ayurveda*, Part V. New Delhi: Concept Publishing Company, 1991.

Deva, B. Chaitanya. *An Introduction to Indian Music*. New Delhi: Ministry of Information and Broadcasting, 1973.

Dossey, Larry. *Meaning and Medicine*. New York: Bantam Books, 1991.

Dossey, Larry. *Space, Time and Medicine*. Boston: Shambhala, 1982.

Frawley, David. *Ayurvedic Healing*. Salt Lake City: Passage Press, 1989.

Goleman, Daniel. *Emotional Intelligence*. New York: Bantam Books, 1995.

Goleman, Daniel, and Joel Gurin. *Mind/Body Medicine*. Yonkers: Consumer Reports Books, 1993.

Hampton, Aubrey. *Natural Organic Hair and Skin Care*. Tampa: Organica Press, 1987.

Hawking, Stephen. *A Brief History of Time*. Toronto: Bantam Books, 1988.

Huang, Chungliang Al, and Jerry Lynch. *Thinking Body, Dancing Mind*. New York: Bantam Books, 1992.

Johari, Harish. *Dhanwantari*. Calcutta: Rupavlo, 1986.

Kinsley, David. *The Goddesses' Mirror: Visions of the Divine from East and West*. Albany: SUNY Press, 1988.

Keville, Kathi, and Mindy Green. *Aroma-therapy: A Complete Guide to the Healing Art.* Freedom: The Crossing Press, 1995.

Lad, Vasant. Ayurveda, *The Science of Self Healing.* Santa Fe: Lotus Press, 1984.

Lad, Vasant, and David Frawley. *The Yoga of Herbs.* Santa Fe: Lotus Press, 1986.

Lambert, Ellen Zetzel. *The Face of Love: Feminism and the Beauty Question.* Boston: Beacon Press, 1995.

Lonsdorf, Nancy, Veronica Butler, and Melanie Brown. *A Woman's Best Medicine.* New York: Jeremy P. Tarcher / Putnam, 1993.

Maharishi Mahesh Yogi. *On the Bhagavad-Gita.* London: Penguin, 1967.

Maharishi Mahesh Yogi. *The Science of Being and Art of Living.* Livingston Manor: MIU Press, 1966.

Massey, Reginald, and Jamila Massey. *The Music of India.* New York: Crescendo Publications, 1976.

Montagu, Ashley. *Growing Young.* Granby: Bergin & Garvey Publishers, Inc., 1989.

Montagu, Ashley. *Touching: The Human Significance of the Skin.* New York: Harper & Row, Publishers, 1986.

Moore, Thomas. *The Re-Enchantment of Everyday Life.* New York: Harper-Collins Publishers, 1996.

Nuernberger, Phil. *Freedom From Stress.* Honesdale, Pa.: The Himalayan International Institute of Yoga Science and Philosophy of U.S.A., 1981.

Ojha, Divakar, and Ashok Kumar. *Panchkarma Therapy In Ayurveda.* Varanasi: Chaukhamba Amarabharati Prakashan, 1978.

Padus, Emrika. *The Complete Guide to Emotions and Your Health.* Emmaus, Pa.: Rodale Press, 1992.

Pitre, Veena. *A Professional Guide to Hair Dressing and Beauty Therapy.* Pune: Veena Pitre, 1988.

Pugliese, Peter T. *Advanced Professional Skin Care.* Bernville, Pa.: APSC Publishing, 1991.

Sachs, Melanie. *Ayurvedic Beauty Care.* Twin Lakes: Lotus Press, 1994.

Sharma, Hari. *Freedom from Disease.* Toronto: Veda Publishing, 1993.

Sharma, R. K., and Bhagwan Dash. *Caraka Samhita.* Varanasi: Chowkhamba Sanskrit Series Office, Vol. I, Vol II.

Sivananda, Swami. *Tantra Yoga, Nada Yoga, and Kriya Yoga.* U.P. Himalayas, India: The Divine Life Society Publication, 1994.

Tiwari, Maya. *Ayurveda: A Life of Balance.* Rochester, Vermont: Healing Arts Press, 1995.

Tiwari, Maya. *Ayurveda Secrets of Healing.* Twin Lakes, Wisconsin: Lotus Press, 1995.

Wallace, Robert Keith. *The Physiology of Consciousness.* Fairfield: MIU Press, 1993.

Weil, Andrew. *Spontaneous Healing.* New York: Alfred A. Knopf, 1995.

Zizmor, Jonathan, and John Foreman. *Super Skin.* New York: Thomas Y. Crowell Company, 1976.